Benchmark Papers in Geology Series

Editor: Rhodes W. Fairbridge, Columbia University

A selection from the published volumes in this series

A complete listing of volumes published in this series begins on page 347.

CONTINENTAL RIFTS

Edited by
A. M. QUENNELL
Consultant Geologist
Bristol, England

A Hutchinson Ross Publication

VAN NOSTRAND REINHOLD COMPANY
New York

Benchmark Papers in Geology, Volume 90
Library of Congress Catalog Card Number: 85-699
ISBN: 0-442-27661-3

Manufactured in the United States of America.

Published by Van Nostrand Reinhold Company Inc.
115 Fifth Avenue
New York, New York 10003

Van Nostrand Reinhold Company Limited
Molly Millars Lane
Wokingham, Berkshire RG11 2PY, England

Van Nostrand Reinhold
480 Latrobe Street
Melbourne, Victoria 3000, Australia

Macmillan of Canada
Division of Gage Publishing Limited
164 Commander Boulevard
Agincourt, Ontario MIS 3C7, Canada

15 14 13 12 11 10 9 8 7 6 5 4 3 2 1

Library of Congress Cataloging in Publication Data
Main entry under title:
Continental rifts.
(Benchmark papers in geology; v. 90)
"A Hutchinson Ross Benchmark book."
Bibliography: p.
Includes indexes.
1. Rifts (Geology)—Addresses, essays, lectures.
2. Continents—Addresses, essays, lectures. I. Quennell, Albert Mathieson. II. Series.
QE606.C68 1985 551.8'7 85-699
ISBN 0-442-27661-3

CONTENTS

PART III: EUROPE

PART IV: ASIA

PART V: NORTH AMERICA

SERIES EDITOR'S FOREWORD

The philosophy behind the Benchmark Papers in Geology is one of collection, sifting, and rediffusion. Scientific literature today is so vast, so dispersed, and, in the case of old papers, so inaccessible for readers not in the immediate neighborhood of major libraries that much valuable information has been ignored by default. It has become just so difficult, or so time consuming, to search out the key papers in any basic area of research that one can hardly blame a busy person for skimping on some of his or her "homework."

This series of volumes has been devised, therefore, as a practical solution to this critical problem. The geologist, perhaps even more than any other scientist, often suffers from twin difficulties—isolation from central library resources and immensely diffused sources of material. New colleges and industrial libraries simply cannot afford to purchase complete runs of all the world's earth science literature. Specialists simply cannot locate reprints or copies of all their principal reference materials. So it is that we are now making a concerted effort to gather into single volumes the critical materials needed to reconstruct the background of any and every major topic of our discipline.

We are interpreting "geology" in its broadest sense: the fundamental science of the planet Earth, its materials, its history, and its dynamics. Because of training in "earthy" materials, we also take in astrogeology, the corresponding aspect of the planetary sciences. Besides the classical core disciplines such as mineralogy, petrology, structure, geomorphology, paleontology, and stratigraphy, we embrace the newer fields of geophysics and geochemistry, applied also to oceanography, geochronology, and paleoecology. We recognize the work of the mining geologists, the petroleum geologists, the hydrologists, and the engineering and environmental geologists. Each specialist needs a working library. We are endeavoring to make the task of compiling such a library a little easier.

Each volume in the series contains an introduction prepared by a specialist (the volume editor)—a "state of the art" opening or a summary of the object and content of the volume. The articles, usually some twenty to fifty reproduced either in their entirety or in significant extracts, are selected in an attempt to cover the field, from the key papers of the last century to fairly recent work. Where the original works are in foreign languages, we have endeavored to locate or commission translations. Geologists, because of their global subject, are often acutely aware of the oneness of our world. The selections cannot therefore be restricted to any one country, and whenever possible an attempt is made to scan the world literature.

To each article, or group of kindred articles, some sort of "highlight commentary" is usually supplied by the volume editor. This commentary should serve to bring that article into historical perspective and to emphasize its particular role in the growth of the field. References, or citations, wherever possible, will be reproduced in their entirety—for by this means the observant reader can assess the background material available to that particular author, or, if desired, he or she too can double check the earlier sources.

A "benchmark," in surveyor's terminology, is an established point on the ground that is recorded on our maps. It is usually anything that is a vantage point, from a modest hill to a mountain peak. From the historical viewpoint, these benchmarks are the bricks of our scientific edifice.

RHODES W. FAIRBRIDGE

PREFACE

The current scientific revolution includes a questioning of the inductive approach to scientific inquiry. As recently as a decade ago, a volume concerned with continental rifts would have focused mainly on objective information on the geology of rift valleys and its correlation with the corresponding geochemistry and geophysics. Indeed, that is what the author set out to do in Benchmark Volume 60, *Rift Valleys:Afro-Arabian*. However, as mentioned in the Introduction, the paths taken by "rift and rifting" investigators have to follow that dictated by the scientific methodology that has accompanied the revolution. The selection and treatment of reprinted articles has to conform to the new approach. Objective published information and standard models constructed from outdated information are not good enough.

In general, and within the limitations imposed by the existing state of the art, it is inevitable that information gathering will take place within designated frameworks and will be tested by subjection to constraints.

The articles of Part I are chiefly concerned with the modern framework of plate tectonics and the Wilson cycle, although some are by authors who worked with earlier earth models. The subsequent parts are defined geographically, by continents; some articles are objective, imparting basic information, whereas others see the information as related to a postulated framework.

For our purposes continental environments are separated into *cratonic* (anorogenic) and *orogenic*, a useful distinction for most of the continents (for example, the Andes and the shields of South America).

In this volume, the terms *rift* and *rifting* are used to cover not only the usual equating of "rift" with "graben" as a rift valley (i.e., a topographic landform), but also all the rifting consequent upon extensional stress and first-order magnitude faulting with slips ranging from dip slip to strike slip.

Advances in geological field mapping in some areas of interest have recently slowed down, but this trend has been offset by the improvements in geophysical and geochemical techniques as well as in remote sensing and landsat imagery mapping. Today, many geophysicists and others accept that, as S. Thomas Crough pointed out, "the ultimate test of any tectonic model must be its consistence with the known geological record." Some leaders in the geophysical field, however, have yet to appreciate this, and in particular to take notice of the geomorphology.

The East African rifts are not treated in this volume because the papers of Benchmark Volume 60 covered the subject adequately (that is, to about 1980). A list of references to important recent articles has been included here.

In addition, two or three papers have been reprinted because of their classical and historical significance. Some others have been selected because they are not currently accessible to many readers. It is hoped that this selection of articles, heterogeneous as it is, will be of value, at least as a source of information, during the International Lithosphere Program.

I acknowledge a debt of gratitude to Dr. Paul L. Hancock of the University of Bristol, for critical reading of the Introduction and for helpful discussions; Professor David A. Brown of Australian National University, Canberra, for assistance with Australian literature and contacts; to Dr. R. W. Girdler of the School of Physics, University of Newcastle-upon-Tyne, for valuable discussion and encouragement. As in the case of Volume 60, Donald Holloway gave valuable assistance during the initial stages in procuring material to be reprinted. I am especially grateful to the authors of reproduced articles for their ready cooperation.

A. M. QUENNELL

[*Publisher's Note:* We regret to inform the reader that Mr. Quennell died on April 15, 1985, during the final stages of production of his book.]

CONTENTS BY AUTHOR

CONTINENTAL RIFTS

INTRODUCTION

The term *rift* in recent years has been restricted by some writers to a major normal fault or fault zone caused by wholly extensional stress at right angles to the strike of the fault, or more particularly to a pair of faults with throws of opposed sense that has resulted in a down-dropped block or blocks, forming a major graben. In this study, however, the term *rift* also embraces the geomorphological meaning it had before it became an abbreviation for "rift valley."

"Rift" can thus be used as an adjective (for example, to qualify "valley"); as a verb, to denote failure under stress; and as a noun, both for the fault zone or "plane" resulting from rifting and for the geomorphic feature. This faulting does not have to be purely extensional, and the displacement is not necessarily purely dip-slip. It can be oblique-slip or strike-slip. In our opinion "rift" should be employed only for features on a continental scale, and their penetration in depth into the lithosphere, whether as simple fault planes or as listric faults, would be as great as is mechanically possible.

The term rift has been used for the major strike-slip faults or megashears (Spencer, 1969, pp. 113–132; Carey, 1958), whether transcurrent or transform. Later strike-slip displacement of formerly extensional grabens was recognized by, among others, Illies (Paper 3) for the Rhine graben, by Billings (1954) for the San Andreas fault, by Sherman (1978) for the Baikal rift zone, and by Quennell (1959, 1984, in press) for the Dead Sea rift.

It appears that some rifts may have evolved from grabens and dip-slip faults through oblique-slip to strike-slip rifts with change in the azimuth of the principal horizontal stress. The dip of the fault plane is no criterion. Strike-slip fault planes are generally steeply inclined or vertical but can have inherited the dip of an initial dip-slip fault.

The depth reached by a rift fault is the order of 15 km (Sibson, 1983) or even greater, as based on seismic information, and is related to the depth to which continental rocks have elastic properties. Where a dip-slip fault reaches an optimum depth, the dip can flatten with conversion to a listric fault. This depends on factors such as the spacing of faults (width of graben).

The serious study of continental rifts and rifting on a project basis began with the International Council of Scientific Unions' 1964 endorsement of the Upper Mantle Committee Project (UMP) and its proposals, which included study of the world rift system. With assistance from UNESCO, the first seminar held was in 1965, on the East African rift system (see publication 12 in the Bibliography). Continental and oceanic rifts were treated separately from the start. Subsequent scientific reports of the UMP dealt chiefly with the Rhine graben. In 1972 the UMP formally ended with publication 7 in the Bibliography. However, the Inter-Union Commission on Geodynamics, through the International Geodynamics Program set up in 1970, preserved continuity of interest under Working Group 4 (publication 20 in the Bibliography). A number of scientific reports commencing in 1974 dealt chiefly with the Rhine graben, the Afar junction, the Baikal rift zone, and finally the symposium edited by Illies (publication 12 in the Bibliography). The final reports, other symposia, and collected works appeared, culminating in Morgan and Baker (1983, publication 21 in the Bibliography).

The Bibliography includes only collected works and symposia brought together in single volumes, but at least as much data and research on continental rifts is dispersed among isolated papers appearing in journals and other channels. A consecutive perusal of the papers listed will show the steady advance in knowledge of continental rifts—geological, geophysical, and geochemical—and the increasing recognition of the importance of an interdisciplinary approach. A history of rift studies by Drake and Girdler, is found in the WG4 final report, *Continental and Oceanic Rifts* (publication 20 in the Bibliography) which appeared in 1981.

The recent intensification of rift studies has taken place within the context of a more general scientific revolution. The papers presented during this period show an increasing independence of the traditional philosophy of science, in that the inductive method is increasingly set aside. The traditional view, which was followed by the earlier papers in the Bibliography, is replaced by the approach advocated by Popper (Magee, 1982), Hallam (1973), Kuhn (1970), and Kitts (1977). Although there are differences in approach and in nomenclature, these authors are following parallel-flowing streams.

Because the scientific revolution is accompanied by changes in methodology and because the later works in the Bibliography (and indeed in those yet to appear) involve these changes, it is important to understand what the revolution is about.

Inductive reasoning involving "development by accumulation" has been generally abandoned, following the experience that it can ultimately lead up blind alleys and is, in any case, logically unsound. Even the concept of uniformitarianism, when expressed as the assumption that because events have followed a particular course in the past they are more likely to do so again, must be reappraised. It is the historians and philosophers of science rather than the scientists themselves who have brought about this new understanding.

The complexity of the tectonics of continental rifts and rift systems is compounded by the wide range of disciplines involved—geomorphological, geological, geochemical, and geophysical. Only geomorphology, structure, and some geochemical properties as seen close to or at the earth's surface are directly observable. For rift studies involving endogenic processes, the shift in scientific method away from inductive reasoning appears to be essential.

When convening their conference on "Processes of Planetary Rifting," Morgan and Baker (1983, p. 1 publication 21 in the Bibliography) deliberately set out their intention:

> to bring together theoretical modellers, geologists, geochemists and geophysicists to give the modellers direct exposure to the constraints imposed by some of the available data. Similarly, workers primarily involved in data collection were exposed to the implications of models to data interpretation and speculation. We attempted to advance the understanding of the processes of rifting by encouraging a multidisciplinary approach to the problem.

In the present study the most direct observations concern tectonic landforms that result from crustal deformation immediately related to rifting. They cannot be explained purely in terms of exogenic processes. Full explanation can only be developed by study of endogenic processes beyond the limits of simple observation, that is, by geophysical and geochemical research in depth eventually reaching the crust-mantle and lithosphere-asthenosphere boundaries. On the other hand, Crough (Morgan and Baker, 1983, p. 35) states "but the ultimate test of any tectonic model must be its consistence with the known geological record."

The revolutionary movement in modern earth-process modeling is away from a static toward a mobile view (Wilson, 1968, 1969). Under plate tectonics, rifts and rifting were first recognized on the ocean floor and the mid-oceanic rises with their axial rift valleys and transform faults. Indeed the crucial chapter in the revolutionary transition from continental drift to plate tectonics was the recognition of seafloor spreading.

Continental rifting is observed chiefly in four principal settings.

First, there is the rifting of cratonic margins, with formation of aulacogens parallel to the separating margins and penetrating aulacogens derived from triple junctions. Then, within the continents there are intracratonic rifts, where rift valleys and rift faults form by intraplate strain. Third, in localized association domal uplift, volcanism and rifting become better understood as the effect of thermal activity and "hot spots" by heat transfer from deep-seated sources. Fourth, the collision of continents gives rise to localized extensional conditions that produce rift freatures, such as grabens and strike-slip structures.

A further development that enriched the original plate tectonic model and solved some important anomalies was the discovery of the Wilson cycle. This concept continues to yield fruitful developments. Logically, if one ocean is opening, another ocean must be closing. Wilson (1968) first enunciated this principle and later (1969) in a William Smith lecture and expanded it in Jacobs, Russell, and Wilson (1974) as a feature of the mobilistic picture of the earth that had to be accepted along with the plate tectonic model. The Wilson cycle is a natural outgrowth of—in fact, a paradigm inherent in—plate tectonics.

Wilson (in Jacobs et al., 1974) describes six stages. The first is the rifting of a cratonic continental plate. This then widens to become a rift sea, a continental separation with rifted margins (e.g., the Red Sea model). With a further development of sea-floor spreading, this becomes a rift ocean with mid-oceanic ridge traversed by transform faults and rift valleys (e.g., the Atlantic). During the separation, further continental rifting may occur with the style of rifting as before. The fifth stage involves imminent plate impact when trenches and troughs develop. Extensional conditions are localized. The sixth stage involves orogeny, fold-mountain building, and its accompaniments of subduction and obduction following continental collision. Localized rifting, especially strike-slip, will result from overthrusting of continental slabs (Burke, 1980). This sixth stage received further attention from Oliver et al. (1983), who, from investigations by Consortium for Continental Reflection Profiling (COCORP) reveal the layered overlapping nature of continental margins where conditions favor overriding and décollement.

There is no synchronism of the Wilson cycle stages on a global basis. Nor is impact simultaneous along the converging margins. Rather it is diachronous, which causes localized differences in strain, giving rise to rifting (Dewey and Burke, 1974). According to Oliver et al. (1983), "At the beginning of the next Wilson cycle, i.e., as the ocean reopened, extensional tectonism produced a new set of normal faults in the overthrust sheet. They became listric at depth into the pre-existing detachment." There will be differing morphotectonic expressions of rifting at the different stages.

Turning to individual rifts and rift zones, processes range between two end-members: "active" rifting, in which rifting is a result of a thermal upwelling of the asthenosphere, and "passive" rifting, in which rifting is a passive response to a regional stress field (Morgan and Baker, 1983; also Şengör and Burke, Paper 1; Turcotte and Emerman, 1983).

In active rifting the determining parameters are the nature and physical state of the outer layers of the earth (crust-mantle or lithosphere-asthenosphere) and the transference of heat. This may be by diapirism, whereby hotter material from the lower layers is injected into the upper layers, or by conduction, with or without thinning of the crust. An extensional stress field (deviatoric) is generated in the brittle upper layers, and this induces rifting. Orientation of rift faults depends on a number of factors. The active process is likely to result in graben rifting. It can give rise to trisection of the domal uplift associated with heat transfer, and eventually to continental separation with triple-point aulacogens.

Passive rifting results from "horizontal movement of plates, their interaction giving rise to differential stresses within the lithosphere that result in rifting" (Şengör and Burke, Paper 1). The passive process, while producing some graben (dip-slip) rifts, is likely to produce oblique-slip and strike-slip rifts. When continents fragment they are not necessarily pulled apart directly but often obliquely (Ballance and Reading, 1980, publication 22 in the Bibliography). Continental collisions can generate localized extensional stress fields (Molnar and Tapponnier, 1975), and strike-slip faulting (Şengör, 1976; Sherman, 1978).

Because the surface manifestation of a young rift fault or faults is a tectonic landform, geomorphology plays a primary role in constraining the modeling of rifts. This is shown by the following eleven points, recorded from published works and from personal field work.

(1) The swells within the continents are, according to some workers, the favored topographic environments for generation of rifting. However, the relationship is not consistent. There are swells in Africa and other continents that have no rifting processes associated with them, and there are also previously undisturbed planation surfaces that have been rift faulted.

(2) The epeirogenic uplift and resulting extension of the surface of swells generally provides insufficient stretching to give rise to new rifts. However, where the superficial extension is most intense and curvature greatest, as on the flanks of the swells, localized extension may be adequate to cause rifting.

(3) Domal uplift—that is, more intense uplift of smaller dimensions than that of swells (Bott, 1981)—appears usually to be accompa-

nied by volcanism. A triple-splayed pattern of active rifting faulting can be associated with it. Controversy continues about the sequence in which domal uplift (arching), volcanism, and rift faulting occur.

(4) Downwarp along the future fault zone, a synform impressed on the doming, can precede faulting. Shoulders are then convex upward.

(5) Rifts of passive origin can lie in terrain that is stressed but has suffered no previous distortion. This can be a planation surface epeirogenetically (uniformly) uplifted. The rifts can occupy the central parts of such swells.

(6) The planation surfaces in which rifts are set do not generally reveal associated concave "rise to the rift," an essential element in the wedge subsidence model advocated by Vening Meinesz (1950) and Bott (1981). The alleged slopes are difficult to identify in the field and can arise from incorrect interpretation of erosional features.

(7) Although topographically a "rise to the rift" is not generally the case, angular unconformities in sedimentary successions that make up the shoulders indicate some outward tilting and arching before or during sedimentation and before erosion under the penultimate cycle (e.g., Upper Rhine graben).

(8) In some cases along the rift margins there are uniformly uplifted (not tilted) remnants of former planation cycles, indicating that the setting of the rift faulting is in morphostructural zones of persistent uplift (Western Rift System of East Africa).

(9) Asymmetry in graben formation is the general rule. Rarely are the opposing faults equal in throw. In some cases one fault will die out along the strike. Models not subjected to geological constraints too often are idealized with respect to symmetry.

(10) Rifting in epeirogenetically uplifted planation surfaces appears to belong to the diastrophism that will have terminated the erosion cycle. The subject of morphotectonic history following termination of an erosion cycle is complex. Usually it shows a period of diastrophism that results in uncompleted partial cycles, rejuvenation of faults, warping and localized uplift (for example, East Africa "end-Tertiary cycle"). It appears that eventually quiescence allows an erosion cycle to resume, giving rise to a further planation surface, which may later be uplifted epeirogenetically.

(11) The form and uplift history of a recognizable planation surface subjected to rift faulting give important information on the timing and magnitude of the faulting. In a similar manner sedimentation within the rifts and the presence of lakes reveals the history of subsidence of rift floor and uplift of flanks.

The reprinted articles do not dwell on the geomorphological aspects of rifts and rift valleys. Reference should be made to Benchmark

Volume 60 (Quennell, 1982) for more specific information, especially the paper by Douglas Johnson. A valuable article by Mohr, "Musings on Continental Rifts," appears in Geodynamics; 1981, Series, Volume 8 (publication 20 in the Bibliography).

REFERENCES

Billings, M. P., 1954, *Structural Geology,* 2nd ed., Prentice-Hall, New York, 514p.

Bott, M. H. P., 1981, Crustal doming and the mechanism of continental rifting, *Tectonophysics* **73:**1-8.

Carey, S. W., 1958, The tectonic approach to continental drift, in *Continental Drift—a Symposium,* Geol. Dept. Univ. Tasmania, Hobart, pp. 177-355.

Dewey, J. F., and K. Burke, 1974, Hot spots and continental breakup: implications for collisional orogeny, *Geology* **2:**57-60.

Hallam, A., 1973, *A Revolution in the Earth Sciences—from Continental Drift to Plate Tectonics,* Clarendon Press, Oxford, 127p.

Illies, J. H., 1978, Two stages of Rhinegraben rifting, in *Tectonics and Geophysics of Continental Rifting,* I. B. Ramberg and E. R. Neumann, eds., D. Reidel, Dordrecht, pp. 63-71.

Illies, J. H., 1979, Preface, mechanism of graben formation, *Tectonophysics* **73:**vii.

Jacobs, J., R. D. Russell, and J. T. Wilson, 1974, *Physics and Geology,* 2nd ed., McGraw Hill, New York, 622p.

Kitts, D. P., 1977, *The Structure of Geology,* Southern Methodist University Press, Dallas, 180p.

Kuhn, T. S., 1970, The structure of scientific revolutions, in *International Encyclopedia of Unified Science,* vol. 2, no. 2, 2nd ed., University of Chicago Press, Chicago, 210p.

Magee, B., 1982, *Popper,* Fontana Paperbacks, Glasgow, 111p.

Mohr, P., 1981, Musings on continental rifts, in *Continental and Oceanic Rifts,* G. Palmason, ed., pp. 293-309.

Molnar, P., and P. Tapponnier, 1975, Cenozoic tectonics of Asia: effects of a continental collision, *Science* **189:**419-426.

Oliver, J., F. Cook, and L. Brown, 1983, COCORP and the continental crust, *Jour. Geophys. Research* **88:**3329-3347.

Quennell, A. M., 1959, Tectonics of the Dead Sea rift, *Assoc. African Geol. Surv. 20th Session Internat. Geol. Cong.,* Mexico, 1956, pp. 385-405.

Quennell, A. M., 1984, Western Arabia rift system, in *Geological Evolution of the Eastern Mediterranean,* Geol. Soc. London Spec. Pub. 17, pp. 587-600.

Quennell, A. M., in press, Rift valleys, in *Encyclopedia of Structural Geology and Plate Tectonics,* Van Nostrand Reinhold, Stroudsburg, Pa.

Şengör, A. M. C., 1976, Collision of irregular continental margins: implications for foreland deformation of Alpine-type orogens, *Geology* **4:**799-782.

Sherman, S. I., 1978, Faults of the Baikal rift zone, *Tectonophysics* **45:**31-40.

Sibson, R. H., 1983, Continental fault structure and the shallow earthquake source, *Geol. Soc. London Jour.* **140:**741-767.

Spencer, E. W., 1969, *Introduction to the Structure of the Earth,* McGraw-Hill, New York, 597p.

Turcotte, D. L., and S. H. Emerman, 1983, Mechanisms of active and passive rifting, *Tectonophysics* **94:**39–50.

Vening Meinesz, F. A., 1950, Les Graben Africains, result de compression ou de tension dans la croute terrestre, *Bull. Inst. R. Colon Belge* **21:**539–552.

Wilson, J. T., 1968, Static or mobile earth: the current scientific revolution, in *Gondwanaland Revisited: New Evidence for Continental Drift,* G. Piel, ed., Am. Philos. Soc. Proc. **112:**309–320.

Wilson, J. T., 1969, Static or mobile earth: the current scientific revolution (abstract), *Tectonophysics* **7:**599–601.

Part I

THEORY AND GENERALIZATIONS

Editor's Comments on Papers 1, 2, and 3

1 **ŞENGÖR AND BURKE**
Relative Timing of Rifting and Volcanism on Earth and Its Tectonic Implications

2 **MILANOVSKY**
Some Problems of Rifting Development in the Earth's History

3 **ILLIES**
Mechanism of Graben Formation

The first paper, by Şengör and Burke, recognizes the distinction between the two basic modes of rifting, end members of a series, active and passive. "Active" refers to thermal activity in the asthenosphere and lithosphere preceding rifting of the crust. "Passive" indicates that the asthenosphere initially plays no role, stresses inducing intraplate rifting, which generally is followed by associated volcanism (Turcotte and Emmerman, 1983). This paper covers other aspects, including the concept of rift clusters (in contrast to isolated rifts). The authors consider the mechanism by which rifts and rift systems have come into being; they believe it is virtually independent of whether the tectonics of the outer layers of the earth involve either vertical or horizontal movement. This view initiated a fresh approach to continental rifting (Morgan, 1982; Baker and Morgan, 1981). Şengör is now at the Technical University of Istanbul, Turkey, and Burke is at the Lunar and Planetary Institute in Houston, Texas.

The second paper is by E. E. Milanovsky, a distinguished professor at the University of Moscow, whose contributions to empirical classification and structural history of rifts and rifting, and correlation with other tectonic processes, extend over two decades. However, his approach is wholly outside the recognition of plate tectonics and the Wilson cycle models. In other papers Milanovsky (1972, 1980, and 1983) states his view that global expansion of median zones of oceanic areas takes place without sea-floor spreading or tangential movement of the continents, with pulsating evolution of the earth. This intermediate view of widening of oceans with stability of continents is referred to by Heezen (1960), who subsequently abandoned it in favor of sea-floor spreading (1969). This paper is of value as containing a classification of rifts, their geological history, and correlations with orogeny and volcanism.

The third paper, by the late Professor Henning Illies of the University of Karlsruhe, West Germany, is the Epilogue to a special issue of *Tectonophysics* devoted to an ICG Symposium, 17th IUGG General Assembly, Canberra, 1979, of which he was editor (publication 12 in Bibliography). In the Preface to the volume Illies defined and explained the term graben, which he emphasized is only partly synonymous with the term rift, to which he gives a wider connotation, even including the results of "extensional shear," for which he cites the San Andreas fault and the Dead Sea rift. A clear explanation of the relationship of graben structure to thermal uplift is given. In another paper (1978) in the case of the Rhine Graben, he defines the essential circumstances that governed its formation: basement control, mantle control, stress-field control. In the present paper he expands this idea and extends its application (see also Illies and Greiner, 1979). The principles stated, although more specifically related to the Rhine graben, are capable of general application.

[*Editor's Note:* The reader is referred to the list following the comments on Papers 4 and 5 (pages **49** and **50**) for the references cited in these comments.]

1

Reprinted from *Geophys. Res. Lett.* **5**(6):419–421 (1978)

RELATIVE TIMING OF RIFTING AND VOLCANISM ON EARTH AND ITS TECTONIC IMPLICATIONS

A.M. Celal Şengör and Kevin Burke

Abstract. Rifts occur in diverse tectonic environments that result from the continuous two-dimensional evolution of the multi-plate mosaic of the Earth and also from the interaction between the mantle processes and the overlying lithosphere. Most rifts have associated volcanics that are mainly basaltic. In continental rifts these basalts are predominantly alkaline and the relative timing of rifting and volcanism is variable; however, despite this variation there seem to be two basic types of rifting-volcanism relative timing: in one, volcanism and usually local doming predates major rift formation whereas, in the other, rifts form first and volcanism (? and doming) follow thereafter. These two basic types of rifting-volcanism relative timing may be related to two basic modes of rifting. In the first the mantle plays an active role, convection 'plumes' dome up and crack the lithosphere, whereas, in the second, the horizontal movements of plates give rise to extension of the lithosphere and induce rifting. In this latter case, the mantle is passive. Numerous local conditions complicate this simple pattern and result in extremely complicated rifting-volcanism relationships that make geophysical/geochemical modelling difficult. Petrologic/geochemical studies of rift volcanics do not provide unique solutions for our understanding of rift environments. Detailed stratigraphic/structural analysis of individual rifts are still the best methods for rift analysis.

Introduction

Rifts, elongate depressions beneath which the entire thickness of the lithosphere has ruptured under extension, occur in diverse tectonic enviroments that result from the continuous two-dimensional evolution of the multi-plate mosaic of the Earth and also from the interaction between mantle processes and the overlying lithosphere. There is a great variation in the sizes and length-to-width ratios of individual rifts, as well as in their mode of occurrence. Some rifts are seemingly isolated such as that of Lake Baikal (*Logatchev and Florensov,* 1978), others are aligned along relatively narrow belts such as those of East Africa (*Burke and Whiteman,* 1973), and yet others are clustered together, in nearly parallel arrangement, in roughly equidimensional areas such as those of the Basin and Range Province of the western United States (*Atwater,* 1970). Although the genetic significance of this variation is far from being clearly understood, it is an empiric conclusion that rifts that occur along 'rift belts' are associated with areas of primary extension, whereas 'rift clusters' are more characteristic of wide, imperfect strike-slip regimes. Isolated rifts occur in both environments.

A feature common to most rifts is basaltic volcanism. In continental rifts, with which we are here concerned, these basalts are mainly alkaline. The relative timing of basaltic volcanism with respect to the associated rifting shows considerable variation. Currently there are two main hypotheses attempting to explain the origin of basaltic magmas beneath rifts: one favors the preliminary cracking of the lithosphere due to differential stresses resulting from two-dimensional plate evolution (e.g., membrane stresses: *Turcotte and Oxburgh,* 1973; *Turcotte,* 1974; stresses due to the collision of continents: *Molnar and Tapponnier,* 1975; Şengör, 1976; stresses resulting from the propagation of existing accreting plate margins into continents: *McKenzie and Weiss,* 1975) that upsets the T/P balance of the underlying mantle resulting in its partial melting. This results in a volume increase in the partially melted area and may induce a post-rifting uplift of the overlying lithosphere as happened in the Upper Rhine Graben (*Şengör et al.,* 1978). The other view holds a complicated convection pattern in the mantle responsible for doming and cracking the lithosphere thereby giving rise to extensional fractures and eventually to rifts (*Burke and Whiteman,* 1973; *Burke and Dewey,* 1973). In the former view, the expected sequence of events is rifting – (?uplifting) – volcanism, whereas in the latter it is doming-volcanism-rifting (Fig. 1). It is our contention that both hypotheses are compatible with the available data, but are applicable to fundamentally different tectonic environments. The first hypothesis is applicable to regions where rifting is the result of horizontal movements of plates and their interaction in which mantle plays a passive role. The second hypothesis seems to be valid where (?small scale) convection in the mantle (*McKenzie and Weiss,* 1975) directly affects the overlying lithosphere and induces rifting as a combined result of primary vertical tectonics (uplift) and (?later) horizontal motion (initial spreading). It appears that this latter mode of rifting occurs mainly on plates that are fixed with respect to the underlying mantle, as Africa has been suggested to have been since 25 m.y. ago (*Burke and Wilson,* 1972). At present, the former mode of rifting is by far the more widespread of the two.

Passive Mantle Hypothesis

The horizontal movement of plates and their interaction give rise to differential stresses within the lithosphere that result in rifting. *Turcotte and Oxburgh* (1973) and *Turcotte* (1974) argued that membrane stresses can be generated within a plate moving longitudinally on the surface of the Earth, due to the ellipticity of the Earth, and result in rifting. *Burke and Dewey* (1974), however, indicated that the elastic stresses thus generated cannot be stored for long periods of time and therefore may not be as important in generating rifts as *Turcotte and Oxburgh* (1973) suggested. Moreover, the supporting evidence that *Oxburgh and Turcotte* (1974) used to apply this idea to the East African rifts consisted of incorrect K/Ar data and misinterpretation of palaeomagnetic results.*

Molnar and Tapponnier (1975) and *Şengör* (1976) pointed out that continental collision, especially of irregular margins, also generates extension within continental plates resulting in rifting more or less perpendicular to the direction of convergence; if collision-induced strike-slip faulting also causes rift formation, as is the case for the Baikal rift (*Sherman,* 1978), such rifts are not likely to have any systematic orientation with respect to the convergence direction. In the foreland of the Alps in Europe, for example, rifting events are well-correlated with the Mesoalpine collision (*Trümpy,* 1973; see *Şengör,* 1976, table 1). Here rifting appears to have predated major basaltic volcanicity. In the case of the Upper Rhine Graben, rifting can be shown to have started during medial Eocene with some minor volcanicity of 42 m.y. age along the master faults (*Şengör et al.,* 1978). Major basaltic volcanic activity began at the northern end of the graben about 25 m.y. ago (*Lotze,* 1974). This volcanic center, the Vogelsberg, appears to have continued its activity until very recently. In the south, well within the major graben trough, the Kaiserstuhl volcano began its activity about 18 m.y. ago; this however, unlike that of the Vogelsberg, was a short-lived event. In the Lower Rhine Graben there are few volcanics, but the maars of Eifel appear to be related to Plio-Pleistocene NW-trending faults (*Greiner and Illes,* 1977). In Bohemia, both along the Thuringian disturbance and in the grabens between the Bohemian and the Thuringian blocks (*Schollen*) volcanicity began during the Oligocene and lasted through the Pliocene; rifting along these lines had begun at the beginning of the Tertiary and accelerated during late Eocene (*Lotze,* 1974). It appears that in Europe it is this kind of intra-plate but inter-*Scholle* activity that caused basaltic volcanism that is largely of olivine-nepheline bearing alkaline type.

Western Turkey and the Basin and Range Province of the western United States are areas of extensive rift development, characterized by numerous sub-parallel grabens in 'rift clusters'. Both areas are related to large, imperfect strike-slip plate boundary zones (western Turkey: *Dewey and Şengör,* 1978; western U.S.: *Atwater,* 1970). In western Turkey rifting began during the late Miocene and is currently active. Most rifts

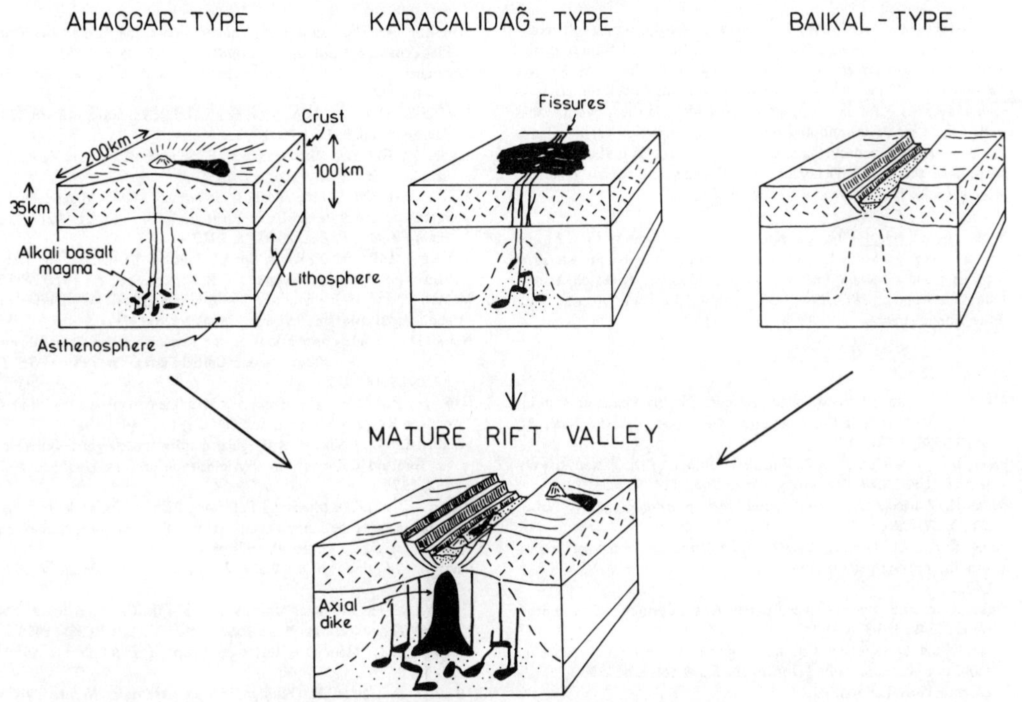

Fig. 1. Idealized and simplified block diagrams showing the evolution of a 'mature rift valley' from three different initial configurations. In Ahaggar-type rift valley evolution mantle is active and induces uplift and subsequent rifting. In this case volcanism is likely to predate rifting. In the Baikal-type mantle is passive and there is no pre-rifting doming or volcanism. Karacalidag-type was added to the diagram to illustrate a complexity: horizontal extension forms fissures through which magma wells up prior to major downfaulting and rift formation. Hence, in the geological record the sequence of events in the formation of a Karacalidag-type rift will look like those of an Ahaggar-type rift, although the mode of rifting was that of a Baikal-type rift. The end products of all rift processes are likely to look very similar.

here are devoid of volcanics with the exception of the Gediz and Simav grabens (*Dewey and Şengör,* 1978). On the northern shoulder of the Gediz graben, an area, 50 km. long and 20 km. wide, is covered with Pliocene to recent alkaline basalts that are nepheline, leucite, and hornblende bearing (kulaite, *Washington,* 1894; *Erinc,* 1970). These basalts emanated from fissures that can be shown to be controlled by faults related to the Gediz graben. *Zeschke* (1954) documented a similar history for the Simav basalts.

In the Basin and Range Province major rifting related to the present regime began about 18 m.y. ago (*Noble,* 1972) and was closely followed by basaltic/rhyolitic bimodal volcanism (*McKee et al.,* 1970). The geographic extent and the close temporal association with a strike-slip regime (*Hamilton,* 1970) of the western Siberian rift system of Triassic age (*Logatchev,* 1977) suggest that it might be a fossil analog of the Basin and Range-type rift regimes, where generally basaltic volcanism postdates rifting.

Active Mantle Hypothesis

Rifts active on the African Plate can be interpreted as products of relatively simple mantle-lithosphere interaction. The control appears to be the lack of relative motion between the African Plate and underlying convective circulation over the last 25 m.y. (*Burke and Wilson,* 1972). The interaction beneath the African Plate shows itself in several phenomena: the distinctive basin and swell structure; the occurrence of intraplate volcanism (largely on swell crests); the development of rifts and the evolution (in the Red Sea and in the Gulf of Aden) of rifts into oceans (see *Burke,* 1977 for a review).

Throughout southern Africa swell crests carry no volcanoes and rifts are mainly not on swells but are reactivated old structures. This indicates that a sequential development from volcanoes on swells to rifts such as *Burke and Whiteman* (1973) distinguished is not universally recognizable. Further evidence that rift development is complex can be seen in the active rifts of East Africa. Volcanism, there, is very unevenly distributed along the length of the rifts and in some areas (for example, near Addis Ababa) is dominantly tholeiitic, whereas in other areas it is dominantly alkaline (see, for example, *Baker et al.,* 1972). The implication is that the petrology of the igneous rocks is a very poor indicator of rift style compared with more direct structural/stratigraphic features such as topography, faulting, and sediment fill. The general absence of signs of interaction between magma and continental crust in the rift igneous rocks is perhaps the most significant feature and is to be expected in a regime where axial dikes and extension dominate.

Older episodes of rifting induced by similar mantle interaction are hard to identify but the lack of motion between Africa and the spin axis during the break-up of Pangaea has been taken as evidence that African rifting during that time was similarly induced (*Burke and Dewey,* 1973). The Pangaean-rupturing rifts on either side of the Atlantic formed just before that ocean opened provide further examples of the diversity of styles of volcanism. Some rifts appear to be without volcanic material as are the majority of the rifts of western Turkey; others are associated with extensive pre-rift igneous activity and yet others show igneous activity only when the rifts are well-developed. As in East Africa now, compositional diversity is the rule. In some areas carbonatites and alkaline syenites abound (e.g., Los, Bagbe, Songo), whereas in other tholeiitic basalts are the only igneous rocks.

Conclusions

Rift studies are now at a very exciting stage. Tentative classifications of rifts and rift systems based on global tectonic hypotheses such as that in this paper are being made and a major need is for comprehensive study of all aspects of selected rifts. They are exceedingly complex structures and timing of events in rifts is a difficult task (see Fig. 1 for a simplified sketch of rifting-volcanism relative timing types and a few of the involved complexities). Some rifts are particularly well-suited for the study of some properties – for example, igneous petrology in the Keweenawan – while others are better adopted for the study of the properties such as subsidence and sediment fill (e.g., the North Sea rifts). A unified approach recognizing that these phenomena are all related to similarly induced processes is likely to prove most rewarding.

Acknowledgements. This paper constitutes Contribution No. 12 of the Basaltic Volcanism Study Project, which is organized and administered by the Lunar and Planetary Institute/Universities Space Research Association under Contract NSR 09-051-001 with the National Aeronautics and Space Administration.

References

Atwater, T., Implications of plate tectonics for the Cenozoic tectonic evolution of western North America, *Geol. Soc. America Bull., 81*, 3513-3536, 1970.

Baker, B.H., P.A. Mohr, L.A.J. Williams, Geology of the eastern rift system of Africa, *Geol. Soc. America Spec. Pap., 136*, 67 p., 1972.

Burke, K., Aulacogens and continental breakup, *Ann. Rev. Earth Planet. Sci., 5*, 371-396, 1977.

Burke, K. and J.F. Dewey, Plume-generated triple junctions: key indicators in applying plate tectonics to old rocks, *J. Geol., 81*, 406-433, 1973.

Burke, K. and J.F. Dewey, Two plates in Africa during the Cretaceous?, *Nature, 249*, 313-316, 1974.

Burke, K. and A. Whiteman, Uplift, rifting and the break-up of Africa, in: *Implications of Continental Drift to the Earth Sciences, 2*, 735-755, Academic Press, London, 1973.

Burke, K. and J.T. Wilson, Is the African Plate stationary?, *Nature, 239*, 387-390, 1972.

Dewey, J.F. and A.M.C. Şengör, Aegean and surrounding regions: complex multiplate and continuum tectonics in a convergent zone, *Geol. Soc. America Bull.*, in press, 1978.

Erinc, S., Kula ve Adala arasinda genc, volkan reliyefi, *Istanbul Üniv. Cogr. Enst. Dergisi, 9*, 7-31, 1970.

Greiner, G. and J.H. Illies, Central Europe: active or residual tectonic stresses, *Pure and Appl. Geophys., 115*, 11-26, 1977.

Hamilton, W., The Uralides and the motions of the Russian and Siberain Platforms, *Geol. Soc. America Bull., 81*, 2553-2576, 1970.

Logatchev, N.A., Rift systems in East Siberia, *Nytt fra Oslofeltgruppen, 6*, 44, 1977.

Logatchev, N.A. and N.A. Florensov, The Baikal system of rift valleys, *Tectonophysics, 45*, 1-13, 1978.

Lotze, F., *Geologie Mitteleuropas*, E. Schweizerbart'sche Verlagsbuchhandlung, Stuttgart, 491p., 1974.

McKee, E.H., D.C. Noble and M.L. Silberman, Middle Miocene hiatus in volcanic activity in the Great Basin area of the western United States, *Earth Planet. Sci. Let., 8*, 93-96, 1970.

McKenzie, D.P. and N. Weiss, Speculations on the thermal and tectonic history of the Earth, *Geophys. J. R. Astron. Soc., 42*, 131-174, 1975.

Molnar, P. and P. Tapponnier, Cenozoic tectonics of Asia: effects of a continental collision, *Science, 189*, 419-426, 1975.

Noble, D.L., Some observations on the Cenozioc volcano-tectonic evolution of the Great Basin, western United States, *Earth Planet. Sci. Let., 17*, 142-150, 1972.

Oxburgh, E.R. and D.L. Turcotte, Membrane tectonics and the East African Rift, *Earth Planet. Sci. Let., 22*, 133-140, 1974.

Şengör, A.M.C., Collision of irregular continental margins: implications for foreland deformation of Alpine-type orogens, *Geology, 4*, 779-782, 1976.

Şengör, A.M.C., K. Burke and J.F. Dewey, Rifts at high angles to orogenic belts: tests for their origin and the Upper Rhine Graben as an example, *Am. Jour. Sci., 278*, 24-40, 1978.

Sherman, S.I., Faults of the Baikal Rift Zone, *Tectonophysics, 45*, 31-39, 1978.

Trümpy, R., The timing of orogenic events in the Central Alps, in: *Gravity and Tectonics*, John Wiley & Sons, New York, 229-251, 1973.

Turcotte, D.L., Membrane tectonics, *Geophys. J. R. Astron. Soc., 36*, 33-42, 1974.

Turcotte, D.L. and E.R. Oxburgh, Mid-plate tectonics, *Nature, 244*, 337-339, 1973.

Washington, H.S., *The volcanoes of the Kula Basin in Lydia*, Robert Drummond, New York, 65 p., 1894.

Zeschke, G., Der Simav-Graben und seine Gestein, *Tür. Jeol. Kur. Bül.*, 5, 1954.

(Received March 15, 1978;
revised April 7, 1978;
accepted April 18, 1978.)

2

Reprinted from pages 385-399 of *Tectonics and Geophysics of Continental Rifts,* I. B. Ramberg and E.-R. Neumann, eds., D. Reidel, Dordrecht, 1978, 444p.

SOME PROBLEMS OF RIFTING DEVELOPMENT IN THE EARTH'S HISTORY

E.E. Milanovsky

Moscow State University, Moscow, USSR

1. INTRODUCTION

During the last 15-20 years our knowledge of the Earth's rift zones has widened and become more profound. It has become evident that rifting is a global geotectonic process which, at least in the Cenozoic, has equalled the geosynclinal process in importance. The theory of rifting is developing rapidly. Some general regularities of the structure and development of rift zones have been revealed, and a structural-genetic classification has been proposed [1,2]. However, a number of important rifting problems remain and is the subject of considerable debate. Some of these problems, in particular those of rifting development in the Earth's history and their correlations with volcanic history and geosynclinal processes are briefly considered in this communication. It must be emphasized that this review is based upon the study of continental rift zones as the structure and geological history of these are much better known and understood than that of oceanic rift zones. Before we turn to the main subject a few words must be said about the rift zone classification accepted by the author and terminology used by him.

According to the character of the deep structure of the Earth's crust the rift zones of the Earth may be divided into four basic categories:

1) continental (or intracontinental) rift zones with crust of continental type, although somewhat thinned, fragmented, anomalously heated and underlain by a lens of the upper mantle material with lower density;

2) intercontinental where the axial part of the rift has a crust similar to that of oceanic rift zones, the peripheral parts have somewhat thinned and reworked continental crust, and the shoulders have typical continental crust (rift zones of Red Sea, Gulfs of Aden and California);

3) oceanic (or intraoceanic) where both the axial rift valley and the flanks of the rift zone have crust similar to oceanic crust, but are underlain by a "pillow" of anomalously heated mantle material with lower density;

4) pericontinental (or perioceanic) rift zones differing from all the preceding types in having sharply asymmetric structures: such rift zones are bounded by very different areas as they have continental type crust on one side and oceanic type on the other. At present such rift zones exist mainly in a passive buried state (along the periphery of continents bordering the Atlantic and Indian oceans), they were active during the Mesozoic and early Cenozoic.

Intercontinental rift zones are evidently the result of the development of continental rifting to a point where total severance of the continental crust has occurred. Only some oceanic rift zones (e.g. the Mid-Atlantic rift belt) may be assumed to result from continued development of an intercontinental rift zone. Others (e.g. the East-Pacific rift belt) are younger than the oceanic space containing them and were undoubtedly initiated in the oceanic crust.

Pericontinental rift zones occur in the process of horizontal extension at an early stage of oceanic trough formation. However, the wide-spread view that these zones are "halves" of continental rifts separated far from each other by the "opening" of the ocean seems unlikely. It is more probable that such zones developed on opposite flanks of intercontinental (subsequently oceanic) rift belts on the margins of continental blocks framing them with fragmented thinned and partially reworked crust [3].

The following basic criteria must be taken into consideration when classifying modern continental rift zones; a) The peculiarities of tectonic position, basement structure and preceding geological history specific to the area which became the scene of rifting; b) the character of the tectonic structures generated in the process of rifting and the regularity of their formation and c) the role, scope and character of magmatic processes accompanying, or sometimes preceding rifting.

On the basis of (a), continental and intercontinental rift belts and zones may be divided into two groups; 1) rift belts and zones of platforms (or epiplatform) where rifting started after a fairly prolonged (200-500 m.y. or more) stage of platform development and 2) rift belts and zones of young folded structures (or epiorogenic) where rifting immediately succeeded the completion of the geosynclinal development (i.e. the orogenic stage) or was even combined with phenomena peculiar to epigeosynclinal orogenesis. The epiplatform rift belts are characterized by rift zones with large single axial grabens and subalkaline or alkaline products of accompanying volcanism. Sometimes alkaline ultrabasites and carbonatites are present. In contrast, epiorogenic rift belts and zones are characterized by combinations of numerous narrow grabens, horsts and unilateral blocks (e.g. Basin-and-Range Province in Cordillera, USA). Volcanic formations in them are of calc-alkali type.

Continental epiplatform rift zones occur mainly on outcrops of folded platform basement which have been subjected to continuous stable uplift before rifting. These zones are seldom found in areas of platform cover development. They are mostly associated with areas of Late-Proterozoic folding or tectono-magmatic regeneration, being superimposed on, or "adjusted" to, previous structural inhomogeneities of the substratum such as longitudinal faults or strike direction of fold zones. As a rule they evade areas of Archean consolidation with development of granite-gneiss domes and other isometric structures. The latter serve as "the frame" of rift belts or form median massifs within them (as for example the Victoria massif in the southern part of the African-Arabian rift belt). During the period of platform development "weakened" zones of the fundament may be repeatedly activated. They serve either as zones of increased permeability facilitating the rise of magmatic melts and emplacement of intrusions, or as zones of fault and graben development. Two extreme types may be distinguished among epiplatform rift zones. They differ significantly in the character of their structures, the relative role of volcanism and formation history. Several transitional types between the two extremes have been recognised.

a) Rift zones of the arch-volcanic type (e.g. Kenyan and Ethiopian rift zones in Africa) are characterized by intense and long sub-aerial volcanic activity. Volcanism appearance is preceded by steady growth of an oval arch-uplift. Subsequently a rather shallow axial graben develops in the "weakened" axial zone.

b) Rift zones of the crevice type are characterized by grabens of greater depth, that is 3-4 km (the Upper Rhine rift) and even 5-7 km (the Baikal rift). The zones of marginal uplifts are here much narrower, and may be discontinuous or totally absent. Within the rift, however, inner narrow zones of uplift sometimes occur (Ruwenzori in the Tanganyika rift zone). Volcanism in crevice zones is either relatively weak, local, and of episodic nature or totally absent. The formation of crevice zones begins with the formation of long, narrow grabens. Marginal and inner uplifts appear later, creating the illusion of arch collapse. Considerable heat leakage seems to be associated with the grabens of crevice rift zones.

Table I. Main categories and types of rift zones (R.Z.) and rift belts (R.b.)

<table>
<tr><th rowspan="3">categories R.Z. and R.b. according to their deep structure</th><th colspan="5">epiplatform R.Z.</th><th colspan="3">epiorogenic R.Z.</th></tr>
<tr><th colspan="2" rowspan="2">types of R.Z.</th><th rowspan="2">predominant structures and movements</th><th colspan="2">volcanism</th><th rowspan="2">types of R.Z.</th><th rowspan="2">predominant structures and movements</th><th rowspan="2">volcanism</th></tr>
<tr><th>intensity</th><th>composition</th></tr>
<tr><td rowspan="3">continental (intra-continental) R.Z</td><td colspan="2">arch-volcanic R.Z.</td><td>oval arch uplifts complicated by relatively shallow axial, sometimes also transverse grabens. Very weak extension of the crust.</td><td>very intense preceding and accompanying all stages of rifting</td><td>volcanic and subvolcanic rocks of subalkaline and alkaline series (from ultrabasic to acid, predominantly basic) and carbonatites.</td><td rowspan="3">continental epiorogenic R.Z. (in the inner or marginal parts of continents).</td><td rowspan="3">very long linear grabens and horsts, "keyboards" of blocks, systems of one-side tilted blocks. General considerable "diffuse" extension and attenuation of the crus</td><td rowspan="3">usually powerful wide-spread subaerial volcanism mainly on the early stages of rifting. Volcanic rocks of calc-alkali series (acid and later basic).</td></tr>
<tr><td rowspan="2">R.Z. of crevice type</td><td>in the areas of general uplift</td><td>deep grabens often bordered from one or two sides by marginal uplifts and sometimes complicated by inner uplifts. Relatively greater extension of the crust.</td><td>weak, sporadic, episodic volcanism, on the different stages of rifting (little volcanic R.Z.).</td><td rowspan="2">volcanism is absent (non-volcanic R.Z.)</td></tr>
<tr><td>in the areas of general subsidence</td><td>deep grabens without the marginal uplifts</td><td></td></tr>
<tr><td>inter-continental R.Z.</td><td colspan="2">intercontinental R.Z. with the platform "frame" (inter-platform R.Z.).</td><td>very deep and wide "step-like grabens" with the separation of the blocks of continental crust in the axial zone. Very considerable extension.</td><td>on the flanks of R.Z.-relatively weak and sporadic, in the axial zone--more intense(?) volcanism, hydrotherms and high heat flow.</td><td>basic rocks of calc-alkali series in the axial zone, subalkaline and alkaline on its ends and flanks.</td><td>intercontinental R.Z. with the frame of young folded structures (interorogenic R.Z and R.b.).</td><td>the same as in the intercontinental R.Z. with the platform "frame".</td><td>calc-acid volcanism.</td></tr>
<tr><td>pericontinental R.Z. and R.b.</td><td colspan="2" rowspan="2"></td><td colspan="6" rowspan="2">subdivision is not given</td></tr>
<tr><td>oceanic (intraoceanic R.Z. and R.b.)</td></tr>
</table>

2. DEVELOPMENT OF RIFTING PROCESSES WITH GEOLOGICAL TIME

The earliest "roots" of rifting extend into the distant past; structures similar to recent continental rift zones have existed since the Late Proterozoic, while related but not identical structures appeared in the Early-Middle Proterozoic period. However, in the Mesozoic and particularly in the Cenozoic, the rifting activity increased markedly, and rifting phenomena were much more widespread, whereas there was a corresponding decrease in the total area of geosynclinal belts and formation of "secondary" "secondary" oceanic depressions. The relative importance of the various categories and types of rift zones and the tectonic environments in which they originated and developed have also changed considerably in the course of geological history (Fig. 1).

In Late Proterozoic the most ancient structures morphologically similar to the Cenozoic continental epiplatform crevice rift zones, the so-called aulacogenes, began to appear. But unlike the Cenozoic continental rift zones many Riphean aulacogenes were somehow tectonically related to the Baikal geosynclinal belts surrounding these platforms. Most of them were their offsets on, or parallel extensions to their margins. Some of the most "mature" aulacogenes lay between platforms and miogeosynclinal troughs and were subjected to folding of moderate intensity. Aulacogenes are not recorded from the Early and Middle Proterozoic periods, but at this time within the consolidated cores of some future platforms the so-called paleo-aulacogenes started forming. These were narrow superimposed graben-like troughs whose development was accompanied by magmatic activity and crowned by folding. Many Phanerozoic rifting areas notable for their peculiarities of development have been subjected to processes of tectonomagmatic activization and repeated magmatism since the Early of Late Proterozoic.

In the Paleozoic the tectonic conditions under which rift zones were formed became more varied. In the Middle Paleozoic (Devonian) some aulacogenes on the ancient platforms were reactivated.

In Early and especially in Middle Paleozoic (Cambrian and Devonian) the first epiorogenic rift zones overlying the young folded epigeosynclinal areas, came into being (for examples in the Baikal folded belt of East Siberia and in the Caledonian folded belt in the North Atlantic). On the platforms of some southern continents the first rift zones appeared in Early-Middle Paleozoic (Australia) and Late Paleozoic (India, Madagascar, S.E. Africa). Their formation seems to be connected with the early stages of Gondwana supercontinent destruction.

In the Triassic-Jurassic the rifting areas continued to

widen and rifting intensity increased. Simultaneously a long epoch of strong trappean magmatism began on the continents, and from the Middle Jurassic an even greater activity of basaltic volcanism may be traced in the oceans. Continental rift zones "opening" towards the Indian Ocean trough and pericontinental rift zones parallel to its margins began to develop on the Hindustanian, Australian and African platforms.

In the Late Triassic the epiorogenic rift system on the western shore of the Atlantic came into being. It is superimposed on the Hercynian folded area of the Appalachians. In the Jurassic powerful basaltic eruptions took place in the marginal zones of the Central Atlantic Ocean. In the Early Triassic a grandiose trappean volcanism took place on the ancient Siberian platform, in the Taimyr and in the Paleozoic folded area of West Siberia. The epiorogenic West Siberian rift system possibly extending as far north as the Arctic Ocean was formed in the Triassic-Lias. During the Mesozoic a system of grabens confined to zones of long repeated tectono-magmatic activity developed in Zabaikalye. Both these rift systems were subjected to some compressional deformations at the end of their development.

In the Cretaceous-Paleogene the rifting and trappean magmatism spread over many areas adjacent to the South and North-Atlantic and continued to spread along the periphery of the Indian Ocean.

Trappean eruptions occurred on the Hindustanian platform during the Cretaceous period, being especially intensive in the Early Paleogene. Later the Cambay-Bombay rift zone was formed in its western margin. An early stage of the Aden rift zone formation is assigned to the Jurassic-Cretaceous, while the origin, or rather regeneration of the Red Sea rift zone may be dated to the Late Cretaceous-Paleogene period. In the Late Cretaceous-Paleogene powerful eruptions of alkaline basalts occurred on Ethiopian territory, which was subjected to epiplatform rifting in the Neogene. During the Early Cretaceous numerous continental and pericontinental rift zones were formed on the margins of the South American and African platforms adjoining the South Atlantic. Basaltic eruption took place on the bottom of the marginal zones of the Southern and Central Atlantic.

The intercontinental rift system of Baffin Bay was formed (or rather regenerated), and the intracontinental rift system of St. Laurence in East Canada became active in the Cretaceous. In the Paleogene the northernmost part of the Atlantic Ocean became an area of voluminous basalt eruptions. Simultaneously the Rhine-Liberyan rift zones appeared. These crossed both the platforms and the Mediterranean geosynclinal belt and continued to develop in the Neogene-Anthropogene.

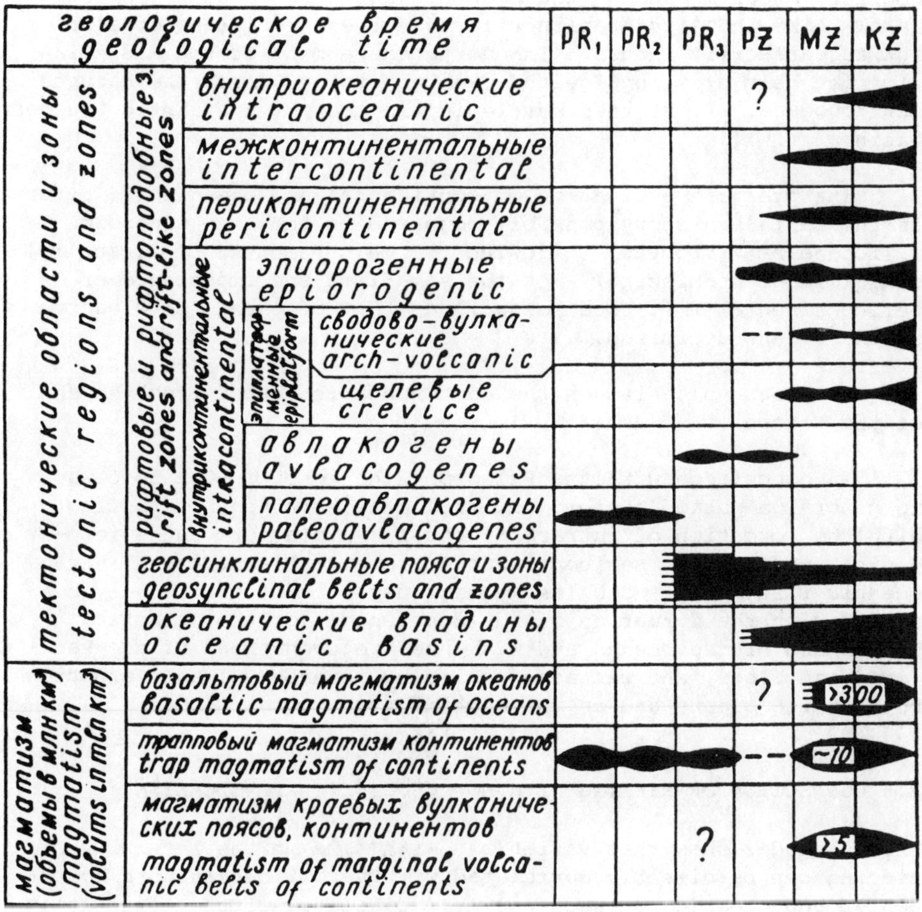

Fig. 1. Distribution of some types of tectonic regions and zones, and magmatic processes in the Earth's history.

In the Late Paleogene-Neogene-Anthropogene the African Arabian rift belt was formed. This included a number of intracontinental epiplatform as well as the two intercontinental rift zones. At the same time the isolated Baikal epiplatform rift belt appeared in Siberia and an epiorogenic rift belt appeared in western North America, superimposed on the Mesozoic folded area of the Cordillera. Like the Afro-Arabian rift belt these represent offsets of intraoceanic rift ranges. The northern segment of the Mid-Atlantic rift belt crossing Iceland was not formed until the Neogene-Anthropogene. It started developing in the North Atlantic trappean basin.

The greatest activation of continental and oceanic (?) rifting in the Earth's history possibly occurred in the Late Cenozoic (Pliocene-Anthropogene) following the main paroxysm of alpine folding (Oligocene-Miocene). At the same time the importance of pericontinental rift zones as well as that of basaltic volcanism in the oceans decreased.

Trappean volcanism on the continents is absent from the end of the Eocene.

In comparison with the Paleozoic the Mesozoic-Cenozoic was in general a period of more powerful manifestations of rifting, with the formation of intracontinental (epiplatform and epiorogenic) and numerous pericontinental, intercontinental and intraoceanic rift zones and belts. Their development was closely connected with the formation of the depressions of the Atlantic, Indian and Arctic oceans and with the "rejuvenation" of the Pacific Ocean floor, and was accompanied by intense volcanic eruptions both in the oceans and on the continents.

3. CONNECTION OF RIFTING WITH NON-GEOSYNCLINAL VOLCANISM

Figs. 1 and 2 show that variations with time in the intensity of rifting can be directly correlated with variations in both continental and oceanic non-geosynclinal volcanism. Such a correlation may be naturally assumed a priori, since both phenomena take place under similar geodynamic conditions - horizontal extension of the Earth's crust. The most important types of non-geosynclinal volcanism are as follows: 1) trappean volcanism of platforms, 2) alkaline-basic volcanism of tectonomagmatic activation areas, 3) essentially intermediate-acid sialic volcanism of marginal volcanic belts and, finally, 4) basaltic volcanism of oceans which in volume exceeds significantly all the other types of volcanism on the Earth.

If we confine ourselves to consideration of the Phanerozoic volcanism we shall see that the first three types appear on the Earth almost exclusively in the Meso-Cenozoic. They are almost unknown in the Paleozoic, being recorded only in the Devonian which was also characterized by a noticeable upsurge of epiplatform and epiorogenic continental rifting. However, the scale of continental volcanism in the Devonian was much lower than that in the Mesozoic or Cenozoic.

The total volume of products of trappean Meso-Cenozoic volcanism of platforms comprises $10\text{-}15 \cdot 10^6$ km^3, in marginal volcanic belts it is not less than $10 \cdot 10^6$ km^3. The volume of products of alkaline-basic volcanism of activated areas, including that in epiplatform rift zones, is several millions of km^3. The total volume of all products of non-geosynclinal Meso-Cenozoic volcanism on continents is therefore less than $25\text{-}30 \cdot 10^6$ km^3.

All the basaltic volcanism in oceans known so far occurred in the Mesozoic (Jurassic, Cretaceous) and Cenozoic. The total volume of the products may be estimated to be minimum $300 \cdot 10^6$ km^3, the average thickness of the second seismic layer in the oceans being about 1.5 km. This means that this volume exceeds that of synchronous non-geosynclinal volcanism of the continents by at least one order of magnitude. The scale of Meso-Cenozoic rifting in the oceans is thus many times greater than that on the continents.

If, during the Phanerozoic, there has been direct correlation between continental rifting intensity and continental non-geosynclinal volcanism, it is natural to assume a similar correlation to have existed in the oceans as well. Therefore we may assume that the intensity of rifting and volcanism within the oceans also increased sharply in the Mesozoic and Cenozoic (as compared to the Paleozoic) when the formation of the Atlantic, the Indian and the Arctic Ocean basins with their rift belts and the "rejuvenation" of the Pacific ocean floor took place. The sharp increase in non-geosynclinal volcanism on the continents and, probably also in the oceans, and the formation of the latter in the Mesozoic and Cenozoic undoubtedly indicate a radical redistribution of energy resources over the Earth's surface. It probably also signifies an essential change in the total energy regime of our planet in the Mesozoic and Cenozoic in comparison with the Paleozoic.

4. CORRELATIONS BETWEEN RIFTING AND GEOSYNCLINAL PROCESS

During the Cenozoic rifting and geosynclinal processes are represented virtually directly opposing mechanism. Rifts and geosynclinal belts are zones of maximum tectonic and magmatic activity. In rift belts deep material rises to the surface and horizontal extension of the Earth's crust takes place. In geosynclinal belts there is submersion of material and horizontal compression,

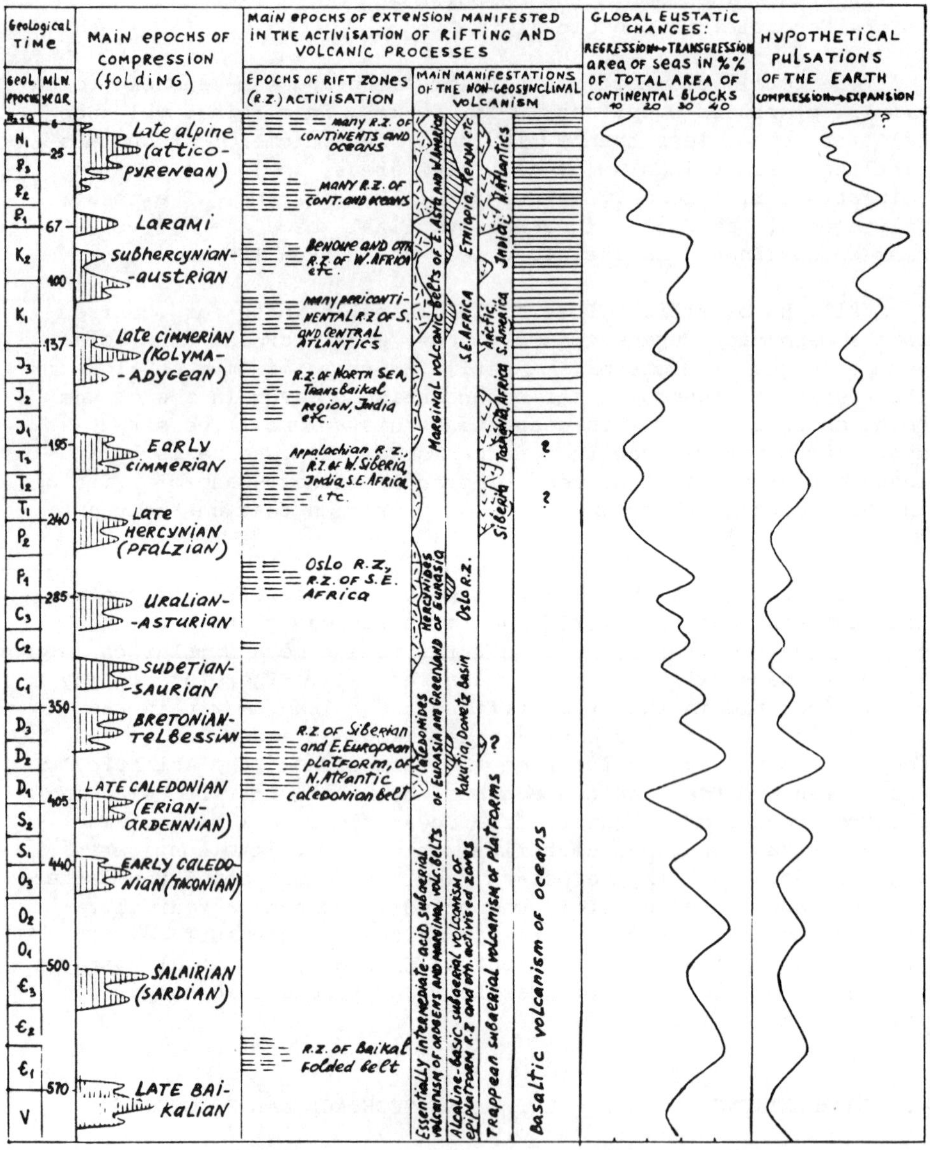

Fig. 2. Correlation of the global epochs of tectonic deformations and volcanism, global eustatic cycles and hypothetical pulsations of the Earth during Phanerozoic. Compiled by E.E. Milanovsky, 1977.

contraction of the Earth's crust prevails. On a global scale these processes approximately compensate each other. In the past, however, such a compensation cannot have taken place since the relative importance of the geosynclinal process has decreased successively in the course of the Earth's history, whereas that of rifting has increased. If we trace the geological history backwards we also see that the radical differences between rifts and geosynclinal zones grow gradually weaker and finally disappear. It must be noted that horizontal extension phenomena are not confined to rift zones alone, but are also observed in geosynclinal zones. The converse is also true, since compression phenomena have been identified in rifts as well as in geosynclinal zones. In the development of all geosynclinal belts, and in eugeosynclines in particular, phases of essential extension with generation of oceanic type crust have taken place. Such phases are recorded especially in the so-called ophiolite zones of eugeosynclines. Later, but long before the end of the geosynclinal "cycle", the latter were subjected to the strongest compression and crushing with formation of tectonic mélange. Essential extension may also occur during the late orogenic phases of development of geosynclinal belts (e.g. in the Mediterranen belt during the Pliocene-Anthropogene). On the other hand, many pre-Cenozoic rift zones underwent compression phases with the formation of linear folds and overthrust structures. Examples of such deformations can be found in the Cretaceous rift of Benoue in Africa, the Mesozoic (Jurassic, Cretaceous) grabens of Transbaikal area, the Triassic grabens of Western Siberia, the Middle Paleozoic Donetsk rift (or aulacogene), subjected to folding late in the Paleozoic, the Devonian graben of Spitsbergen, the Cambrian grabens of the Baikal folded area, etc. There was also a compression phase in the history of the Oslo graben at the very beginning of the Devonian. Compression phases are also thought to have taken place in the history of development of oceanic rift belts, in the Mid-Atlantic, in particular [4]. Upper Proterozoic and, partially, even Paleozoic continental rift zones (the so-called aulacogenes) were structures intermediate in nature between the platform and miogeosynclinal depressions. The largest and most mature among them branched off from geosynclinal belts and intruded into the body of adjacent platforms. These inherited the position and strike of the Proterozoic folded zones in their basement. Still closer (related) to geosynclinal zones are the so-called paleoaulacogenes - narrow linear volcanically active depressions of Early and Middle Proterozoic age - were revealed in the structure of the Siberian and East European platform. Thus, the most ancient - Proterozoic analogs of continental rift zones are structures close to underdeveloped epicraton geosynclinal depressions. This indicates common historic "roots" for geosynclinal and rift zones. The situation and orientation of most Proterozoic and Phanerozoic rift zones has been determined by the structural pattern of their consolidated epigeosynclinal substratum. This also confirms the relationship.

In an earlier time - in the Archean - there were evidently no rift zones differing in principle from geosynclinal zones. Numerous but relatively narrow protogeosynclinal zones in this epoch of the Earth's history were separated not by vast platforms but by relatively small consolidated massifs similar to the median massifs in the geosynclinal belts of the Phanerozoic. Therefore those tectonic tendencies which were more "polarized" later on into geosynclinal and rift areas were also typical of the Archean and Early Proterozoic mobile protogeosynclinal belts.

5. PERIODICITY OF RIFT PROCESS DEVELOPMENT AND ITS CORRELATIONS WITH PERIODICITY OF GEOSYNCLINAL DEVELOPMENT

As the study of rift zones and belts has progressed, it has become more and more evident that their development is a complicated and long process. The duration may be comparable with the development of geosynclinal belts. At the same time it is now clearer that this development is not regular in time. There are relatively short phases of strong rifting activation alternating with periods of "weakened" activity. Such an irregularity of rift zones development in time was demonstrated recently by Pilger and Rösler [5]. They used the Afro-Arabian rift system as an example while Kazmin [6] described such a periodicity of rifting as a more general phenomenon. This irregularity in development of rift zones in time may be observed clearly in the Rhine rift system, the rift zones of Iceland, etc. It must be emphasized that the variations are not only in intensity of horizontal extension (which at certain moments may be even replaced by some compression), but also in orientation of the dynamic field. The latter type of variation was well described by Illies and Greiner [7] for the Rhine rift system.

The author's research shows that the phases of extension are not restricted to particular rift zones but common for whole rift belts and evidently for the whole rift system. Thus periodicity in rift belt development is a phenomenon similar (though opposite in its main kinetic tendency) to that of geosynclinal belt development. The latter is indicated by the existence of global phases and epochs of folding as stated by Stille [8]. In this case the so-called folding (compression) phases in geosynclinal areas, and extension (taphrogenesis) phases in rift zones represent different examples of the same general global irregularity and periodicity of geodynamic activity in the course of the Earth's development. If this is true a question arises - are the phases of folding (horizontal compression) in geosynclinal belts simultaneous with those of taphrogenesis (horizontal extension) in rift belts or, do they, on the contrary, alternate with each other. Many researchers consider these phenomena to be simultaneous (e.g. [9]). This view agrees well with the "plate tectonics" concept assuming mutual relation and kinematic "equilibrium" between spreading in some zones of the Earth and collision of lithospheric plates (subduction, obduction, etc.) in others. The assumption that the Earth's radius is constant is a natural consequence of the "plate tectonics" concept [10]. There are,however, a number of facts which do not agree with the idea of temporal coincidence between compression and extension paroxysms in the different tectonically active zones of the Earth. It has for example already been stated that, with the exception of recent ones, rift zones of different age demonstrate that periods of extension (prevailing on the whole) alternate with periods of compression. Periods of slight compression or no extension in rift zones correlate in time with folding phases in geosynclinal belts. It has also been established that quite significant crustal extensions have occurred in geosynclinal areas. In particular, these are phases of regeneration of the oceanic type crust in ophiolite zones like that in the Alpine Mediterranean belt in the middle of the Mesozoic. Another example is the Pliocene-Quaternary epoch of extension in the Southern Mediterranean part of the Alpine geosynclinal belt which followed the epoch of intensive fold-overthrust deformations in this belt in the Miocene. At the same time the Pliocene-Quaternary epoch was characterized by a sharp intensification of rifting process in numerous rift zones, continental (Baikal and others), intercontinental (the Red Sea rift) and oceanic. Thus, according to Steiner [11] the extension rate in Median rift zones of the Atlantic, Indian and Pacific oceans in the Pliocene increased considerably as compared to the Miocene. Epiorogenic rift zones occur within the young folded epigeosynclinal areas as a result of extension. They were subjected to this extension after considerable compression which created their folded structure. Extension phases in rift zones coincide in time with world wide phases of volcanism activity. The latter took place in the Devonian, Mesozoic and Cenozoic. The whole Meso-Cenozoic epoch of rifting intensification is an epoch of marked reactivation of non-geosynclinal continental and evidently oceanic volcanism.

An attempt to correlate directly in time the main global epochs of folding (compression) increase in geosynclinal belts and extension in rift zones is made in Table 2 which illustrates the lack of coincidence and marked tendency to alteration of these phenomena. On the whole extension weakening in rift zones (which may even invert to slight compression) corresponds to phases of intensification of compression (folding) in geosynclinal belts. Conversely, strong extension activity in rift zones is accompanied by compression decrease in geosynclinal belts and even by the appearance of extension structures in them, an example being many orogenic belts in the Pliocene-Anthropogene. Global epochs of extension intensification are synchronized with the largest positive phases of sea level eustatic fluctuations ("world transgressions"), while the epochs of marked compression correlate with the negative phases of the latter ("world regressions").

The concept briefly stated above [12] naturally requires the assumption of the alternation of general compression and extension periods and epochs in the course of the Earth's development. This brings us back to the idea of the existence of global pulsations of the Earth. Already 40 years ago that was expressed by an American geologist,Bucher [13], and later by the Soviet geologists Obrutchev [13] and Ussov [15]. Periods of general compression on the Earth are global epochs of folding intensification and rifting when volcanism decrease. General extension periods are periods of folding attenuation, rifting activation and massive volcanism. The rate of pulsation of the Earth during the Phanerozoic is approximately 40-50 m.y. Together with these rhytmic oscillations it is possible to detect a general trend of expansion of the Earth in the Mesozoic and Cenozoic compared with the Paleozoic. It is indicated by the sharp increase in rifting and volcanism that involved a large part of the planet's surface at that time, by the formation of modern oceans and by the successively diminishing role played by geosynclinal belts in the general structure of the Earth.

The idea of some expansion of the Earth in the Phanerozoic put forward originally by Hilgenberg and Holm in the thirties looks rather likely from a geological point of view. It is therefore not accidental that it has acquired ever growing support of present day geologists (B. Heezen, S. Carey, B. Cirič, J. Steiner, N.P. Kropotkin, M.V. Muratov, Yu. V. Chudinov and others). According to the data on volcanism and rifting development the main epoch of Earth expansion covers the Mesozoic and Paleogene. Apparently it was suspended or even stopped in the Oligocene and, especially, in the Miocene but renewed again in the Pliocene-Anthropogene.

REFERENCES

1. E.E. Milanovsky, Tectonophysics, 15, 65, 1972.
2. E.E. Milanovsky, Rift zones of continents, Nedra, Moscow, 1975 (in Russian).
3. W.L. Van der Linden, Mar. geophys. Res., 3, 209, 1977.
4. A.V. Peive, The tectonics of the Middle Atlantic ridge, Geotectonica, 5, 3, Moscow, 1975 (in Russian).
5. A. Pilger and A. Rösler, Temporal relationships in the tectonic evolution of the Afar depression (Ethiopia) and the adjacent Afro-Arabian rift system, in: Afar between continental and oceanic rifting, ed. by A. Pilger and A. Rösler, E. Schweizerbart'sche, Stuttgart, 1976.
6. V.Q. Kazmin, Epochs of riftogenesis and some problems of origin of rift structures, Geol. Geophiz., 9, 3, Novosibirsk, 1975 (in Russian).

7. H. Illies and G. Greiner, Oberrhein. Geol. Abh., 25, 1, 1976.
8. H. Stille, Abh. Preuss. Geol. Landesanstalt, N.F.H. 95, 110, 1925.
9. W. Schwan, Clausthales geol. Abh., 17, 1, 1974.
10. X. Le Pichon, I. Francheteau and I. Bonnin, Plate tectonics, Elsevier Sci. Publ. Comp., Amsterdam, 1973.
11. I. Steiner, Geology, 5, 312, 1977.
12. E.E. Milanovsky, Some regularities in tectonic development and volcanism of the Earth during phanerozoic (The problems of pulsations and expansion of the Earth), Geotectonica, 5, Moscow, 1978 (in Russian).
13. W.H. Bucher, The deformation of the Earth crust, Princeton Univ. Press, 1938.
14. V.A. Obructhev, The pulsation theory of geotectonics, Izvestia Akad. Nauk USSR ser. geol., 1, 12, 1940, (in Russian).
15. M.A. Ussov, Structural geology, Gosgeolizdat. Moscow, 1940, (in Russian).

3

Reprinted from *Tectonophysics* **73:**249–266 (1981)

MECHANISM OF GRABEN FORMATION

J. HENNING ILLIES

Geologisches Institut, Universität Karlsruhe, Karlsruhe (F.R. Germany)

The tectonic and morphologic features of continental grabens, in general, are outlined here against other structural phenomena by considering a number of specific attributes. All grabens are framed by two convergent dip-slip faults, which separate downthrown wedge-blocks. The framing crustal units are often upwarped towards the graben. The frames as a whole form a dome-like uparching, whilst the graben generally splits the dome along its crestal line. In many cases, the wedge-block is disintegrated by numerous listric dip-slip faults along which tilt-blocks rotated, commonly antithetically. Subsided wedge-blocks and tilted blocks fill the lateral space that has been created by horizontal extension normal to the rift axis. The parallelity of the master faults in cases where a zigzag configuration of the rims has been observed marks the wood-cut-like traces on geological maps. The crust, attenuated by rifting, is often penetrated by volcanic extrusions of a specific chemical composition. And the open fractures associated with rifts enable the hydrothermal convection and positive heat flow anomalies in these regions. The composition of the sediment-fill in the fault-troughs is mostly controlled by specific environmental conditions and short-distance transport from the shoulders towards the basin. Lacustrine or lagoonal facies are prevailingly observed while the fauna in the isolated basins, recent and fossilized, is of an endemic character.

In spite of these and other common trends considered to be typical for graben structures, there are other pecularities that are totally different in the individual segments. The affected tectonic stockworks vary considerably. Larger grabens, like the Baikal and the Rhinegraben rifts, have their roots in the upper mantle or asthenosphere. On the other hand, in the domain of salt tectonics it may even be the basement that remained unaffected by the observed near-surface rifting.

Kinematically, rifting may be controlled by a lateral extension normal to the graben axis, as it is the case in the East African rift system. Or it may be an extensional shear which governed rifting and rift valley propagation; the Dead Sea rift may serve as an example for this phenomenon. Of a very different extent is the volcanic activity associated with rifts. Some features like the Afar depression in Ethiopia were nearly flooded by volcanic extrusions.

In others, like the Rhinegraben, the quantity of volcanic products remained relatively scarce.

One may observe other significant differences when taking into consideration the regional tectonic setting of graben structures. Features like the Rhinegraben or the Pantelleria rift are found nearby and normal to collision fronts of the Alpine system. Contrary to such foreland structures, the East African grabens indicate an ongoing process of further break-up of Gondwanaland. Such rifts are coming nearer to real plate boundaries than the typical intraplate features of the first category. The rift-like structures on both ends of the Basin-and-Range Province along Owens river and Rio Grande may be referred to tensional forces as a consequence of plate interaction between the Pacific plate and stable North America.

Whatever the local conditions of graben formation may be, splittting of the continental crust has effected a series of subsequent processes that had evolved commonly in a congruent way under different prerequisites. Whether that may be considered as coincidence or not, a taphrogenic cycle came into existence whose stages and appearances may allow a comparative study of the graben phenomenon as a whole. By means of a series of diagrams,

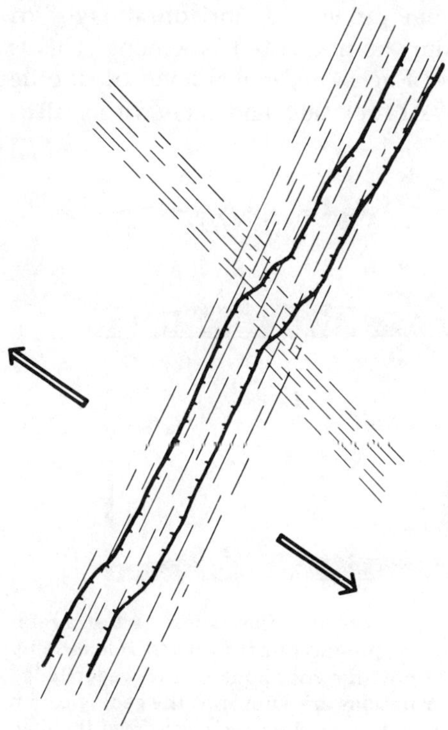

Fig. 1. Grabens, in the majority of cases, are basement controlled. If there are tensional forces, the splitting of the crust will make use of pre-existent weakness zones as far as their trends fit about the direction of maximum tension.

sometimes pertaining to already mentioned examples, it will be attempted to explain how the mechanism of graben formation has taken place and what implications are involved.

First in the pre-rift domain some congruences may be observed. It can be shown that the majority of graben structures follow subsequently preexisting weakness zones in the basement (Fig. 1). Many Cenozoic grabens in East Africa, the Red Sea rift, Lake Baikal rift or Rhinegraben are oriented parallel to mylonite zones, shear elements or dike swarms of Precambrian or Hercynian age. Rifting has followed given weakness zones along which the tensile strength of the lithosphere had diminished relative to the adjoining plate unit.

When the first rift event had started, a feature was formed with a specific individual width (e.g., for the Rhinegraben it is 36 km on average). The specific width has evolved from near-parallelity of the framing master faults. It may range from a few decimeters at some micro-grabens up to more than 100 km at the aulacogens on the East European platform. Grabens form a wedge-block, its convergent master faults display dips between 60° and 65° on an average. Following this concept, the average width of a graben appears to be a function of the depth of a triangular wedge-block (Fig. 2a). Since rifting is considered to be an extensional process, a horizontal layer of decoupling separated the rifted crust from the non-rifted basement. This is required for the wedge-block subsidence. Such a reference layer of ductile material may be realized by intercalations of rock salts and clays with sealed-

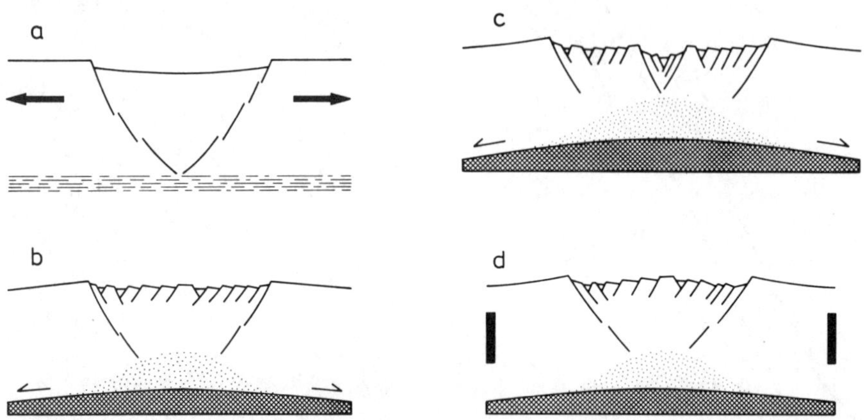

Fig. 2. a. Ideal graben structures are defined as wedge-blocks, their apex start at a reference layer of mechanical decoupling. b. The ongoing process of rifting leads to mantle uplift, rise of a thermal dome and partial resorption of the root zone of the wedge-block. External shoulder uplift and internal tilt-block rotations are known as the geological implications. c. A rising asthenolith will support gravity slide, sideward away from the mantle bulge. Rift-in-rift features are involved in further crustal spreading. d. In case the tectonic boundary conditions impede gravity slide, consequent regional uplift of the graben floor will be observed.

in porewaters or, in larger grabens, by a magmatic behavior of the concerned crust or mantle layers. For the Rhinegraben, a depth of nearly 31 km of the primary wedge-block apex may be calculated, i.e. an approximate depth of the Moho presumed during a pre-rift stage in that area.

Wedge-block subsidence and disintegration of the wedge into a mosaic of rotating tilt-blocks indicates sideward crustal spreading. Horizontal extension involves vertical attenuation of the crust. For the underlying upper mantle, necking of the crust will cause a subsequent unloading effect. Unloading will trigger phase transformations and reduction in density of the affected mineral material. A body of low-density mantle material will grow up to create a widespanned, uninterrupted subcrustal dome underneath the physiographic rift phenomenon (Fig. 2b).

The ascending mantle bulge makes it possible for the blocks of the upwarped graben shoulders to slide gravitationally sidewards away from the crest of the mantle bulge. During this stage, gravitational stresses superimpose and, in case of congruence, increase the regional stress conditions. This will trigger longitudinally further rift valley propagation and transversally crustal spreading. As an additional effect, external wedge-block subsidence and internal tilt-block rotations will follow. Volcanic activity, fed from the upsurging low-density mantle material, will penetrate open fractures associated with rift propagation. The graben shoulders will be passively upwarped in response to an isostatic readjustment to the expanding asthenolith of hot low-density mantle material. Consequently, shoulder upwarping will often follow in a bilateral-symmetric way on the opposite flanks of the rift valley. Self-acceleration and self-propagation by the mutual triggering of gravitational stresses and tectonothermal processes typify the present-day stage over wide parts of the Kenya rift valley.

It might be possible that the regional tectonic setting does not permit further sideward yielding of the block units framing the rift valley on both sides (Fig. 2d). In that case, the regional updoming will additionally affect the graben floor; its fill sediments will be subjected to fluvial erosion. In case a discrete rift valley has not yet been formed, plateau uplift or upthrown horst blocks will substitute for fault trough subsidence. As a relevant example, the Mount Ruwenzori as a basement block up to 5119 m elevated interrupts a chain of down-thrown grabens in the western branch of the East African rift system. Such an uplift of intra-rift block units is an autonomous process without any need for balancing it by a corresponding fault trough subsidence in the surrounding region. No reciprocating action has been observed between the mantle and thermal controlled processes of crustal updoming and the depth of wedge-block subsidence which is primarily dependent on the rate of crustal spreading.

On the contrary, if the stress conditions favor further gravitational slide, crustal spreading by tilt block rotations and fissure eruptions will be supported. A rift-in-rift feature will appear over the rising mantle body (Fig. 2c). During this stage, extreme tensile stress conditions on top of the mantle

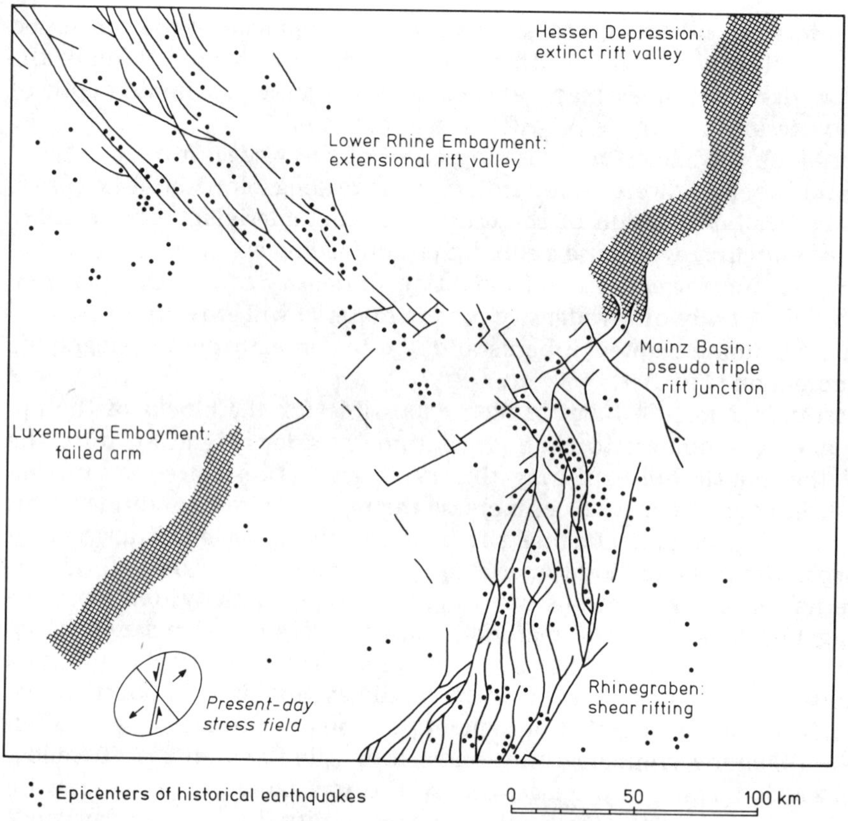

Fig. 3. The Cenozoic rift pattern of Central Europe appears complex. It is a consequence of the rotation of the peri-Alpine regional stress conditions during geologic time. Active rifting is observed in the Rhinegraben controlled by a sinistral shear motion. In the Lower Rhine Embayment, the same stress pattern provokes extensional rifting.

bulge will favor the development of open fissures and related dyke injections in the inner portion of the rift valley. The Lake Hannington depression in the Kenya rift may serve as an example for this phenomenon. This stage of rifting precedes the semi-oceanic conditions as observed in the Afar Depression of Ethiopia.

Last but not least, rifting is a stress-field controlled process. Tensional forces, with a relative maximum component of horizontal stress parallel to the rift axis, are required to start rift faulting and wedge-block subsidence. The Lower Rhine Embayment (Fig. 3) exhibits present-day dip-slip faulting as controlled by an active tensile stress regime; its average direction of σ_1 fits the trend of active fault scarps. On the other hand, Quaternary tectonics and seismotectonic activity of the Rhinegraben are governed by a sinistral shear strain. Here the average direction of σ_1 trends obliquely to the axis of the rift valley, and the corresponding strain release is that of a left-lateral strike-slip motion. In this case, primarily an extensionally formed graben has been remodelled during the Pliocene period into a sinistral shear zone as a consequence to the rotation of the regional stress conditions in course of geologic time. First generation normal faults were overprinted by horizontal slickensides of the second generation strike-slip faulting.

The Eocene to Lower Miocene extensional faulting of the Rhinegraben has caused a slight zigzag configuration of the rift valley (Fig. 4). Therefore, the functional alteration from extension to shear has developed compression shear for the central segment of the graben. Here, a primary rift valley has converted in a ramp valley, and erosional landforms as well as high rates of present-day uplift are observed. For the northernmost segment of the rift valley the same shear vector, but then under a deviatory trend of the given local rift axis, has caused extension shear with high rates of Pliocene to Recent subsidence. The kinematics of shear motion was mainly guided by the given fault pattern which has been inherited from the 1st generation extensional rifting. No discrete 1st order shear planes were formed. Instead of them, 2nd order or Riedel shears have been observed (Fig. 5). Pleistocene to Recent fault activity along the Riedel fractures seen on the graben floor is restricted to distinct segments which form a scattered array in harmony with the characteristic pattern of feather jointing (Fig. 6). Subordinated to major

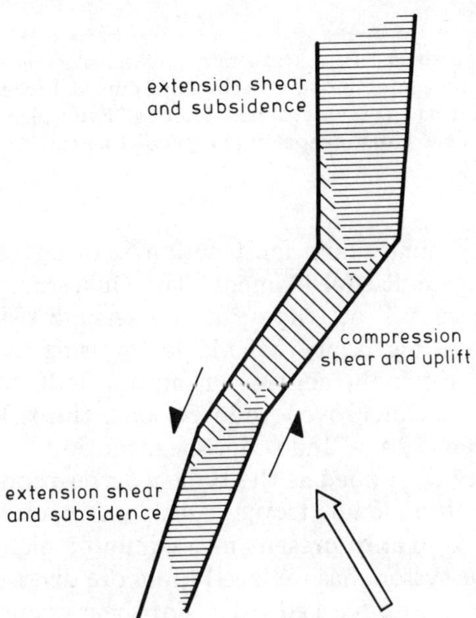

Fig. 4. Cartoon to illustrate the present-day horizontal and vertical motions along Rhinegraben rift. The given crooked course of the primordially extensional rift, under sinistral shear conditions, causes compression and uplift in its central segment, but extension and subsidence in the outer parts.

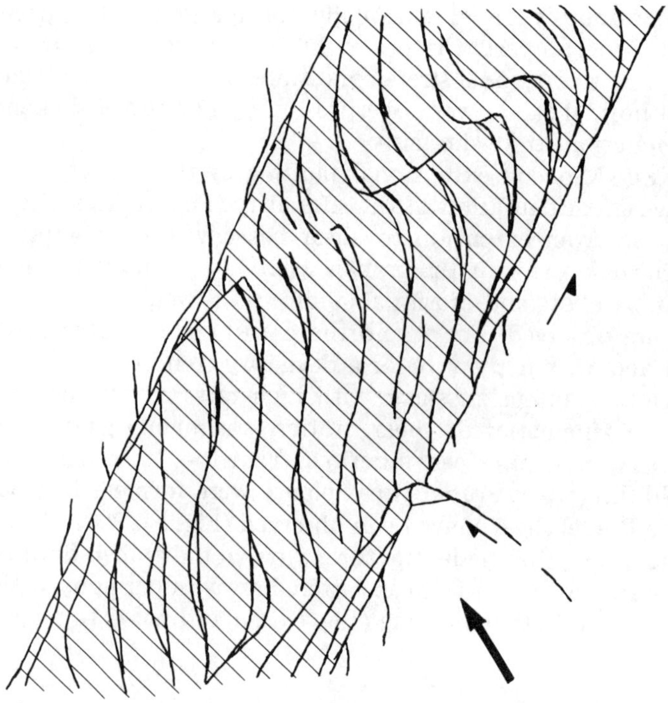

Fig. 5. Idealized pattern of Rhinegraben-type shear rifting. Low-rate sinistral shear motion along a primary extensional graben is dissipated along en echelon arranged Riedel fractures. The shear strain is released by near-surface extensional faulting. Fault-plane solutions of earthquakes reveal 2nd order shear along deeper faults parallel to surficial faulting.

en echelon faults local E—W trending quaternary fault action is observed which may be interpreted as a conjugate Riedel element. The Quaternary active faults often exhibit historical seismic activity. But the seismic risk associated with recent Rhinegraben faulting is not calculable by using the 310 km length of the graben structure. Since the continuous master faults of 1st generation extensional rifting are extinct over wide regions, the risk depends mainly upon the complex kinematics of 2nd order shear motion.

En echelon faults of Riedel shears are developed as the typical pattern controlled by a low rate of axial shear motion. Fault trough subsidence caused by 1st order strike-slip displacement, is mainly present in medium or high-rate shear zones, where a discrete shear system has replaced the more diverse fault pattern of Riedel shear. In this case, dog-legged offsets of shear planes initiated a graben-like subsidence during shear progression (Fig. 7).

Even if the regional stress conditions are adequate for rifting, the strain response to the stress regime may not initiate the formation of a rift. The concerned rock unit in a strain release should be of a lithology favorable to

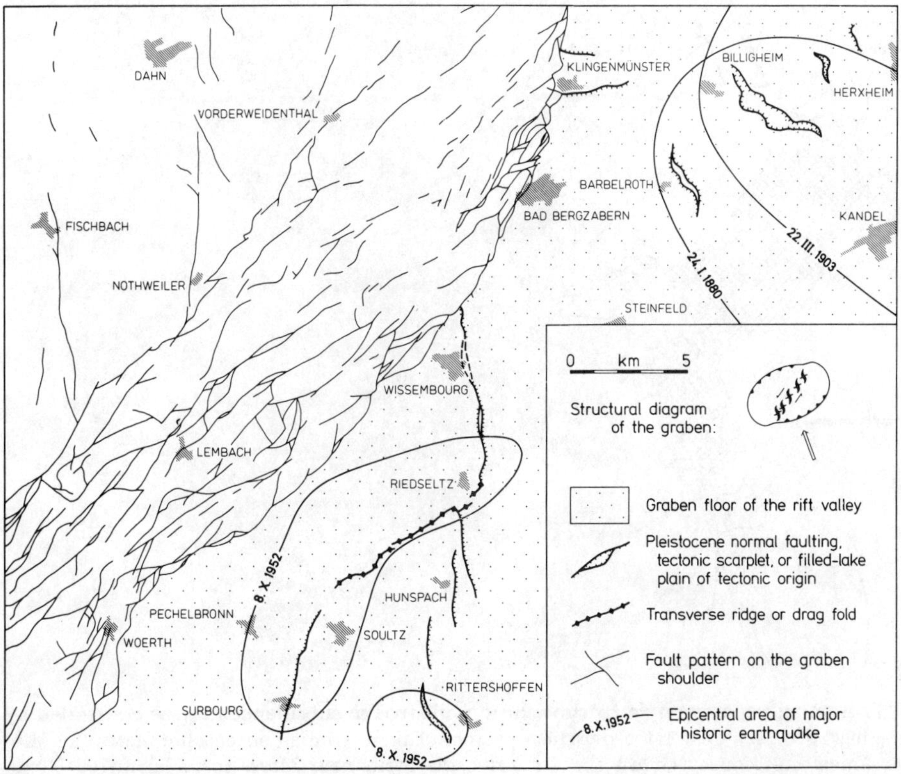

Fig. 6. Surface faulting at the knee of the western margin of the central segment of the Rhinegraben (west of the city of Karlsruhe) is distinctive of the present-day sinistral shear. Features of Pleistocene tectonics on the graben floor, often seismically active, comply with the mechanical principle of Riedel shear and conjugate Riedel shear.

tectonically competent behavior. Otherwise, in cases rocks being incompetent, a ductile response will impede brittle reactions like rifting (Fig. 8). An example for rift development as dependent on the lithology is given by the Rhenish Schiefergebirge. Kinematically, this unit acts as a hinge between shear rifting along the Rhinegraben and extensional rifting at the Lower Rhine Embayment (Fig. 14). Due to an incompetent "slaty" behavior of the upper crust of this Hercynian unit, physiographic rift features are absent over wide parts of the Schiefergebirge. Theoretically, in its regional tectonic setting, in its stress pattern, and in its seismotectonics, a rift segment should have been traversing this unit to complete the rift belt between Rhinegraben and Lower Rhine Embayment. In this segment, instead of physiographic rifting a Quaternary plateau uplift is observed which had forced Rhine and Mosel rivers to form deep antecedent river valleys. Plateau uplift, in this case, is thought to be a consequence of ductile crustal extension and a subsequent

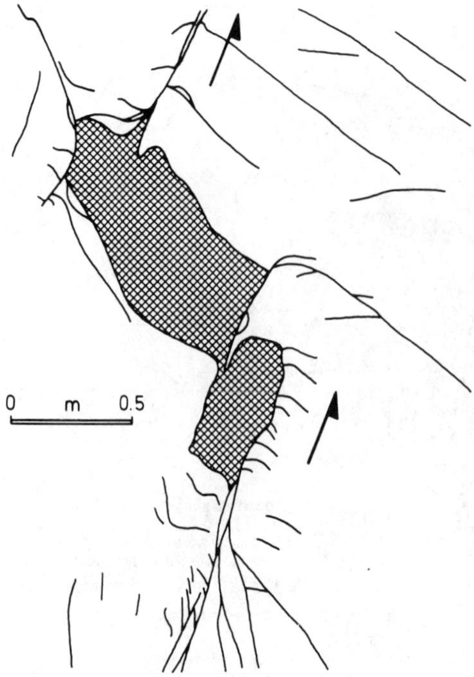

Fig. 7. A small-scale example to show how fault-trough subsidence may be controlled by a dog-legged offset of a 1st order shear system. Taken from a neotectonic feature in Miocene limestones, observed on the wave-cut platform near Marsalforn, Island of Gozo, Malta.

formation and rise of a low-density mantle (Fig. 9).

The ductile strain response of the Rhenish Schiefergebirge additionally caused a specific modification of extensional rifting in the northward adjacent graben unit of the Lower Rhine Embayment. This segment exhibits brittle rotational tilt-blocks with a fan-shaped arrangement and increasing spreading rates towards the NW (Figs. 3 and 14). The absence of a rotational shear pattern to compensate such a divergent spreading mechanism can be explained by the ductile response of the crust south of the hinge of the fan-shaped feature.

These and other interdependencies between the crust affected by rifting and the development, trend and width of the individual rift segments may indicate how relatively shallow-seated the boundary conditions were located that influenced the initial stages of graben formation. This may not imply that the early stages of rifting were only controlled by factors hidden out in the crust. Along all major graben elements, rift-type volcanism started long time before the appearance of physiographic features related to grabens. As an example the volcanic activity was initiated about 100 my ago in the Rhinegraben, about 50 m.y. earlier than the oldest fill sediments had been

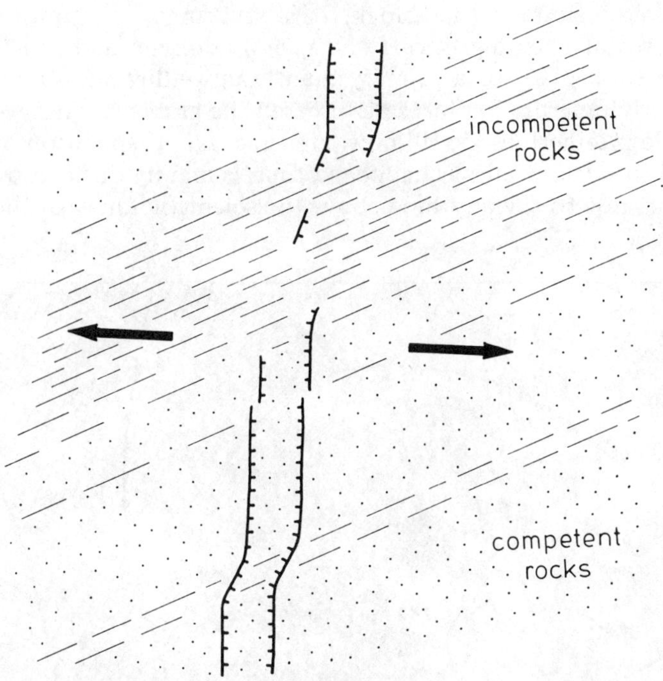

Fig. 8. Rift phenomenon demands a specific ruptural reaction of the concerned rock units. If a ductile strain release may evolve, i.e. an incompetent rock behavior will be given, graben structures remain absent, in spite of the action of tensile stresses.

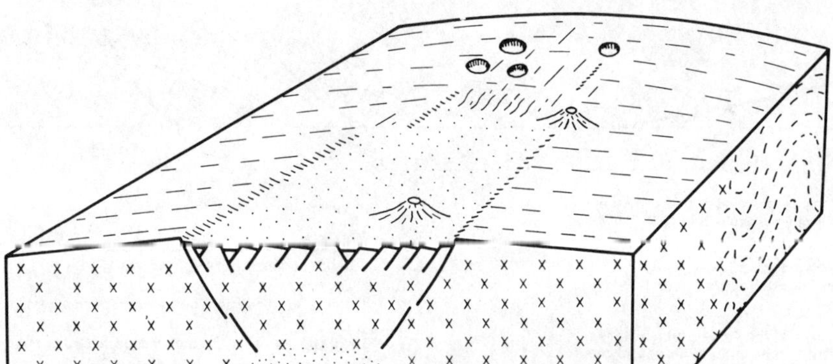

Fig. 9. The cross-sections of the diagram depict crustal rocks of competent and incompetent behavior. In the competent unit a discrete rift valley has evolved framed by upwarped shoulders. Under incompetent rock conditions, the graben is replaced by an uparched plateau. Both shoulder upwarping and plateau uplift are thought to be a consequence of crustal attenuation and a subsequent rise of low-density mantle material.

deposited on top of a subsiding wedge-block. It is still an open question whether the ductile crustal thinning during the pre-rift domain and subsequent mantle reaction took place or a primary mantle upwelling and subsequent break of the overlying crust to pieces has created the grabens. What we observe, is crust-mantle interaction. Up till now, it seems that a separation of causes and effects and to settle what has happened first, is mainly decided by the personal viewpoint, not to say anything above the scientific fancy of the writer in charge.

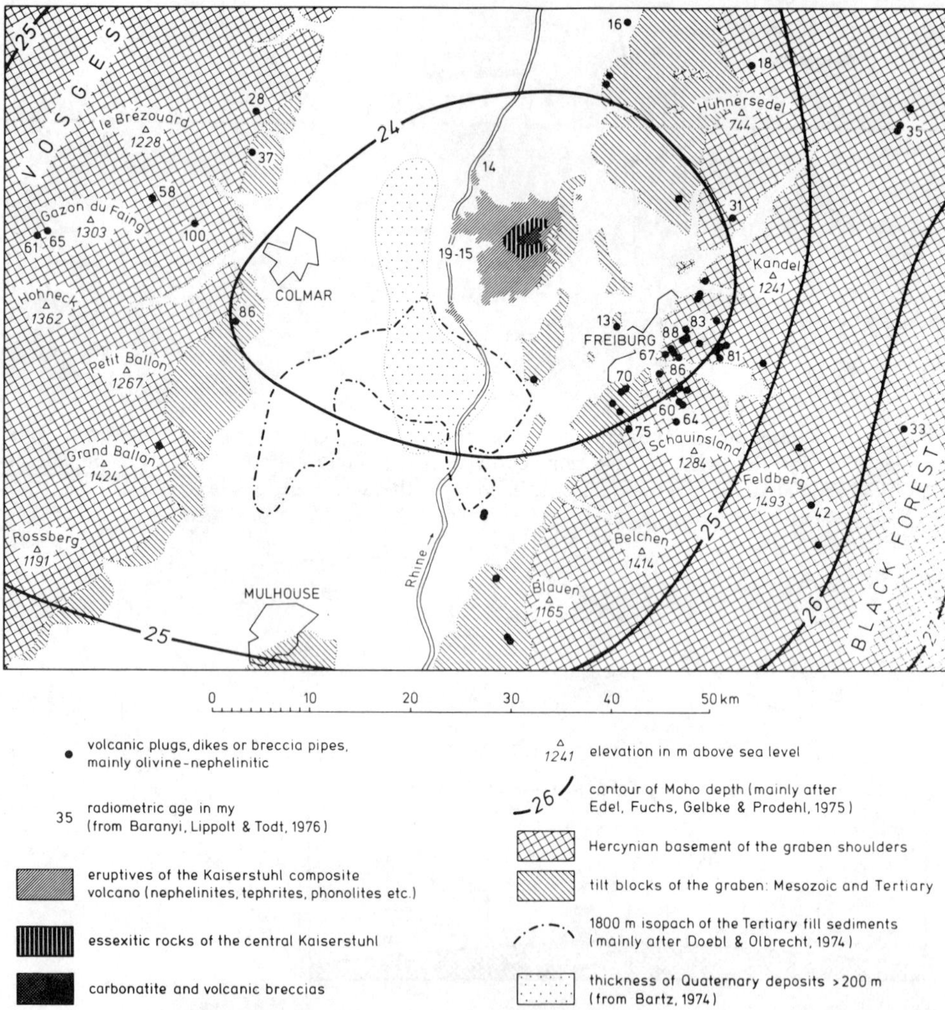

Fig.10. The Kaiserstuhl mantle asthenolith or hot spot acted as the primary spreading center of Rhinegraben rifting. Volcanic activity of olivine-nephelinitic composition started 100 m.y. ago, nearly 50 m.y. earlier than surficial rifting set in during the mid-Eocene. Shoulder upwarping of the Black Forest and Vosges coincides with the configuration of the subcrustal mantle bulge.

Nevertheless, the hot spot under the southern part of the Rhinegraben which is centered by the Miocene composite volcano of the Kaiserstuhl, had feeded a nearly uninterrupted volcanic activity between 100 and 13 m.y. (Fig. 10). Since the Upper Miocene, no volcanic eruptions came off. The observed present-day subsurface temperatures are nearly normal in this area (Fig. 13) and are not related to the distribution of the Moho depth. Active tectonics with a dense pattern of seismic epicenters and high rates of Pleistocene to Recent time tectonics swings around westward the Kaiserstuhl which acts now as a hard inclusion within the mobile shear belt of the graben. The primary hot spot had turned over to a cold spot with a tectonically competent behavior of the remaining asthenolith. Mantle rise, shoulder upwarping and thermal conduction were acting as constituents of the first stage of rifting. Axial shear, local shoulder downwarping and hydrothermal convection result from the stress/strain behavior of this rift valley during Recent time.

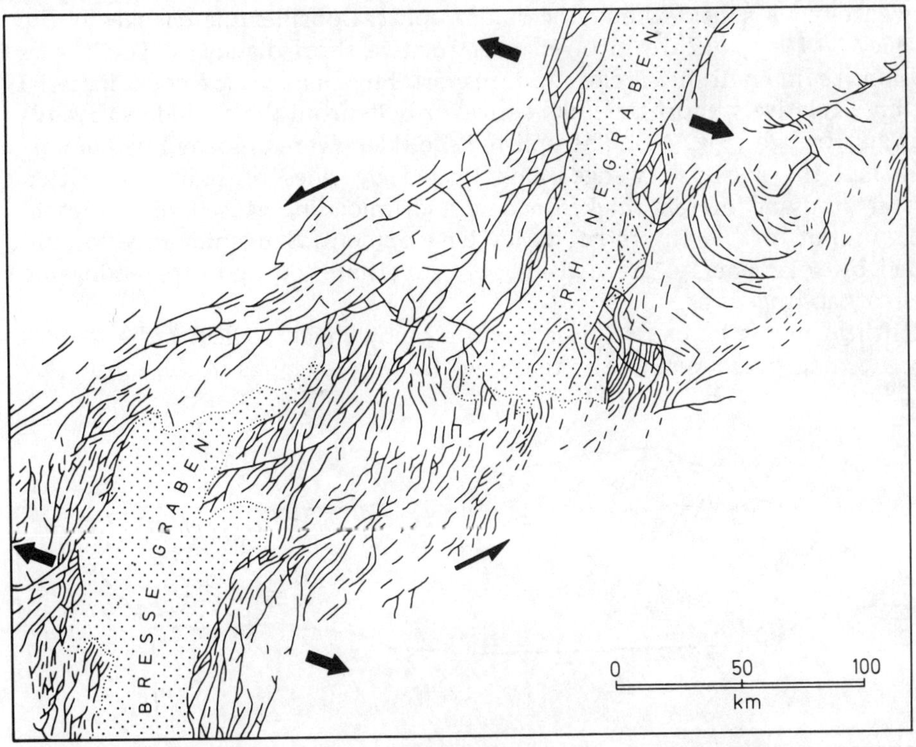

Fig. 11. Discrete transform faults are only seldom observed in the continental domain. Instead of them, lower rates of crustal spreading and the pre-existent crustal anisotropy produce a pattern of en echelon faults. They are acting to release the shear strain, resulting from extension along the two grabens at a rift-rift offset. The fracture pattern combining Rhinegraben and Bresse graben illustrates this mechanism.

In the oceanic domain, transform faults are the most conspicuous characteristics of the mid-oceanic ridge-rift systems. Their functional operation is to release the shear strain arising from the antagonistic shift of the rift frames due to ocean floor spreading along discontiguous rift segments. Lateral offsets along rift belts and the constraint to compensate the shear strain belonging to them exist in the continental sphere as well. But the spreading rates starting from the rift are mostly several times lower than observed on the ocean floor. And the crustal anisotropy molded by the older orogenic history of the basement is more distinctive as in the oceanic realm. It is the oceanic lithosphere which approximates the ideal plates as defined by the plate tectonics hypothesis. Consequently, in most cases no discrete transform faults have been served to intervene rift-rift offsets of a continental provenance. Instead of them, the shear strain will be commonly released by en echelon Riedel fractures (Fig. 11); their pattern may be mostly superimposed by the reactivation of the given framework of pre-existent fault zones in the basement.

The physiographic effect of continental rifting is mainly vertical because of wedge-block subsidence and shoulder uplift. Considerable changes in the base-level of erosion are involved over relative short distances. This causes erosion—sedimentation cycles to transport huge masses of rock material within a relatively short time like conveyor-belts from the shoulders towards the fault trough (Fig. 12). The uplifted shoulders were unloaded by denudation, and the subsided wedge-block loaded by piles of sediments, often several thousand meters thick. Loading and unloading caused an additional energy input for the future rift valley development. Exogenic processes, impelled by solar energy, are considerably superimposed upon the endogenic cycle of taphrogenesis.

Rifting, may it be caused by extension or by shear, indicates the temporary opening of deep-seated extensional fracture systems in the crust. The

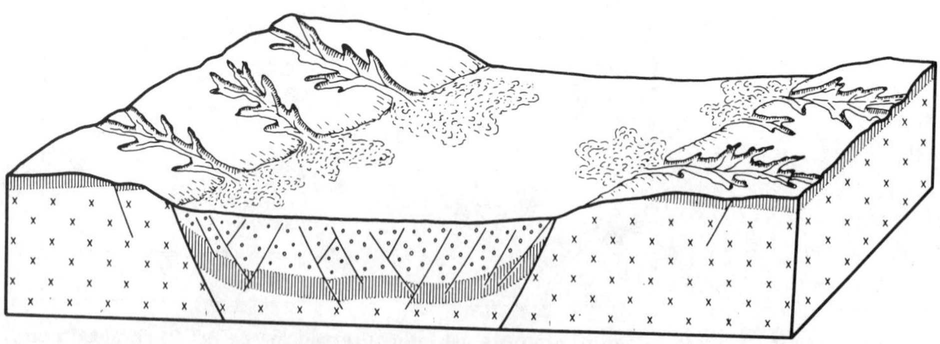

Fig. 12. Shoulder uplift and graben subsidence achieve erosion—sedimentation cycles over only short distances. The conveyor-belt mechanism effects unloading of the shoulders and loading of the graben floor, both acting as an additional energy input for the evolution of crustal rifting.

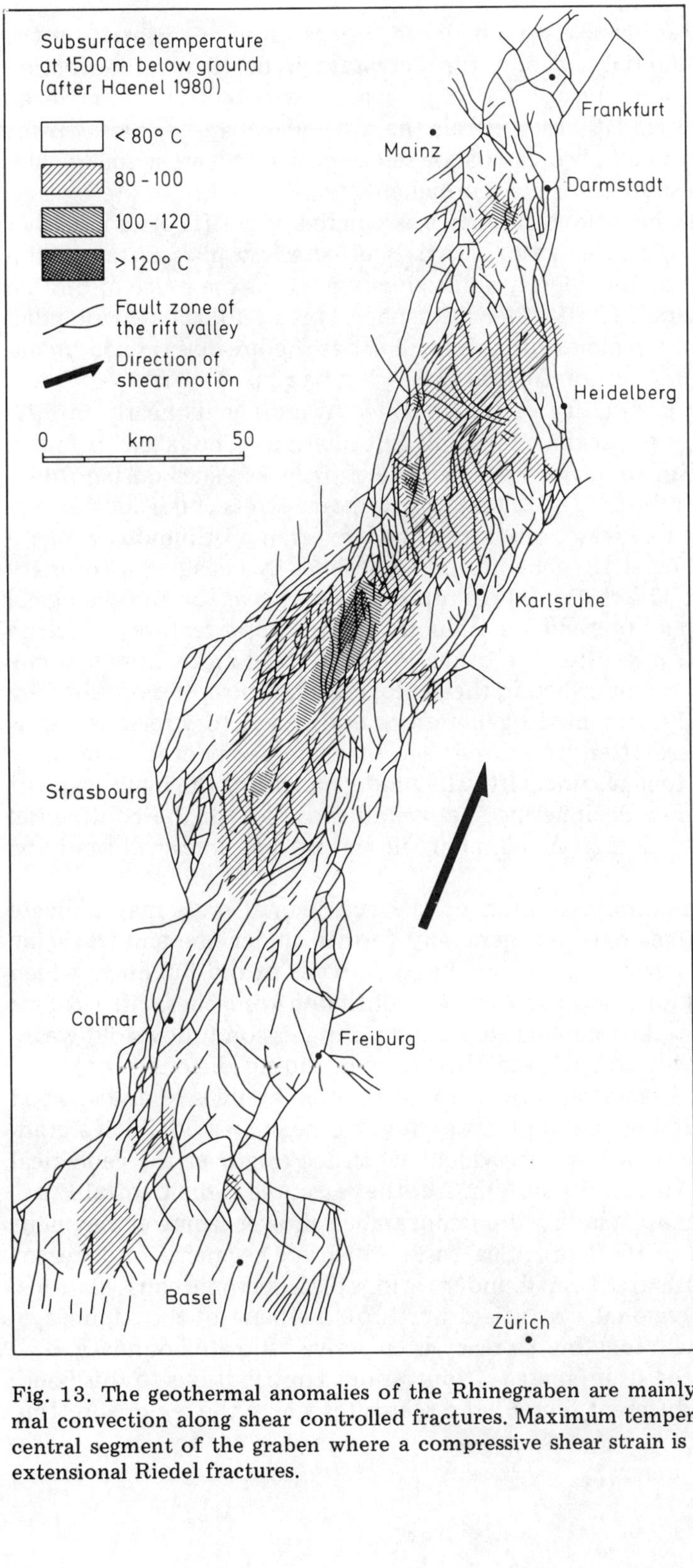

Fig. 13. The geothermal anomalies of the Rhinegraben are mainly caused by hydrothermal convection along shear controlled fractures. Maximum temperatures are found in the central segment of the graben where a compressive shear strain is released by fissuring of extensional Riedel fractures.

temperature increases regularly with depth and is often augmented by the rise of hot mantle material which, in turn, creates a hydrothermal convection within the fissures. Hot springs and high heat flow values characterize all active rift valleys (Fig. 13). In general, the rift valleys are well-known for their sources of petroleum, brown-coal or salt deposits, and are actually gaining importance as resources for geothermal energy.

Tectonically, high heat-flow anomalies associated with rifts offer a further prerequisite for crustal attenuation, as it is effected by piles of unconsolidated sediment fill, by the densely splitted basement, or the bulge of molten asthenospheric material. If a further lithospheric break-up will be demanded by the global tectonic regime, rift belts will act as the pre-determined breaking points along which the continental plates will be ripped apart.

Another specific peculiarity of the continental crust is its nearly infinite memory for all stages of previous tectonic deformations. The fabric of faults, joints, metamorphism or magmatism survive mostly as relics during subsequent tectonic revolutions. The fluctuating regional stress conditions of successive orogenic events create structures that are often posthumously reactivated features of older strain generations, sometimes by changing its primary kinematic function. The structural palimpsest, as it appears on the geological map, occasionally may pretend simultaneity of structural features which in reality were formed one after the other. After looking more closely, some triple rift junctions, as described in the geological literature, were formed by the superposition of different rift generations under deviatory regional stress fields. Failed arms are often of another age than the shoulder joint or body and may develop into an extinct rift valley independent of the main segment of the rift system. For deciphering the overall evolution of the continental rift pattern, a precise dating of sediment fill and rift volcanics will firstly be required.

The complex structural evolution of the continental crust may indicate that major rift systems have not generally formed under one and the same paramount tectonic regime. Like the chinks in walls of old buildings widen gradually by joining one another to open continuous breaches, rift systems are built up by multiple segments. Rift valley progression follows old weakness zones, links fault troughs together formed during different stages of crustal evolution and reactivates buried rifts covered by piles of undisturbed sediments till the present coherent rift system has been created. Such a gradual stage-by-stage development is evident when looking over the geological history of the East African rift system. Another example is the Central European rift belt (Fig. 14), including the Rhinegraben, Lower Rhine embayment and Central Graben of the North Sea basin which are segments of different age and structure, primary formed under various plate tectonic constellations. But it is the active regional stress field north of the bend of the Alpine system which forced them together to react as an active subplate boundary.

Continental grabens, as it emerges from various contributions to this issue, were formed under different tectonic prerequisites where the regional setting

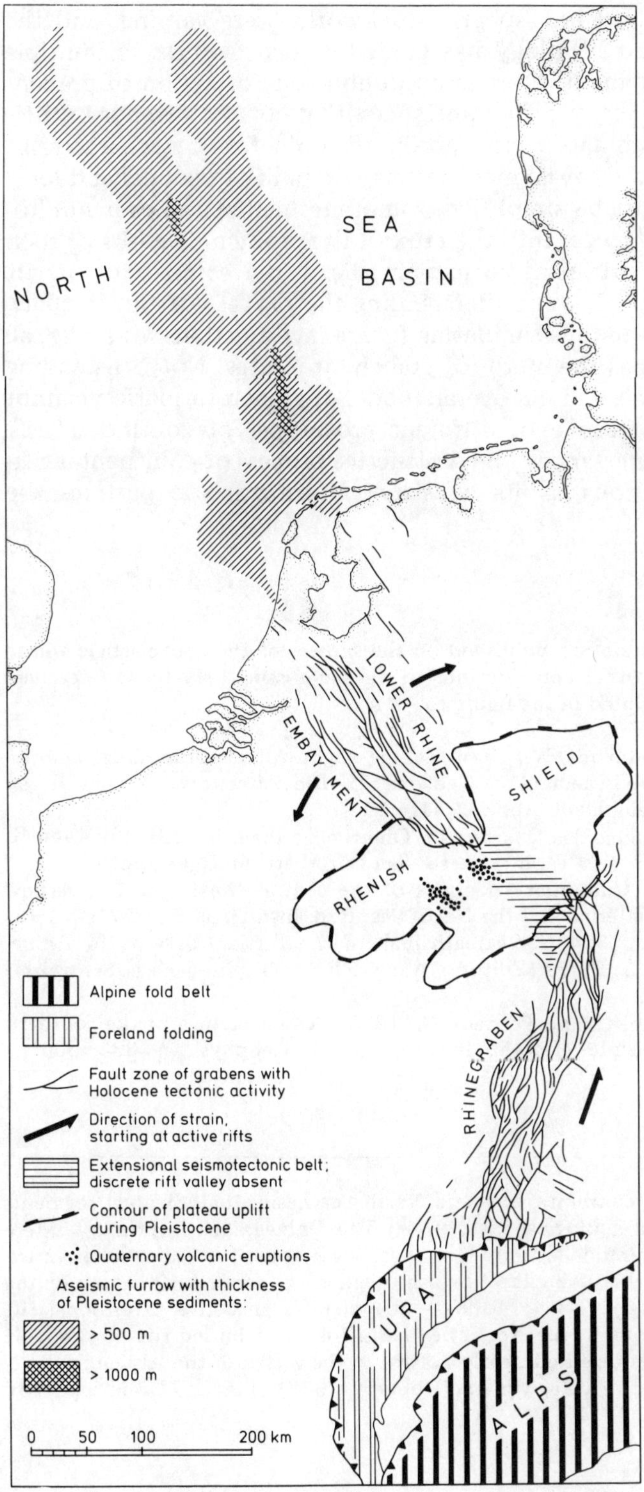
NORTH
SEA
BASIN
LOWER RHINE
EMBAYMENT
RHENISH
SHIELD
RHINEGRABEN
JURA
ALPS
Alpine fold belt
Foreland folding
Fault zone of grabens with
Holocene tectonic activity
Direction of strain,
starting at active rifts
Extensional seismotectonic belt;
discrete rift valley absent
Contour of plateau uplift
during Pleistocene
Quaternary volcanic eruptions
Aseismic furrow with thickness
of Pleistocene sediments:
> 500 m
> 1000 m
0 50 100 200 km

was manifold, the involved lithospheric stockworks were varying, and the pre-existing structure and lithology has guided graben features to multiple variations. On the other hand, there is no doubt that formation of grabens has been governed by their own circumstances, that common trends in their structural evolution dominate, and certain tectonic features characterize them. Trying to model the causal mechanisms, the bulk of regional and local boundary conditions must be simplified, sometimes perhaps an oversimplification must be accepted to get into the crux of the problem. Results of such a procedure cannot be sufficient to explain the whole complexity of rift valley formation. But it helps in understanding the process better, in specifying possible principles and in stimulating future investigations. In-so-far, all kind of modelling enriches the study of graben structures. Notwithstanding the superiority of this method, an overall model of graben formation cannot be found because of presence of different prototypes of continental rift valleys. Under such circumstances, the destructive process of continental rifting appears as multivarious as its constructive antagonistic partner, the orogeny.

REFERENCES

Because of the bulk of literature published on the subject of the above article and to avoid cross references with other contributions to the same issue, only those references are listed below which are quoted in the figure captions.

Baranyi, I., Lippolt, H.J. and Todt, W., 1976. Kalium-Argon-Altersbestimmungen an tertiären Vulkaniten des Oberrheingraben-Gebietes. II. Die Alterstraverse vom Hegau nach Lothringen. Oberrhein. Geol. Abh., 25: 41—62.

Bartz, J., 1974. Die Mächtigkeit des Quartärs im Oberrheingraben. In: J.H. Illies and K. Fuchs (Editors), Approaches to Taphrogenesis. Schweizerbart, Stuttgart, pp. 78—87.

Caston, V.N.D., 1977. The Quaternary deposits of the central North Sea 1,2. A new isopachyte map of the Quaternary of the North Sea. Rep. Inst. Geol. Sci., 77/11: 1—8.

Doebl, F. and Olbrecht, W., 1974. An isobath map of the Tertiary base in the Rhinegraben. In: J.H. Illies and K. Fuchs (Editors), Approaches to Taphrogenesis. Schweizerbart, Stuttgart, pp. 71—72.

Edel, J.B., Fuchs, K., Gelbke, C. and Prodehl, C., 1975. Deep structure of the southern Rhinegraben area from seismic-refraction investigations. Z. Geophys., 41: 333—356.

Fig. 14. An active rift belt combines Alps and North Sea basin. Its individual segments have different structure and geological history. (a) The Rhinegraben, a primary extensional rift valley, has been remodelled into a sinistral shear zone. (b) The zone of extensional seismotectonic activity crossing the Rhenish shield is a concealed rift caused by the ductile rock behavior. (c) The Lower Rhine embayment is an active extensional rift valley. (d) The aseismic furrow in the North Sea basin follows a buried rift of Mesozoic age. The whole rift system is oriented nearly normal to the vertex of the Alpine fold arc. Data mainly from Caston (1977), Zagwijn and Doppert (1978), Illies and Greiner (1979), Illies et al. (1979).

Haenel, R. (Editor), 1980. Atlas of Subsurface Temperatures in the European Community. Comm. Europ. Commun., Hannover, 36 pp., 43 maps.

Illies, J.H. and Greiner, G., 1979. Holocene movements and state of stress in the Rhinegraben rift system. In: C.A. Whitten, R. Green and B.K. Meade (Editors), Recent Crustal Movements, 1977. Tectonophysics, 52: 349—359.

Illies, J.H., Prodehl, C., Schmincke, H.E. and Semmel, A., 1979. The Quaternary uplift of the Rhenish shield in Germany. In: T.R. McGetchin and R.B. Merrill (Editors), Plateau Uplift: Mode and Mechanism. Tectonophysics, 61: 197—225.

Zagwijn, W.H. and Doppert, J.W.Chr., 1978. Upper Cenozoic of the southern North Sea Basin: palaeoclimatic and palaeogeographic evolution. In: A.J. van Loon (Editor), Key-notes of the MEGS-II (Amsterdam, 1978). Geol. Mijnbouw, 57: 577—588.

Editor's Comments on Papers 4 and 5

4 MILANOVSKY
Excerpts from *Aulacogens of Ancient Platforms: Problems of Their Origin and Tectonic Development*

5 ŞENGÖR, BURKE, and DEWEY
Rifts at High Angles to Orogenic Belts: Tests for Their Origin and the Upper Rhine Graben as an Example

The term *aulacogen* was defined in the USSR some forty years ago, but it does not have quite the same range of meaning for most Soviet geologists as it does for those of the West. Shatsky (1946*a,b;* 1961) recognized in Europe and Asia "ancient, buried, narrow and extensive grabens of the platforms." Milanovsky, in Paper 4, regards them as generally Paleozoic, Proterozoic, and older, and, in one reference as "ancient, 'dead' buried tectonic elements of the platforms" (1974). He recognizes three types—"through, penetrating, and inner or blind." On the other hand most Western geologists, including authors of Paper 5, use the term in a single sense only, corresponding to the "penetrating" type of Milanovsky.

In spite of these differences in concept and definition—largely the outcome of differing tectonic concepts (fixism versus mobilism) —Milanovsky's writings offer much that is relevant. The paper has been pruned of what is of local interest, leaving only that which has general application. Information about the distribution of features within the continents and along their margins has been omitted. In any case not all those examples are of equal authenticity. Paper 4 is nevertheless useful for reference purposes.

In Paper 5 Şengör, Burke, and Dewey concentrate on aulacogens and collision rifts. For them and most Western geologists, an aulacogen is the "failed" arm of a rift system at a triple junction, that is, the arm that does not "split" to form part of separating continental margins. In general, the authors recognize only the "penetrating" aulacogen of Milanovsky (see above). Collisional rifts ("impactogens") are those that form by continental collision before the orogenic belt is formed, the Upper Rhine graben being taken as an example. Useful data on this much studied rift are thus made available. Comparisons are offered with other explanations. The chief point is that the Alpine collision is directly responsible for the Rhine graben rifting. We are dealing with rifts that have formed at different stages of the Wilson cycle (Wilson, 1969; Burke, 1980; Burchfiel, 1980). The Atlantic was opening when the Benue aulacogen (West Africa) was forming, whereas the Rhine graben was forming when the Mediterranean was closing (Dewey and Burke, 1974; Şengör, 1976; Burke, 1977).

REFERENCES

Baker, B. H., and P. Morgan, 1981, Continental rifting: progress and outlook, EOS, *Am. Geophys. Union* **62:**585–586.

Beloussov, V. V., 1969, Continental rifts, in *The Earth's Crust and Upper mantle,* Am. Geophys. Union Mon. 13, pp. 539–543.

Burchfiel, B. C., 1980, Plate tectonics and the continents: a review, in *Continental Tectonics,* Panel on Continental Tectonics, National Academy of Sciences, Washington, pp. 15–25.

Burke, K., 1977, Aulacogens and continental breakup, *Ann. Rev. Earth and Planetary Sci.* **5:**371–396.

Burke, K., 1980, Intracontinental rifts and aulacogens, in *Continental Tectonics,* Panel on Continental Tectonics, ed., National Academy of Sciences, Washington, pp. 42–49.

Dewey, J. F., and K. Burke, 1974, Hotspots and continental break-up: implications for collisional orogeny, *Geology* **2:**57–60.

Heezen, B. C., 1960, The rift in the ocean floor, *Sci. American,* Oct. 1960 **203:**98–110.

Heezen, B. C., 1969, The World Rift System: an introduction to the symposium, *Tectonophysics* **8:**269–279.

Hoffman, P., J. F. Dewey, and K. Burke, 1974, Aulacogens and their genetic relation to geosynclines, in *Modern and Ancient Geosynclinal Sedimentation,* R. A. Dott, ed., Society of Economic Paleontol. Mineral., Spec. Pub. 19, pp. 38–45.

Illies, J. H., 1978, Two stages Rhinegraben rifting, in *Tectonics and Geophysics of Continental Rifting,* I. B. Ramberg and E. R. Neumann, eds., D. Reidel, Dordrecht, pp. 63–71.

Illies, J. H., and G. Greiner, 1979, Holocene movements and state of stress in the Rhinegraben rift system, *Tectonophysics* **52:**349–359.

Milanovsky, E. E., 1972, Continental rift zones: their arrangement and development, *Tectonophysics* **15:**65–70.

Milanovsky, E. E., 1974, Rift zones of the geologic past and their associated formations, Report 1, *Int. Geol. Rev.* **18:**515–531.

Milanovsky, E. E., 1980, Problems of the tectonic development of the Earth in the light of concept on its pulsations and expansion, *Rev. Geol. Dyn. et Geogr. Phys.* **22:**15–27.

Milanovsky, E. E., 1983, Major stages of rifting evolution in the earth's history, *Tectonophysics* **94:**599–607.

Morgan, P., 1982, Heat flow in rift zones, in *Continental and Oceanic Rifts* (Rept. of Working Group 4), G. Palmason, ed., American Geophysical Union/Geological Society of America, Washington, D.C., 309p.

Ramberg, I. B., and E.-R. Neumann, 1978, Paleorift systems—Introduction, in *Petrology and Geochemistry of Continental Rifts,* I. B. Ramberg, and E.-R. Neumann, eds., D. Reidel, Dordrecht, 444p.

Şengör, A. M. C., 1976, Collision of irregular continental margins: implications for foreland deformation of Alpine-type orogens, *Geology* **4:**779-782.

Shatski, N. S., 1946*a,* Basic features of the structures and development of the East European platform, *SSSR Akad. Nauk Izv., Geol. Ser.,* No. 1, pp. 5-62 [in Russian].

Shatski, N. S., 1946*b,* The Great Donets Basin and the Wichita system, *SSSR Akad. Nauk Izv., Geol. Ser.,* No. 6, pp. 57-90 [in Russian].

Shatski, N. S., 1961, *Vergleichende Tektonik alter Tafeln* Berlin: Akademie-Verlag, 220p. [in German].

Turcotte, D. L., and S. H. Emmerman, 1983, Mechanisms of active and passive rifting, *Tectonophysics* **94:**39-50.

Wilson, J. T., 1969, Static or mobile earth: the current scientific revolution, *Tectonophysics* **7:**599-601.

4

Reprinted from pages 213-216, 229, and 243-245 of *Tectonophysics* **73**:213-248 (1981)

AULACOGENS OF ANCIENT PLATFORMS: PROBLEMS OF THEIR ORIGIN AND TECTONIC DEVELOPMENT

E.E. MILANOVSKY

Geological Department, Moscow State University, Moscow (U.S.S.R.)

(Received July 31, 1980)

ABSTRACT

Milanovsky, E.E., 1981. Aulacogens of ancient platforms: problems of their origin and tectonic development. In: J.H. Illies (Editor), Mechanism of Graben Formation. Tectonophysics, 73: 213—248.

Riphean and Palaeozoic linear graben-like platform depressions — aulacogens —jointly with linear swell uplifts in a platform cover, intraplatform folded zones and Meso-Cenozoic epi-platform rift zones are different morphological manifestations of a single group of linear platform structures. Aulacogens are characterized by the long duration of their development and ability for regeneration. Their existence is certain, mainly since the Late Proterozoic (Riphean). Usually they inherit more ancient linear attenuated tectonic zones of platform basement or "adjust" to them. Aulacogens pass through common stages of development forming a single "cycle" or in many cases, even several similar though not identical "cycles". The "cycle" starts with the incipience (or regeneration) of a linear graben-like depression limited by normal faults and after a more or less deep and long subsidence, accompanied often by basic or alkaline-basic volcanism, it is completed in many cases by an inversion stage. At this stage horst-like uplifts appear limited by thrust and overthrusts in the aulacogen basement, while linear swell-like uplifts or their systems and even linear folded structures occur in its cover. The early stages of an aulacogen evolution occur under horizontal extension conditions, whereas the inversion stage takes place under a compressional regime. "Cycles" of aulacogen development resemble (in attenuated, miniature form) the development "cycle" of adjacent geosynclinal areas, coincide with them in time and many "through" and "penetrating" aulacogens are their direct offsets.

The Late Proterozoic "cycle" of aulacogen evolution corresponds to a "pre-plate" stage of the platform history ("aulacogen stage" by Bogdanov) and is characterized by the widest development of a network of differently oriented aulacogens over the whole area of ancient platforms which have been subjected to some general horizontal extension ("crawling away") and only partially compensated by compression during the inversion stage.

The subsequent aulacogen development in Palaeozoic and on Laurasian platforms — partly in Meso-Cenozoic — was spatially and genetically connected with the history of geosynclinal (later orogenic) belts bordering these platforms. With regard to Gondwana platforms (and later the North American and Siberian), the Mesozoic saw the development of many aulacogens or rift zones ranging parallel to transversally with respect to the recent margins of platforms. This process was connected with the break up of the Gondwana supercontinent and the formation of "secondary" oceanic depressions. As a rule, it has not been completed by inversion, but took place under extension conditions and probably some general expansion of the earth activated during the Meso-Cenozoic.

INTRODUCTION

On all ancient platforms linear or strip-like structures are widely spread together with sub-isometric large structural forms (shields, anteclises, syneclises, etc.). These structural elements occurred or developed after the platform formation. They are distinguished by the relief of their basement surface and by the deformations of those or other horizons of the platform cover. Morphologically they are very diverse. The following structures refer to linear platform structures: (1) swell-like uplifts formed by deposits of the platform cover (placanticlines by Shatsky, 1945), their chains, bundles and zones of platform folds as well as large flexures. To this category belongs in particular the famous Zhiguli dislocation on the East European platform, its discovery and description by Pavlov in 1887 started a study of structures of ancient platforms; (2) numerous, mainly, Cenozoic and Mesozoic grabens and other linear structures of epi-platform rift zones: they are located mostly on the uplifts of Precambrian basement (shields) of the Gondwana platforms, as well as Siberian and some other platforms of northern continents; (3) Late Proterozoic and Palaeozoic deep linear graben-like depressions revealed over the recent decades on the platform basement surface. They are filled in with thick sediments and volcanics in the lower part of the platform cover (sometimes rather deformed) and usually superimposed and "masked" by its upper horizons. These linear graben-like depressions of ancient platforms were called aulacogens by Shatsky and Bogdanov (1961). Unfortunately, Shatsky had no time to give a comprehensive definition of aulacogens and later there appeared some discrepancies while interpreting this term. The latter were connected to a considerable extent with the morphological diversity of structures of the "aulacogen family" that are at different stages of development. At present, a close connection among all three above-mentioned types of linear platform structures becomes ever more evident. Initially, however, they were distinguished independently and on the basis of different criteria. On the other hand, it was made clear that linear dislocations in the platform cover, in particular swells (placanticlines), their chains, bundles of platform folds (placanticlinoria, megaswells and others) as well as flexures and less frequently large fractures in the platform cover are situated as a rule, above aulacogens, or along their margins. They occur as a result of repeated basement deformations within these aulacogens. The character of vertical block movements in the latter can change up to a complete inversion, that leads to the formation of swells (placanticlines) above older grabens (i.e. Vyatka swell above the Kirov aulacogen on the East European platform). Thus, either the graben depth gets smaller due to the elevation of the whole bottom or a narrower horst uplift occurs within it in the basement. The growth of the latter leads to formation of placanticlines, flexures or fractures in the platform cover. On the other hand an identity of many indices of Meso-Cenozoic epiplatform rift zones and ancient (Upper Proterozoic and partly Palaeozoic) aulacogens has been revealed (Milanovsky, 1976,

1978b, 1979). Due to this reason such ancient graben-like structures of platforms are more frequently called palaeorifts or palaeorift zones (one should certainly not ignore some differences between old and young rift zones of platforms associated with the irreversible evolution of the earth). Finally, during recent time some Cenozoic and Mesozoic epiplatform rift zones have been observed to occur at the place of much older — Palaeozoic or even proterozoic — structures of the rift type. Thus it means the former are inherited from, or spatially "predetermined" by, the latter. For instance, McConnel (1979) has shown that Cenozoic and Mesozoic rift systems of East Africa are located at the area with numerous manifestations of tectonomagmatic activities and epi-craton rifting, with the earliest among them as old as the formation of the "great dyke of Rhodesia" (2600 m.y.) and the Bushveld Intrusive Complex (2000 m.y.). Many aulacogens of the East European and Siberian platform (Kirov, Dnieper—Donetz, Vilyui and others) were stated to be set up in the Riphean, then, after a long pause in their development were regenerated in Devonian and in Late Palaeozoic or Mesozoic times were subjected to new structural transformations up to a complete inversion. Thus, the rather long duration (up to 1500—2500 m.y.) and a multi-phase development of many zones of linear platform structures became clear. They seem to be tectonic forms with exceptional "vitality" and capable of regeneration sometimes accompanied by inversions of the sign of vertical movements at the successive stages of their development. At early stages, these zones are mainly linearly extended graben-like depressions (aulacogens or "palaeorifts" and their recent analogues; Cenozoic rift zones of platforms). During the subsequent stages some swell-like uplifts, horsts, linear folded zones and others may occur within these zones.

Though a close relationship of aulacogens, swells and rift zones of ancient platforms has become nowadays ever more evident, the question of origin and kinematic conditions of this type of platform structure development is treated and interpreted in the literature in very different ways. Whereas the formation of Cenozoic rift zones of platforms is considered by most researchers as a result of more or less significant horizontal extension, the aulacogens, zones of swelllike uplifts, flexures and other types of structures of ancient platforms are explained differently. Up to now, a concept of the leading or even exceptional role of vertical differential basement movements in their genesis, remains predominant, especially among the Russian geologists (for instance, the classical works of Shatsky, 1955). However, with increasing knowledge of some concrete linear platform structures and the relationship between rifts and aulacogens, some geologists are now more inclined to admit the essential significance of lateral extension (Chekunow, 1967), shifts and lateral compression (at a later inversional stage of development) in formation of aulacogens. Most clearly, these considerations were stated in the Russian literature by Kropotkin (1971) and Valeev (1977, 1978) and in an American paper by Burke (1977) who interprets the aulacogens as "Failed rifts" and "blind" branches of main rift (spreading) zones, penetrating into the continents.

This paper deals with general regularities of dislocation, morphology, development stages and tectonic evolution of linear platform structures that can be distinguished on the basis of studying the structure and evolution of concrete aulacogens and rift zones of ancient platforms and their comparative analysis. Particular attention is paid to the linear structures of the East European platform, for which aulacogens and swells were first described and studied in detail; the study also involves some comparative material from other platforms.

[*Editor's Note:* Material has been omitted at this point.]

REGIONAL PECULIARITIES OF DISPOSITION AND DEVELOPMENT OF AULACOGENS ON DIFFERENT PLATFORMS

Some important features of aulacogen arrangements on the ancient platforms have been noted by Bogdanov (1961). He distinguished three types among them: (1) *"through" aulacogens* disecting the platform into two usually unequal parts, for instance, Patom—Vilyui on the Siberian, South Emba and probably, Timan aulacogen on East-European platform; (2) transversal or rather *penetrating aulacogens* entering into the platform body and attenuating within it (for instance, Wichita and Dnieper—Donetz aulacogen *; and (3) *inner or "blind"* aulacogens attenuating at both ends in the platform body (aulacogens of the Middle Russian zone).

Mobility of aulacogens, the rate of their subsidence and the extent of successive inversion decreases consecutively in this order reflecting the reducing connection of aulacogens with geosynclinal, folded areas or oceanic basins framing the platform. "Through" aulacogens and the segments of penetrating aulacogens approaching the platform edges are usually the deepest, and folding occurs often later in sediments filling them. One cannot, in praxis, clearly distinguish them from intracraton geosynclinal depressions. On the contrary, inner, "blind" terminations of penetrating aulacogens and inner "blind" aulacogens are shallower as a rule, and inversion does not always occur in them. If it takes place, it is commonly expressed only in a change of the sign of vertical basement block movements, by the occurrence of inner horsts and unilateral horsts within the latter and by swell-like uplifts in the platform cover.

* K. Burke (1977) use the term "aulacogen" only as regards to aulacogens of this, "entering" or "penetrating" type.

[*Editor's Note:* Material has been omitted at this point.]

CONCLUSION

(1) Riphean and Palaeozoic linear graben-like troughs of platforms — aulacogens, jointly with linear swell-like uplifts in platform cover, intraplatform folded zones and Meso-Cenozoic epiplatform rift zones are various morphological manifestations of a single group of linear platform structures. They differ from each other in age, tectonic position, extent of "maturity", stage of development and depth of erosion.

(2) Aulacogens are as a rule, characterized by the long duration of their development ("vitality") and their ability to regenerate. The existence of aulacogens or structures close to them ("palaeoaulacogens") is fixed from the early stages of platform history, since Early Proterozoic and mainly, since Upper Proterozoic (Riphean). They usually inherit older linear attenuated tectonic zones of the platform basement or adjust themselves to them.

(3) In their development aulacogens pass through common general stages of evolution, forming a single cycle and in many cases even several similar, though not identical "cycles" (Fig. 24). The "cycle" starts with the incipience (or in subsequent "cycles" with regeneration) of a linearly extended graben-like trough (or a system of troughs) limited by normal faults. It is subjected to more or less continuous subsidence often accompanied by basic (in Riphean) or alkaline-basic (in Phanerozoic) volcanism and is concluded by the inversion stage (not always manifested). At this stage horst-like uplifts occur in the aulacogen basement, limited by upthrusts and overthrusts, whereas in its cover—swell-like uplifts, their systems or even linear folded and folded-overthrust structures occur. Early stages of aulacogen development undoubtedly proceed under the conditions of horizontal extension (quite recognized nowadays for their Meso-Cenozoic analogues — rift zones of platforms). The late inversional non-volcanic stage, is characterized by the conditions of some significant horizontal compression.

(4) Complete "cycles" of aulacogen development on platforms resemble (in an attenuated and miniature form) those of the development of miogeosynclinal zones, adjacent to these platforms in the general direction of tec-

tonic evolution, succession of stages and their content. They coincide in time, and many "through" or "penetrating" aulacogens represent their direct branches and are connected with them by gradual transitions.

(5) The Late Proterozoic "cycle" of aulacogen evolution corresponds to the pre-plate stage of history of ancient platforms. It is characterized by the widest development of aulacogen networks (including volcano-active aulacogens) occurring simultaneously and genetically connected with Riphean geosynclines bordering the platforms. Close combinations of differently oriented aulacogens in these systems indicate, that territories of platforms were subjected to some general, differently directed, horizontal extension as if "creeping apart" at the time of their formation. It may be proposed that some general expansion of the earth took place during this stage. The extension of ancient platform bodies was only in part compensated by compression of some aulacogens at the latest inversional stage of their development (at Grenville or Baikal tectonic epochs).

(6) The subsequent development of aulacogens in Palaeozoic and on Eurasian platforms — up to the Palaeogene — was mainly connected with "cycles" of development of geosynclinal (later-orogenic) belts framing the platforms: Caledonian, Hercynian, partly Mesozoic (on the Siberian) and Alpine (on the East European platform) belts. Most widespread and powerful on northern platforms was the Hercynian "cycle" of aulacogen development. In contrast to the Riphean, aulacogen systems in the Phanerozoic did not cover the whole territory of platforms, but only their wide zones adjacent to these belts. Newly formed or regenerated aulacogens stretched subparallel to the boundaries of neighbouring geosynclinal belts and at the place of their sharp convex turns penetrated deep into the platform bodies. Development of many aulacogens (except the smallest) was accompanied by inversion. During the early stages of tectonic "cycles", platform zones adjacent to Phanerozoic geosynclinal belts, were subjected to some horizontal extension (especially effectively manifested in aulacogens of the East European and Siberian platform in Devonian). During late stages, they experienced some horizontal compression sometimes frequently repeated after the main folding in a neighbouring geosynclinal area, in connection with the repeated orogenic processes there.

(7) On Gondwana platforms since Middle Palaeozoic (in Australia) or Late Palaeozoic (in areas adjacent to the Indian Ocean) or Mesozoic (in areas adjacent to the Atlantic Ocean) begins the process of origination (regeneration) and subsidence of numerous aulacogens or rift zones. They strike parallel to the margins of modern platforms (pericontinental rift zones) or transversal to them. In the Mesozoic similar processes covered also the southeastern and northeastern periphery of the North American platform. These processes were undoubtedly, connected with the decline of Gondwana supercontinent and formation of "secundary" oceanic basins. Occurrence and development of this generation of aulacogens took place under the conditions of extension probably connected with the hypothetical expansion of

the earth during the Meso-Cenozoic (Carey, 1976). With very rare exceptions (the Benue zone in Late Cretaceous, the zone of West Spitsbergen in Paleogene) they did not experience inversion or compression. Most intensively fragmentation and general extension ("creeping apart") processes were manifested in the Mesozoic on the Indian and the southeastern part of the African platform and during the Cenozoic (?) in the northern part of the North American platform. In these areas systems of intersecting differently oriented "trough" and "penetrating" grabens occurred. They resembled a network of Riphean aulacogens on the East European platform.

(8) The global process of renewed activization of extension, especially intensified in Pliocene-Anthropogen, involved a number of ancient platforms. It manifested itself most strongly on the Afro-Arabian platform and in some peripheral parts of the North American, Siberian and Indian platforms, where many rift zones appeared or developed both directly associated with the worldwide system of intraoceanic rift ranges and isolated intracontinental ones (Baikal, Kambay and oth.). At the same time, a number of ancient graben-like basins is "revived" on other platforms.

Thus aulacogens (and rift zones) of ancient platforms originated and developed during the process of horizontal extension to which they were subjected jointly with the mobile belts of the earth framing them. With high intensity this process was manifested in the Late Proterozoic, when deformations of extension covered not only peripheral zones of platforms but even their whole territory, being later replaced by less intensive and local compression in some zones. Similar extension and compression processes in the Palaeozoic (and sometimes in the Mesozoic) were manifested in a more scanty way only in some aulacogens of ancient platforms adjacent to active geosynclinal-orogenic belts. However, since the end of the Palaeozoic first southern platforms of the Gondwana group and then peripheral zones of some northern platforms are subjected to extension connected with Gondwana decline, growth of "secondary" oceans and, probably with activisation of the process of some general expansion of the earth (Carey, 1976, Milanovsky, 1978a, b; 1980). It may be supposed that this expansion process in the earth history (similarly with another planet of the Earth group) passed not uniformly and continuously in time, but on the contrary was manifested mainly during several stages of the geological history, namely in Archean (age of the "greenstone belts" formation), in Late Proterozoic (age of Stille's "Umbruch" or Algonkian regeneration, formation of the Neogeikums geosynclinal belts and of the first generation of aulacogens on the ancient platforms) and, finally, during the Meso-Cenozoic time.

[*Editor's Note:* Only the references cited in the preceding excerpts are reprinted here.]

REFERENCES

Bogdanov, A.A., 1961. On some problems of tectonics of Europe. Vestn. Mosk. Univ. Ser. Geol., 1979 (5), (in Russian).

Burke, K., 1977. Aulacogens and continental breakup. Annu. Rev. Earth Planet. Sci., 5: 371—396.

Carey, S.W., 1976. The Expanding Earth. Elsevier, Amsterdam, 488 pp.

Chekunov, A.V., 1967. On formation mechanism of aulacogen type structures (taking the Dnieper-Donetz depression as an example). Geotektonika, 1967 (3): 3—18 (in Russian).

Kropotkin, P.N., 1971. Deep structure and deformations of ancient platforms. In: P.N. Kropotkin, B.M. Valyaev, P.A. Gafarov (Editors), Deep Structure of Ancient Platforms of the Northern Hemisphere. Nauka, Moscow, pp. 321—364 (in Russian).

McConnel, R.B., 1979. A precambrian origin for the proto-rift dislocation belt of Eastern Africa. In: Geodynamic Evolution of the Afro-Arabian System. Int. Symp. Abstr. Acad. Nac. dei Lincei, Rome, pp. 62—65.

Milanovsky, E.E., 1976. Rift Zones of Continents. Nedra, Moscow, 280 pp. (in Russian).

Milanovsky, E.E., 1978a. Some regularities of tectonic development and volcanism of the Earth in Phanerozoic (problems of the Earth pulsations and expansion). Geotectonika, 1978 (6): 3—16. (in Russian).

Milanovsky, E.E., 1978b. Rifting development in the Earth's history. In: J.B. Ramberg and E.R. Neumann (Editors), Tectonics and Geophysics of Continental Rifts. Reidel, Dordrecht, pp. 385—399.

Milanovsky, E.E., 1979. On the problem of origin and development of linear platform structures. Byull. Mosk. Ova. Ispyt. Prir., Otd. Geol., 1979 (6): 29—58. (in Russian).

Milanovsky, E.E., 1980. Problems of the tectonic development of the Earth in the light of the concept on its pulsations and expansion. Rev. Géogr. Phys. Géol. Dyn., 22 (1): 15—27.

Pavlov, A.P., 1887. Samarskaya Luka and Zhiguli. Tr. Geol. Kom., 1887 (2) (5), (in Russian).

Shatsky, N.S., 1945. Essay of tectonics of Volga-Urals oil-bearing region and adjacent parts of the western slope of southern Urals. Izd. Moskovskogo Obshchestva Ispytately Prirody, Moscow (in Russian).

Shatsky, N.S., 1955. On the origin of the Patchelma depression. Byull. Mosk. Ova Ispyt. Prir., Otd. Geol., 1955 (5), (in Russian).

Shatsky, N.S. and Bogdanov, A.A., 1961. A international tectonic map of Europe, scale 1:2.500.000. Izv. Akad. Nauk SSSR, Ser. Geol., 1961 (4), (in Russian).

Valeev, R.N., 1977. East European platform. In: A.I. Suvorov (Editor), Raslomy i Horizontalnye Dvizhenia Platformennykh Oblastei SSSR, Nauka, Moscow, pp. 48—67 (in Russian).

Valeev, R.N., 1978. Aulacogens of the East European platform. Nedra, Moscow, 152 pp. (in Russian).

5

Reprinted from *Am. Jour. Sci.* **278**:24–40 (1978)

RIFTS AT HIGH ANGLES TO OROGENIC BELTS: TESTS FOR THEIR ORIGIN AND THE UPPER RHINE GRABEN AS AN EXAMPLE

A. M. CELÂL ŞENGÖR, KEVIN BURKE, and JOHN F. DEWEY

Department of Geological Sciences, State University of New York at Albany, Albany, New York 12222

ABSTRACT. Rift valleys striking at high angles to orogenic belts are of two main kinds: those originating at ocean opening (aulacogens) and those originating at collision (impactogens). We here consider differences between the geological evolution of the two kinds of rifts that can be helpful in distinguishing between them. We review the Permian to Recent evolution of the Upper Rhine Graben as an example. There is little evidence to support the idea that there was active rifting in the Rhine Graben in Late Triassic times contemporary with the opening of the Alpine Tethys. Major rifting in the graben began during the Eocene at the same time as the Mesoalpine collision. The subsequent development of the Rhine Graben includes strike-slip and compressional faulting and volcanism. The relationship of the Upper Rhine Graben to the Hessen and Lower Rhine Graben appears more complex than that of a simple triple-rift on a dome as Cloos suggested.

INTRODUCTION

Although the association between the world's orogenic belts and grabens that strike into them at high angles was recognized early during this century (Weber, 1921, 1923, 1927; Shatski, 1946a,b, 1947, 1955) and later reemphasized (DeSitter, 1956; J.T. Wilson *in* Jacobs, Russel, and Wilson, 1959: Wilson, 1966a), their genetic relationships remained blurred or were denied altogether (Cloos, 1939; see a summary of various objections to Weber's views *in* Weber, 1927) until the development and wide acceptance of the theory of plate tectonics and its geologic corollaries. Burke and Whiteman (1973) showed that, in East Africa, Neogene rifts commonly developed a three-armed pattern that they compared to the pattern of the rifts that had led to the opening of the South Atlantic Ocean establishing an apparently common sequence as follows: doming, development of three- (in some cases more) armed rift systems on the crests of the domes, continental break-up as two of the three-armed rifts link up and develop into an accreting plate margin, while the third arm is left as a graben or a "failed arm." Burke and Dewey (1973) argued, on the basis of their survey of 52 rift systems throughout the globe, that most of the ocean-opening cycles, both young and old, have been initiated by and evolved through the sequence that Burke and Whiteman (1973) described; further studies at oceanic and fold belt margins seem, generally, to have confirmed their views (for example, Curray and Moore, 1974; Burchfiel and Davis, 1975; R.C.L. Wilson, 1975; Rankin, 1976). Burke and Dewey (1973), Dewey and Burke (1974), and Hoffman, Dewey, and Burke (1974) suggested that, when an ocean closes, the previously "failed arms" become rifts at high angles to the resulting collisional orogen or, in Shatski's

(1946a) terminology, aulacogens.[1] Recently, however, Şengör (1976a,b) proposed that continental collision also produces rift structures very similar to aulacogens in map view and internal geometry, with the important difference that the origin of aulacogens predates (whereas that of collision rifts postdates) ocean closing and the resultant collisional orogeny.

In two recent reviews, Burke (1976, 1977) emphasized the importance of rift structures that strike into mountain belts at high angles, herein called high-angle rifts irrespective of their mode of origin. He pointed out their generally much less severely deformed state when compared with that of the adjacent orogen and indicated that if these rifts predate the orogeny and are related to the initial ocean opening phase they then may preserve the valuable syn-rifting stratigraphic record of the vanished ocean whose early record is generally obliterated by the intense collisional orogeny. If, however, the rifts are collision-induced, their peculiar stratigraphy is unlikely to have any relation to that of the obliterated ocean. It is, therefore, of great interest in tectonics to be able to distinguish between aulacogens and collision grabens.

The purpose of this paper is to outline briefly expected differences, primarily in the stratigraphic and structural evolution, between aulacogens and collision rifts, to point out those differences that may be decisive and those that are likely to be deceptive, and, finally, to describe the evolution of one well-studied example of a high angle rift, the Upper Rhine Graben of Central Europe.

DIFFERENCES IN GEOLOGICAL EVOLUTION BETWEEN AULACOGENS AND COLLISION RIFTS

Burke (1977) suggested that a study of the date of origin of a high-angle rift might reveal its origin, because it would show whether the rift predated or postdated the collisional orogeny. There are, however, many complications in the evolution of a high-angle rift that warrant a closer examination of the life histories of aulacogens and collision rifts. In order to emphasize the point that aulacogens and collision rifts are not the only types of high-angle rifts, we include the discussion of the life history of one kind of "random high-angle rift" (fig. 1E-G) and attempt to show how it differs from the other two.

Figure 1 is a composite diagram that schematically illustrates events during the origin and evolution of an aulacogen, a collision rift, and a rift that randomly originates at or near the continental margin after the initial rifting and opening of the associated ocean. In the case of an aulacogen (fig. 1A-D), rifting is likely to be predated by doming at the future site of formation of an rrr-junction, possibly induced by a mantle plume (Burke and Whiteman, 1973; Burke and Dewey, 1973; Burke and J. T. Wilson, 1976). It may be possible to document and date this doming event

[1] In this paper we follow Burke's (1977) practice of calling aulacogens only those high-angle rifts striking into fold belts that had originated during and in relation to the ocean opening phase. Rifts that originate during a collision could perhaps be called impactogens.

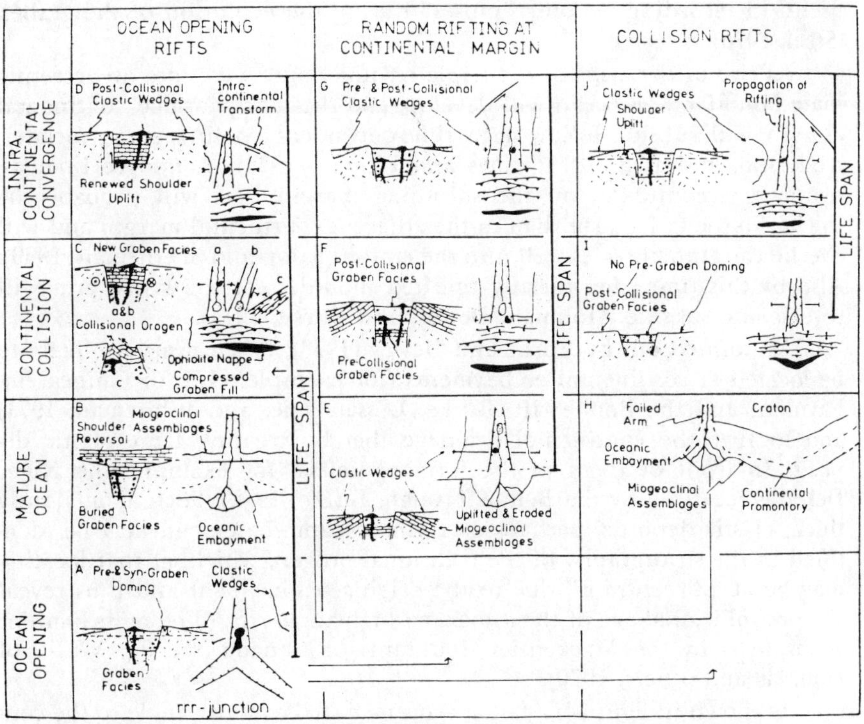

Fig. 1. A schematic illustration of events during the origin and evolution of an aulacogen (A-D), a random rift at high angle to continental margin (E-G), and a collision rift or impactogen. Cross sections show the expected differences in the stratigraphic evolution of the three kinds of high-angle rifts. For more detailed discussion see the text.

by means of peripheral clastic wedges. For example, the Miocene clastics of Western Kenya and Uganda were produced by accelerated erosion during arching over the sites of the future western and eastern rifts during the early Neogene and were lated preserved beneath the younger volcanics (King, 1970). Similarly, R.C.L. Wilson (1975) presented strong evidence for late Jurassic uplift on the site of the northwestern corner of the Iberian Peninsula, where grabens developed at the end of the Jurassic before the opening of North Atlantic and the Bay of Biscay. A doming event may be accompanied by volcanicity (for example, Kenya, King and Chapman, 1972) and eventually followed by crestal rifting (for example, Kenyan and Ethiopian rifts in East Africa, Baker, Mohr, and Williams, 1972). Within the rift, clastics in the form of fanglomerates, arkoses and mudstones, and locally evaporites, the 'graben facies' of Bird and Dewey (1970), may form. The marginal fanglomerates are particularly useful in stratigraphic analysis because they record the successive levels of exposure on the graben shoulders, thereby complementing the syngraben doming record of the clastic wedges outside the graben. Fanglomerates are also useful in recording tectonic "pulses" during the evolution of the graben (Steel, 1976).

As two of the arms of the triple rrr-junction evolve into an accreting plate boundary, an ocean will develop, and the newly-formed continental margin will subside following a time-dependent cooling curve (Sclater, Anderson, and Bell, 1971; Hays and Pitman, 1973). A miogeoclinal assemblage accumulates on the subsiding margin that will probably be thicker in the failed arm than in the adjacent continental margin and will overlie the graben facies well into the craton (Salop and Sheineman, 1969). Also by this time, the initially uplifted shoulders of the failed arm will reverse and subside (Hoffman, Dewey, and Burke, 1974).

As pointed out by Burke and Dewey (1973), failed arms are likely to be located at continental embayments (for example, Gulf of Guinea embayment and the Benue rift, Burke, Dessauvagie, and Whiteman, 1971) and localize the continental drainage thereby creating large clastic deposits in front of them in the form of deltas (for example, the Niger Delta as localized by the Benue Trough, Burke, 1972). Such anomalously thick, clastic deposits, perhaps overlying oceanic basement, may be identified in the stratigraphy of the collisional orogen, which, at that locality, may be at a "reentrant" due to the original embayment and thus reveal the pre-collisional age of the associated high-angle rift. (The most familiar example is in the Mackenzie Mountains of Canada, see fig. VIII-2 of Douglas and others, 1970).

As shown in figure 1C, failed arms may strike at any angle to the continental margin (a,b,c), but not all of them may survive through the collision to become aulacogens: those that are highly oblique to the colliding margins (c) are likely to undergo compressional deformation with minor strike-slip displacement during the collision: an example of this kind of a compressed, oblique aulacogen is represented in the present Timan Mountains (Kraus, 1972; Siedlecka, 1975). Compression due to shoulder reversal may also occur (see above), but it is unlikely to produce strike-slip displacement along the graben axis. Failed arms that strike at high angles to the collision front may be reactivated as rifts with a strike-slip component. Failed arms that are perpendicular to the collision front would be reactivated as purely extensional features. During the collisional reactivation and further evolution during intra-continental convergence (fig. 1C, D), the shoulders of the aulacogen are likely to rise again to provide a post-collisional clastic wedge to the periphery of the resulting arch; in the rift trough, a new graben facies will develop above the preserved miogeoclinal assemblages (fig. 1D).

In sharp contrast to aulacogens, collision grabens do not necessarily have an associated pre-graben doming, neither do they have any stratigraphic relation to the adjacent mountain belt (with the possible exception of the exogeosynclinal sequences, such as a possible correlation between the Middle Pechelbronn beds of the Upper Rhine Graben and the lowest units of the Lower Marine Molasse of the Alps). Their prerifting basement is composed of the rocks of the orogenic foreland, and the overlying graben facies is clearly post-collisional. Although they begin as purely extensional structures (Şengör and Göçmen, in preparation), a subsequent change in the orientation of the convergence vector across the collision zone may impose a late strike-slip component onto the structure as happened to the Upper Rhine Graben during Miocene and, mainly, Pliocene times (Illies, 1974a; see below). Post-rifting doming will give rise to clastic wedges, which, along the correlative fanglomerates of the graben fill, would reveal the age of onset of doming (fig. 1H-J).

As seen from the above discussion, the decisive test as to whether or not a high-angle rift is an aulacogen would be the detailed study of the stratigraphy of the graben fill; however, unless the graben is subsequently deformed, its pre-graben basement and the early sedimentary infill are generally not accessible to field examination. Even if the sedimentary record is revealed, either by drilling or by geophysical methods, it is often non-marine and therefore difficult to date. Therefore, the identification, in the stratigraphic record of the adjacent mountain belt opposite a high-angle rift of an exceptionally large delta and a deflection of the orogen toward the rift at this spot may be used as another, perhaps more practical, criterion from the viewpoint of the field geologist, for identifying the graben as an aulacogen. Early, pre-graben doming and early syn-collisional strike-slip movement along the rift are additional but risky criteria, as they may have other causes and should be used in conjunction with stronger evidence to identify a high-angle rift as an aulacogen.

Random rifting at or near a continental margin after the initial opening either as a result of a mantle plume or strike-slip tectonics parallel to the margin or as a result of membrane stresses (Turcotte and Oxburgh, 1973; Turcotte, 1974) may further complicate the picture (fig. 1E-G). In this case, the pre- (or syn- and/or post-, depending on the mode of origin) rifting doming will result in the erosion of the miogeoclinal strata, and the graben facies will be deposited on a thinner-than-normal miogeoclinal assemblage, as opposed to an aulacogen's thicker-than-normal, and a collision rift's altogether absent miogeoclinal sediment content. Clastic wedges associated with such a rift would be deposited unconformably over the miogeoclinal strata. However, if a failed arm should be subjected to a rejuvenation before the collision, then it may be extremely difficult, if not impossible, to establish its exact date of origin; a line of alkaline intrusives aligned along the graben axis and predating the later rejuvenation may give clues to the age of original rifting.

As a result of the difficulties involved in the study of grabens, with the exception of those that contain economic resources, most rift structures are rather poorly known. Therefore, especially in the case of pre-Phanerozoic high-angle rifts, it may be extremely difficult to deduce the nature of the relation between the graben and the adjacent orogenic belt. In the following sections we describe the evolution of a high-angle rift, the Upper Rhine Graben of Central Europe, which has variously been interpreted as having evolved as a result of doming and key-stone collapse during the Tertiary without relationship to the Alpine Orogeny (Cloos, 1939; Burke and Dewey, 1973); as a collision rift related to the Alpine Orogeny (Illies, 1947a,b; 1975; Şengör, 1976a,b); as a drag feature related to strike-slip faulting (Molnar and Tapponnier, 1975); and as an aulacogen, originated in the Triassic and reactivated by the Alpine Collision (Burke, 1977). The long history of geological investigation and rich economic resources of the Rhine Graben provides an extensive data base that allows us to show it to be collision induced. Recent efforts of German, Swiss, and French geoscientists, under the Upper Mantle and Geodynamics projects, have greatly contributed to our understanding of the evolution and present behavior of the Upper Rhine Graben. Their results are published largely in three recent compendia (Rothé and Sauer, 1967; Illies and Müller, 1970; Illies and Fuchs, 1974), which make up a large portion of our data base.

PERMIAN TO MIDDLE EOCENE DEVELOPMENT OF THE UPPER RHINE GRABEN AREA

If the Upper Rhine Graben is an aulacogen related to the Alpine Orogen, it should have a rifting history that would date back to the initial rifting in the Alpine area during the Triassic. In this section we describe the stratigraphy and gross basement structure of the Rhine Graben area from Permian to the end of the Mesozoic to show that there was no sign of a rift structure, a "failed arm", on the site of the future Rhine Graben.

Throughout its entire length of about 300 km, from the Jura Mountains to the young volcanic center of Vogelsberg, the Upper Rhine Graben is located mainly in the *Rhenoherzynikum* and *Saxothuringikum* zones of the Central European Variscides; the age of deformation in these zones ranges from late Devonian to Stephanian (Lotze, 1974).

The orogenic development of the Central European Variscides was largely completed during the late Pennsylvanian. By Rotliegendes (Early Permian) times, various east-northeast striking basins (for example, Saar-Nahe Basin, Kraichgau Basin) appeared and received largely clastic sediments with occasional intercalated volcanics (Boigk and Schöneich, 1970, 1974; Lotze, 1974; P. A. Ziegler, 1975). It is important to emphasize that all isopachs of the Rotliegendes deposits cross the future site of the Upper Rhine Graben indiscriminately (fig. 2). To the south of the Rhine Graben, in the Jura Mountains, continental Permian rests on Upper Carboniferous with angular unconformity.

The number of control points for the construction of the distribution of Zechstein sediments is unfortunately too few to permit an isopach analysis; however, from the existing data (Boigk and Schöneich, 1970, 1974; Lotze, 1974) and from P. A. Ziegler's (1975) paleogeographic map of the Zechstein, the following conclusions may be drawn: The sea of the Permian Basin in Northern Germany extended a tongue in the direction of the present Rhine Graben, in which marls, carbonates, and evaporites with local salt deposits were laid down. At about this time, the site of the

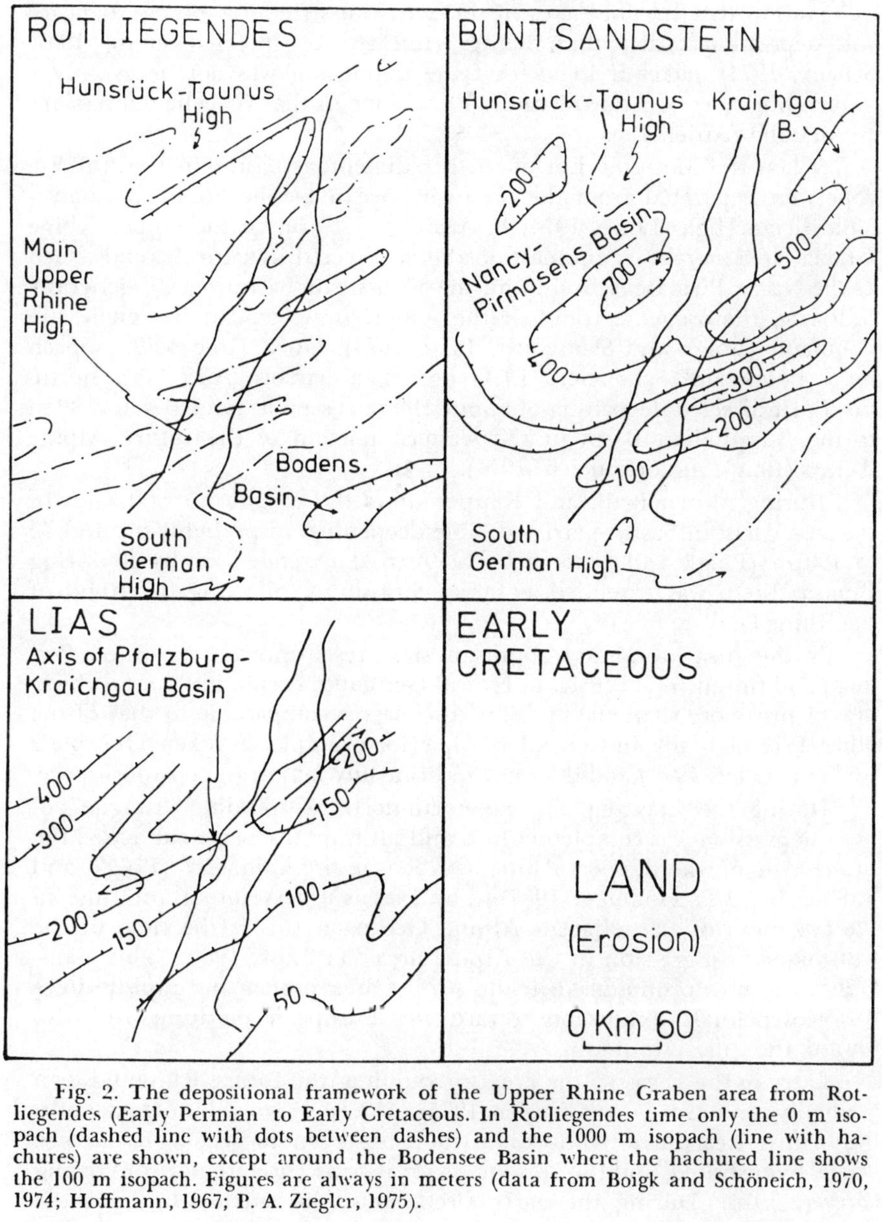

Fig. 2. The depositional framework of the Upper Rhine Graben area from Rotliegendes (Early Permian to Early Cretaceous. In Rotliegendes time only the 0 m isopach (dashed line with dots between dashes) and the 1000 m isopach (line with hachures) are shown, except around the Bodensee Basin where the hachured line shows the 100 m isopach. Figures are always in meters (data from Boigk and Schöneich, 1970, 1974; Hoffmann, 1967; P. A. Ziegler, 1975).

future Hessen Graben appears to be a preferred site of subsidence, although not as large as the future Upper Rhine Graben. Moreover, the observation that the fold axes of the Variscan structures plunge beneath this depression rather than being truncated by it (Hedemann, 1957; Schenk, 1974) makes it likely that the depression was not necessarily a graben but more of a downbending, like some of the Neogene depressions of Anatolia (Ardel, 1965).

When the Mesozoic Era opened, sedimentary basins in Central Europe were separated from the Tethyan Ocean by the South German—Vindelician High (Lotze, 1974; P. A. Ziegler, 1975). In the Upper Rhine Graben area, northeast and east-northeast directed structural trends, such as the Nancy-Pirmasens Basin and the Kraichgau Basin (fig. 2), generally following the orogenic trend of the underlying Variscan basement, are dominant (Boigk and Schöneich, 1970, 1974). On a large scale isopach map (Boigk and Schöneich, 1970, fig. 2), a narrow (~15 km), north-northeast directed deepening of about 100 m is seen around Basel, related to the Burgundian Basin in the western portion of the future Alpine Tethys (Boigk and Schöneich, 1974).

During Muschelkalk and Keuper times the east-northeast trends of the post-Variscan basins persisted. The deepening, of probably around 75 to 100 m (Boigk and Schöneich, 1970), that extended to Basel during Buntsandstein was extended as far as Strassburg following the trend of the Rhine Graben.

In the Jura Mountains, Buntsandstein rests unconformably on Permian and the entire Triassic, in typical Germanic Facies, has a total thickness of just more than 600 m. This thickness is comparable to that of the other Triassic basins in Central Europe (for example, Franken/Oberpfalz ≃845 m; Lotze, 1974) and does not indicate any major graben subsidence.

During the Lias (fig. 2), the north-northeast-trending "tongue" of weak depression was completely lost, and all isopachs again cut across the future site of the Upper Rhine Graben indiscriminately (Boigk and Schöneich, 1970; Hoffman, 1967). The Lias is a very important time in the evolution of the adjacent Alpine Ocean, as this is the time of the first major transgression in the Alpine area (Trümpy, 1960). This transgression is most pronounced in the Alpine area proper and rapidly loses its "geosynclinal" expression toward the European platform and also toward the Jura Mountains.

Later in the Jurassic, the area surrounding the future Rhine Graben gradually shallowed (P. A. Ziegler, 1975). Although syn-sedimentary faulting in this area during Oxfordian times has been suspected, it has not been demonstrated, and the arguments for its existence are not convincing (Breyer, 1974). During the early Cretaceous, the surroundings of the Rhine Graben became dry land (fig. 2) (Umbgrove, 1947; P. A. Ziegler, 1975). This so-called Rhine Shield (Cloos, 1939) was not confined, however, only to the future Upper Rhine Graben area but encompassed a wide region extending from Holland to the Bohemian Massif (P. A. Ziegler, 1975, fig. 16). Toward the end of the Mesozoic Era, volcanism began to affect various parts of Central Europe, including the Vosges, Schwarzwald, Mainz Basin, and the Taunus (Illies, 1947a). A majority of this activity is today preserved in the form of plugs and dikes that give ages in the range of 90 to 100 m.y. (Illies, 1974a). No major uplift, similar in magnitude to those observed in Africa (Burke and Whiteman, 1973), is reported to have accompanied these eruptions, although regional warping, tilting, and jostling of rigid to semi-rigid blocks (*Schollen*), giving rise to local intense deformations (for example, the remarkable Osning overthrust zone, Stille, 1953) have occurred since the Late Jurassic in Central Europe (Lotze, 1953, 1974; Keller, 1976). It is uncertain whether this period of magmatic activity, mainly of olivine-nepheline-bearing basic rocks (Illies, 1974a) can be related to this kind of deformation.

In summary, the Mesozoic Era, when the Alpine Ocean was born and later during Cenomanian began to contract (Dewey and others, 1973; Dietrich, 1976), closed with no indication of any kind of major graben subsidence on the site of the future Rhine Graben. The Mesozoic strata "bear no relation to the graben" (Sittler, 1969, p. 545). They thicken from the south to the north in Europe, probably due to the influence of the Vindelician High. During the early Cretaceous, Central Europe became land, and later in the Cretaceous volcanicity appeared accompanying faster uplift of the northern sector of Central Europe (Sittler, 1969).

LUTETIAN TO RECENT EVOLUTION OF THE UPPER RHINE GRABEN

In this section we present data that indicates that the Upper Rhine Graben originated during the Lutetian as a result of the Mesoalpine collision (Şengör, 1976b). The Tertiary history of the Upper Rhine Graben area is much clearer than the Mesozoic. The initial downfaulting of the present Rhine Graben began in the south and is indicated by the conglomerates of the *Siderolitikum* of probable Lutetian age (Doebl, 1970). This initial downfaulting was accompanied also by mafic volcanicity along the master faults of the rift, which volcanics yield ages at around 48 m.y. (Illies, written commun.). Drilling and geophysics (gravity and magnetics) show that the graben floor inherited the east-northeast-trending trough and sill structure of the post-Variscan basement, and the initial infill of the graben also shows variable thickness and facies within different sub-basins (Sittler, 1969, figs. 3 and 4). By Priabonian (late Eocene) times, subsidence had accelerated in the south (Sittler, 1969; Illies, 1947b), and the Limnea marls of fresh water origin are up to about 900 m thick to the southwest of Freiburg, 500 m thick near Karlsruhe, and wedge out near Mannheim (Illies, verbal commun., 1977; fig. 3B). In early Oligocene times, the Pechelbronn beds, generally of terrestrial origin with the exception of the Middle Pechelbronn strata laid down by a marine incursion, were deposited. The Pechelbronn beds extend into the Hessen Graben, which had become a graben during the Eocene (Illies, 1974b), where they are known as the *Melania* Clay (Lotze, 1974). During Pechelbronn times along the master faults of the Rhine Graben in the south are the so-called *Küstenkonglomerate* (coastal conglomerates) that indicate not only the activity of the faults here, but also the post-rifting uplift of the graben shoulders. Early pebbles of this series contain Jurassic material (Illies, 1967), showing that the major uplift of the graben shoulders occurred after faulting (Illies, 1970).

The area of most rapid subsidence of the graben floor was still in the south during Pechelbronn times, although rifting in general had already reached the Hessen Graben as indicated by the continuity of the Pechelbronn beds and the Melania Clay to the north. The thickness of the Pechelbronn beds reaches 1600 m to the southwest of Freiburg and about 900 m near Karlsruhe (Illies, 1974b; fig. 3B).

During the medial Oligocene, graben subsidence levelled off and the Grey Beds, a marine marl sequence, were deposited with more or less uniform thickness (Doebl, 1970) (fig. 3B). During the late Oligocene, the subsidence slowed down, and the deposits of this age, the Niederroedern Beds, are disconformably overlain (fig. 3B).

With the onset of the Miocene, two important changes occurred in the evolution of the Rhine Graben: first, the center of subsidence shifted to the north, where today the thickest accumulations of Tertiary sediments are found (Illies, 1974a,b; Sittler, 1969; fig. 3A), and later in Miocene times (18 m.y.) volcanicity began within the rift (Illies, 1975).

The Aquitanian section in the Rhine Graben consists of ~1500 m of clastics and carbonates (fig. 3B). However, after close of the Oligocene, the subsidence axis in the Upper Rhine Graben shifted from a north-northeast-trend to almost northwest (Illies, 1974a,b). This is also reflected by the thickness of the late Tertiary/Quaternary fill of the graben (fig. 3A). Likewise, the Aquitanian section was deposited in a "secondary" northwest-trending rift within the main trough of the Upper Rhine Graben (Illies, 1974a).

The Kaiserstuhl volcanic center began its activity about 18 m.y. ago (Illies, 1974a,b, 1975), about 27 m.y. after the Rhine Graben had originated and in the south subsided to more than 2.5 km. The subvolcanic breccias of the Kaiserstuhl (Baranyi, 1974) and the close correspondence of volcanism and graben suggest that faulting controlled the volcanism.

Toward the end of the Tertiary, Rhine Graben tectonism began to change from pure extension normal to the graben axis to sinistral strike-slip (Illies, 1972, 1974a,b; 1975; Illies and Greiner, 1976). Today the graben as a whole is a broad strike-slip zone with associated second-order extensional and compressional features (Ahorner, 1970; Ahorner and Schneider, 1974; fig. 4). The central portion of the graben is no longer a rift but a ramp valley as shown by the high-angle thrust faults near Baden-Baden (Illies and Greiner, 1976, fig. 3) and by seismic first motion studies (Ahorner and Schneider, 1974) (fig. 4).

Detailed knowledge of the structure and stratigraphy of the Upper Rhine Graben trough and its basement enabled Illies (1967) to reconstruct palinspastically the pre-rifting geometry and thereby determine the

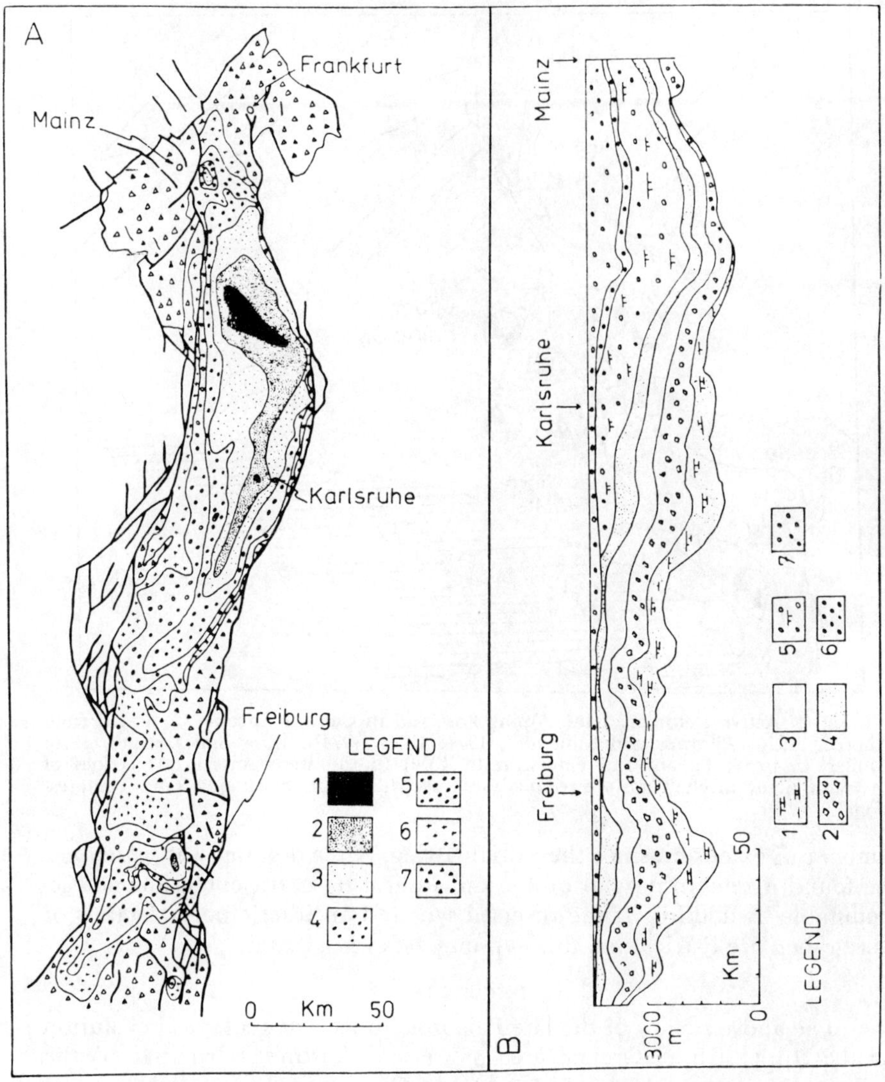

Fig. 3. (A) Distribution of the thicknesses of the Tertiary sediments in the Upper Rhine Graben trough. Heavy lines are faults. Key: 1, > 3000 m; 2, 3000 to 2500 m; 3, 2500 to 2000 m; 4, 2000 to 1500 m; 5, 1500 to 1000 m; 6, 1000 to 500 m; 7, <500 m.

(B) Generalized stratigraphic section along the Upper Rhine Graben trough. Key: 1, Lymnea marls (inclusive of *Siderolitikum* here); 2, Pechelbronn beds; 3, Gray beds; 4, Niederrödern beds; 5, Aquitanian deposits; 6, Upper Miocene; 7, Plio-Pleistocene.

Both (A) and (B) are simplified after Illies, 1974b.

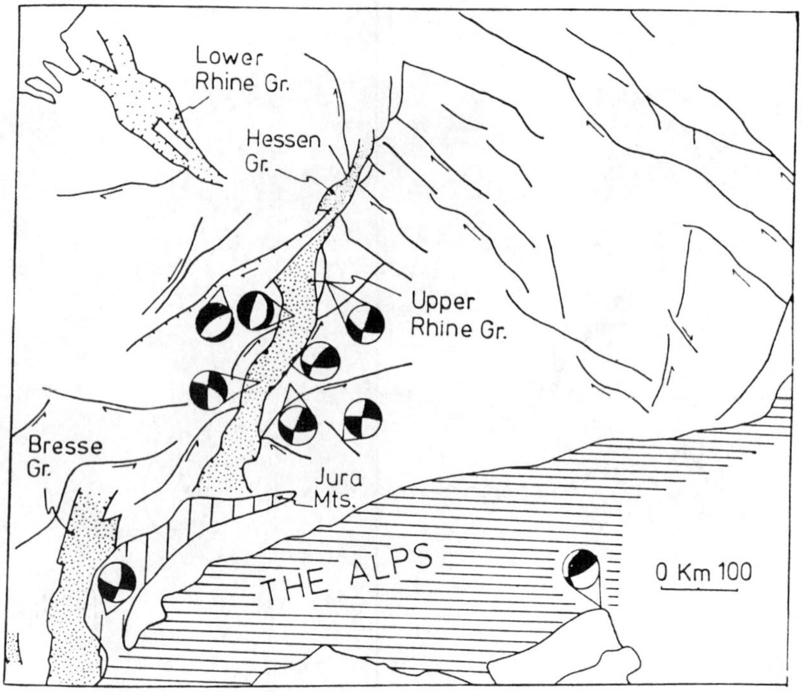

Fig. 4. Active tectonics of the Alpine Foreland in Central Europe. Compiled from Ahorner, 1970; Ahorner and Schneider, 1974; Illies, 1974b; Illies and Greiner, 1976; Müller, in press; Pavoni and Peterschmitt, 1974. In the stereographic projections of the fault plane mechanisms white quadrants are dilational, black quadrants compressional.

amount of extension since the initial rifting. After doming was taken out, he found a remaining gap of 4.8 km, indicating extra extension not accountable by doming. If the reversed syn- and antithetic normal faults of the graben are listric, then this gap may be even greater.

DISCUSSION

The above review of the late Paleozoic to Recent geological evolution of the Upper Rhine Graben area of Central Europe shows that (1) the miogeoclinal assemblages of the Alpine Ocean do not extend into the present graben; (2) during late Palaeozoic and Mesozoic times, the Rhine Graben area did not have a "precursor" of the present graben; (3) before the graben formed during Lutetian time, there had been no pre-graben doming in the area that could have led to a "key stone-drop" to form the Rhine Graben. From this, it is evident that the Upper Graben is not, as suspected by Burke (1977), an aulacogen. Cloos (1939) and Burke and Dewey (1973) suggested that it formed as a result of doming around the Vogelsberg volcano; the Hessen and the Lower Rhine Grabens were also ascribed to the same doming, and the three rifts were suggested to form the Frankfurt triple junction (Burke and Dewey, 1973). However, the data, not only from the Upper Rhine Graben but also from the Lower Rhine Graben and Hessen Graben, do not support the doming-rifting-triple junction formation hypothesis. The northwestern part of the Lower Rhine Graben was active during the Maastrichtian (Teichmüller, 1974), and possibly, was related, originally to the North Sea rift system (Whiteman and others, 1975; P. A. Ziegler, 1975; W. H. Ziegler, 1975). Renewed rifting during Pliocene and Recent times extends from the Rheinische Schiefergebirge to the North Sea and is associated with Quaternary maar explosions and eruptions in Eifel area (Lotze, 1974; Greiner and Illies, in press). This new period of northwest-trending normal faulting coincides in time with sinistral shearing along the north-northeast striking Upper Rhine Graben. It may, therefore, be a second-order extensional feature related to the Upper Rhine Graben shear zone. The Hessen Graben has been largely inactive during late Tertiary times. Cloos' (1939) interpretation of the Upper and Lower Rhine and the Hessen Grabens does not explain the temporal evolution of the three rifts and especially the late Tertiary strike-slip movement of the Upper Rhine Graben. Moreover, the Upper Rhine Graben began its subsidence in the south and propagated northward; the doming-rifting hypothesis of Cloos demands the opposite.

Molnar and Tapponnier's (1975) suggestion that the Upper Rhine Graben originated as a "drag" structure suffers from the lack of a suitable wrench fault at the southern end of the graben. The zone of left-lateral shear connecting the southern end of the Upper Rhine Graben and the northern end of the Fosse Bressan (Contini and Theobald, 1974) appears to be younger than the initial graben formation and was perhaps formed as an intra-continental transform fault between the two rifts (Illies, 1974b).

Illies (1975) and Şengör (1976b) correlated the rifting events in the Upper Rhine Graben with the Alpine collision and later intra-continental convergence in the Western Alps, concluding that the Upper Rhine Graben is the *result* of the Alpine orogeny. Illies (1974a, 1975), however, believes that subduction in the Alpine Realm first gave rise to a subcrustal swelling of the mantle beneath the southern Upper Rhine Graben, which caused or greatly facilitated rifting. However, this view is difficult to reconcile with the southerly dip of the Alpine subduction zone (Dewey and others, 1973; Trümpy, 1975; Hawkesworth, Waters, and Bickle, 1975; Dietrich, 1976) notwithstanding Oxburgh's (1972) flake model. Even if the system involved a three-plate geometry with two subduction zones of opposing polarity, as suggested by Roeder (1976), the conspicuous absence of a subduction related arc and the paucity of andesitic volcanism make it unlikely that the magnitude of northerly and/or southerly subduction was great enough to have given rise to a mantle swell almost 200 km away from the area of the Alpine suture in the Ivrea Zone. The late Miocene collision along the Bitlis suture (Hall, 1976) in southeastern Turkey gave rise without any indication of a crustal uplifting or preliminary dike injection to north-south trending rifting and fissure formation, along which the Plio-Pleistocene alkaline basalts of the Karacalidağ volcanic province erupted (Şengör and Göçmen, in preparation). We therefore follow Şengör's (1976a,b) interpretation that holds the Alpine collision directly responsible for the Rhine Graben rifting.

In conclusion, we emphasize that high-angle rifts may have multiple causes of origin both genetically related to the evolution of the associated mountain belt and also independent of it. Because oceans are likely to open and close several times along roughly the same lines (J. T. Wilson, 1966b), high-angle rifts may be rejuvenated several times during their life history with all the complexities of their pre- and post-collisional evolution, as schematically shown in figure 1. Such multiple events may obscure the origin of a high-angle rift or rift system beyond any reasonable hope of unraveling their history. Geologists working in such high-angle rift areas should be aware of the bewildering complexities of their evolution and hesitate to draw hasty conclusions as to their origin and tectonic significance; for instance, if a collisional orogeny is immediately followed by ocean opening along the same lines it may be especially difficult to decide whether the rift belongs to the collision event and is therefore a collision rift, or to the opening cycle and is therefore a failed arm.

ACKNOWLEDGEMENTS

We thank Professor Henning Illies and Mr. Gerhard Greiner of Universität Karlsruhe, Germany, for long hours of discussion on the evolution of the Upper Rhine Graben and a pre-publication copy of their paper on the present stress regime in Europe. We are also grateful to Professor Sirri Erinç and Dr. Kemâl Göçmen of Istanbul Üniversitesi, Turkey, for discussions on the Karacalidağ basalts in southeast Turkey. We are indebted to Professor Henning Illies for critically reading the manuscript.

REFERENCES

Ahorner, L., 1970, Seismo-tectonic relations between the graben zones of the Upper and Lower Rhine Valley, *in* Illies, J. H. and Mueller, S., eds., Graben Problems: Stuttgart, E. Schweizerbart'sche Verlagsbuch., p. 155-166.

Ahorner, L., and Schneider, G., 1974, Herdmechanismen von Erdbeben im Oberrhein-Graben und in seinen Randgebirgen, *in* Illies, J. H. and Fuchs, K., eds., Approaches to Taphrogenesis: Stuttgart, E. Schweizerbart'sche Verlagsbuch., p. 104-117.

Ardel, A., 1965, Anadolu Havzalarinin Teşekkül ve Tekamülü Hakkinda Düşünceler: Istanbul Üniv. Coğrafya Enstitüsü Dergisi, v. 8, p. 60-73.

Baker, B. H., Mohr, P. A., and Williams, L. A. J., 1972, Geology of the Eastern Rift System of Africa: Geol. Soc. America Spec. Paper 136, 67 p.

Baranyi, I., 1974, The subvolcanic breccias of the Kaiserstuhl Volcano (SW-Germany), *in* Illies, J. H. and Fuchs, K., eds., Approaches to Taphrogenesis: Stuttgart, E. Schweizerbart'sche Verlagsbuch., p. 226-230.

Bird, J. M., and Dewey, J. F., 1970, Lithosphere plate—continental margin tectonics and the evolution of the Appalachian Orogen: Geol. Soc. America Bull., v. 81, p. 1031-1060.

Boigk, H., and Schöneich, H., 1970, Die Tiefenlage der Permbasis im nördlichen Teil des Oberrheingrabens, *in* Illies, H. J. and Mueller, S., eds., Graben Problems: Stuttgart, E. Schweizerbart'sche Verlagsbuch., p. 45-55.

———— 1974, Perm, Trias und älterer Jura im Bereich der südlichen Mittelmeer-Mjösen-Zone und des Rheingrabens, *in* Illies, J. H. and Fuchs, K., eds., Approaches to Taphrogenesis: Stuttgart, E. Schweizerbart'sche Verlagsbuch., p. 60-71.

Breyer, F., 1974, Structure and development of the south part of the Rhine Graben according to geological and geophysical observations, *in* Illies, J. H. and Fuchs, K., eds., Approaches to Taphrogenesis: Stuttgart, E. Schweizerbart'sche Verlagsbuch., p. 145-154.

Burchfiel, B. C., and Davis, G. A., 1975, Nature and controls of Cordilleran Orogensis, western United States: extensions of an earlier synthesis: Am. Jour. Sci., v. 275-A, p. 363-396.

Burke, Kevin, 1972, Longshore drift, submarine canyons, and submarine fans in development of Niger Delta: Am. Assoc. Petroleum Geologists Bull., v. 56, p. 1975-1983.

——— 1976, Development of graben associated with the initial ruptures of the Atlantic Ocean: Tectonophysics, v. 36, p. 93-112.

——— 1977, Aulacogens and continental breakup: Ann. Rev. Earth Planetary Sci., v. 5, p. 371-396.

Burke, K., Dessauvagie, T. F. J., and Whiteman, A. J., 1971, Opening of the Gulf of Guinea and geological history of the Benue Depression and Niger Delta: Nature Phys. Sci., v. 233, p. 51-55.

Burke, K., and Dewey, J. F., 1973, Plume-generated triple junctions: key indicators in applying plate tectonics to old rocks: Jour. Geology, v. 81, p. 406-433.

Burke, K., and Whiteman, A. J., 1973, Uplift, rifting and the break-up of Africa, *in* Tarling, D. H. and Runcorn, S. K., eds., Implications of Continental Drift to the Earth Sciences: London, Acad. Press, p. 735-755.

Burke, K., and Wilson, J. T., 1976, Hot spots on the earth's surface: Sci. Am., v. 235, p. 113-130.

Cloos, H., 1939, Hebung-Spaltung-Vulkanismus: Geol. Rundschau, v. 30, p. 405-527.

Contini, D., and Théobald, N., 1974, Relations entre le Fossé rhénan et le Fossé de la Sâone, Tectonique des régions sous-vosgiennes et prejurassiennes, *in* Illies, J. H. and Fuchs, K., eds., Approaches to Taphrogenesis: Stuttgart, E. Schweizerbart'sche Verlagsbuch., p. 310-321.

Curray, J. R., and Moore, D. G., 1974, Sedimentary and tectonic processes in the Bengal Deep-Sea Fan and Geosyncline, *in* Burk, C. A. and Drake, C. L., eds., The Geology of Continental Margins: Berlin, Springer-Verlag, p. 617-628.

DeSitter, L. U., 1956, Structural Geology: New York, McGraw-Hill Book Co., 552 p.

Dewey, J. F., and Burke, K., 1974, Hot spots and continental break-up: implications for collisional orogeny: Geology, v. 2, p. 57-60.

Dewey, J. F., Pitman, W. C., III, Ryan, W. B. F., and Bonnin, J., 1973, Plate tectonics and the evolution of the Alpine System: Geol. Soc. America Bull., v. 84, p. 3137-3180.

Dietrich, V. J., 1976, Plattentektonik in den Ostalpen. Eine Arbeitshypothese: Geotekt. Forsch., no. 50, p. 1-84.

Doebl, F., 1970, Die Tertiären und Quartären Sedimente des Südlichen Rheingrabens, *in* Illies, J. H. and Mueller, S., eds., Graben Problems: Stuttgart, E. Schweizerbart'sche Verlagsbuch., p. 56-66.

Douglas, R. J. W., Gabrielse, H., Wheeler, J. O., Stott, D. F., and Belyea, H. R., 1970, Geology of western Canada, *in* Douglas, R. J. W., ed., Geology and Economic Minerals of Canada: Ottawa, Canada Geol. Survey, p. 365-488.

Greiner, G., and Illies, J. H., in press, Central Europe: active or residual tectonic stresses: Pure and Appl. Geophysics, v. 115.

Hall, R., 1976, Ophiolite emplacement and the evolution of the Taurus Suture Zone, southeastern Turkey: Geol. Soc. America Bull., v. 87, p. 1078-1088.

Hawkesworth, C. J., Waters, D. J., and Bickle, M. J., 1975, Plate tectonics in the eastern Alps: Earth Planetary Sci. Letters, v. 24, p. 405-413.

Hays, J. D., and Pitman, W. C., III, 1973, Lithosphere plate motion, sea level changes and climatic and ecological consequences: Nature, v. 246, p. 18-22.

Hedemann, H. A., 1956/1957, Die Gewölbestruktur des Sollings und ihre Entstehung: Geol. Jahrb., v. 72, p. 73.

Hoffman, K., 1967, Der Oberrhein-Graben Zwischen Heidelberg und Offenburg in der Lias-Zeit [abs.]: Deutsche Geol. Gesell. Zeitschr., v. 118, p. 236.

Hoffman, P., Dewey, J. F., and Burke, K., 1974, Aulacogens and their genetic relation to geosynclines with a Proterozoic example from Great Slave Lake, Canada, *in* Dott, R. H. and Siever, R. H., eds., Modern and Ancient Geosynclinal Sedimentation: Soc. Econ. Paleontologists Mineralogists Spec. Pub. 19, p. 38-55.

Illies, J. H., 1967, An attempt to model Rheingraben tectonics: Abh. geol. Landesamt Baden-Württemberg, v. 6, p. 7-9.

——— 1970, Graben tectonics as related to crust-mantle interaction, *in* Illies, J. H. and Mueller, S., eds., Graben Problems: Stuttgart, E. Schweizerbart'sche Verlagsbuchhandlung, p. 4-27.

——— 1972, The Rhine Graben Rift System—Plate Tectonics and Transform Faulting: Geophys. Survey, v. 1, p. 27-60.

——— 1974a, Intra-Plattentektonik in Mitteleuropa und der Rheingraben: Oberrhein. Geol. Abh., v. 23, p. 1-24.

——— 1974b, Taphrogenesis and plate tectonics, *in* Illies, J. H. and Fuchs, K., eds., Approaches to Taphrogenesis: Stuttgart, E. Schweizerbart'sche Verlagsbuch., p. 433-460.

——— 1975, Intraplate tectonics in stable Europe as related to plate tectonics in the Alpine System: Geol. Rundschau, v. 64, p. 677-699.

Illies, J. H., and Fuchs, K., eds., 1974, Approaches to Taphrogenesis: Stuttgart, E. Schweizerbart'sche Verlagsbuch., 460 p.

Illies, J. H., and Greiner, G., 1976, Regionales stress-feld und Neotektonik in Mitteleuropa: Oberrhein. geol. Abh., v. 25, p. 1-40.

Illies, J. H., and Mueller, S., eds., 1970, Graben Problems: Stuttgart: E. Schweizerbart' sche Verlagsbuch., 316 p.

Jacobs, J. A., Russel, R. D., and Wilson, J. T., 1959, Physics and Geology: New York, McGraw-Hill Book Co., 424 p.

Keller, G., 1976, Saxonische Tektonik und die Osning-Zone: Deutsche Geol. Gesell. Zeitschr., v. 127, p. 297-307.

King, B. C., 1970, Vulvanicity and rift tectonics in east Africa, *in* Clifford, T. N. and Gass, I. G., eds., African Magmatism and Tectonics: Edinburgh, Oliver and Boyd, p. 263-284.

King, B. C., and Chapman, G. R., 1972, Volcanism of the Kenya Rift Valley: Royal Soc. London Philos. Trans., A., v. 271, p. 185-208.

Kraus, E., 1972, Die Entwicklungsgeschichte der Kontinente und Ozeane: Berlin, Akad. Verlag, 429 p.

Lotze, F., 1953, Einige Probleme der Osningtektonik: Geotekt. Forsch., no. 9/10, p. 7-17.

——— 1974, Geologie Mitteleuropas: Stuttgart, E. Schweizerbart'sche Verlagsbuch., 491 p.

Molnar, P., and Tapponnier, P., 1975, Cenozoic tectonics of Asia: effects of a continental collision: Science, v. 189, p. 419-426.

Müller, G., in press, Fault-plane solution of the earthquake in northern Italy, 6 May 1976, and implications for the tectonics of the eastern Alps: Geophys. Jour., in press.

Oxburgh, E. R., 1972, Flake tectonics and continental collision: Nature, v. 239, p. 202-204.

Pavoni, N., and Peterschmitt, E., 1974, Das Erdbeben von Jeurre vom 21. Juni 1971 und seine Beziehungen zur Tektonik des Faltenjura, *in* Illies, J. H. and Fuchs, K., eds., Approaches to Taphrogenesis: Stuttgart, E. Schweizerbart'sche Verlagsbuch., p. 322-328.

Rankin, D. W., 1976, Appalachian salients and recesses: Late Precambrian continental breakup and the opening of the Iapetus Ocean: Jour. Geophys. Research, v. 8, p. 5605-5619.

Roeder, D. H., 1976, Die Alpen aus Plattentektonischer Sicht: Deutsche Geol. Gesell. Zeitschr., v. 127, p. 87-103.

Rothé, J. P., and Sauer, K., eds., 1967, The Rhinegraben Progress Report 1967: Abh. geol. Landesamt Baden-Württemburg, v. 6, 146 p.

Salop, L. I., and Scheineman, Y. M., 1969, Tectonic history and structure of platforms and shields: Tectonophysics, v. 7, p. 565-597.

Schenk, E., 1974, Die Fortsetzung des Rheingrabens durch Hessen. Ein Beitrag zur tektonischen Analyse der Riftsysteme, *in* Illies, J. H. and Fuchs, K., eds., Approaches to Taphrogenesis: Stuttgart, E. Schweizerbart'sche Verlagsbuch., p. 286-302.

Sclater, J. G., Anderson, R. N., and Bell, M. L., 1971, Elevation of ridges and evolution of the central eastern Pacific: Jour. Geophys. Research, v. 76, p. 7888-7915.

Şengör, A. M. C., 1976a, Rift valley formation as a combined result of parting and uplift [abs.]: Geol. Soc. America Abs. with Programs, v. 8, p. 65.

——— 1976b, Collision of irregular continental margins: implications for foreland deformation of Alpine-type Orogens: Geology, v. 4, p. 427-430.

Shatski, N., 1946a, Basic features of the structures and development of the east European platform: Akad. Nauk SSR Izv. Geol. Ser., no. 1, p. 5-62.

———— 1946b, The Great Donets Basin and the Wichita System: comparative tectonics of ancient platforms: Akad. Nauk SSR Izv. Geol. Ser., no. 6, p. 57-90.

———— 1947, Structural correlations of platforms and geosynclinal folded regions: Akad. Nauk SSR Izv. Geol. Ser., no. 5, p. 37-56.

———— 1955, On the origin of the Pachelma Trough: Mosk. Lyubit. Prir. Byull. Geol. Sec., v. 5, p. 5-26.

Siedlecka, A., 1975, Late Precambrian stratigraphy and structure of the northeastern margin of the Fennoscandian Shield (East Finnmark-Timan Region): Norges geol. undersøk., no. 136, p. 313-348.

Sittler, C., 1969, The sedimentary trough of the Rhine Graben: Tectonophysics, v. 8, p. 543-560.

Steel, R. J., 1976, Devonian basins of western Norway—sedimentary response to tectonism and to varying tectonic context: Tectonophysics, v. 35, p. 207-224.

Stille, H., 1953, Zur Geschichte der Osningforschung: Geotekt. Forsch., no. 9/10, p. 1-6.

Teichmüller, R., 1974, Die tektonische Entwicklung der Niederrheinischen Bucht, *in* Illies, J. H. and Fuchs, K., eds., Approaches to Taphrogenesis: Stuttgart, E. Schweizerbart'sche Verlagsbuch., p. 269-285.

Turcotte, D., 1974, Membrane tectonics: Royal Astron. Soc. Geophys. Jour., v. 36, p. 33-42.

Turcotte, D., and Oxburgh, E. R., 1973, Mid-plate tectonics: Nature, v. 244, p. 337-339.

Trümpy, R., 1960, Paleotectonic evolution of the central and western Alps: Geol. Soc. America Bull., v. 71, p. 843-908.

———— 1975, Penninic-Austroalpine boundary in the Swiss Alps: a presumed former continental margin and its problems: Am. Jour. Sci., v. 275-A, p. 209-238.

Umbrove, J. H. F., 1947, The pulse of the earth: The Hague, Martinus Nijhoff, 358 p.

Weber, M., 1921, Zum Problem der Grabenbildung: Deutsche Geol. Gesell. Zeitschr., v. 73, p. 238-291.

———— 1923, Bemerkungen zur Brunchtektonik: Deutsche Geol. Gesell. Zeitschr., v. 75, p. 184-192.

———— 1927, Faltengebirge und Vorlandbrüche: Centralbl. Mineralogie, Geologie, Paläontologie, Abt. 5, p. 235-245.

Whiteman, A. J., Rees, G., Naylor, D., and Pegrum, R. M., 1975, North Sea Troughs and Plate Tectonics: Norges geol. undersøk., nr. 136, p. 137-161.

Wilson, J. T., 1966a, Some rules for continental drift, *in* Gardland, G. D., ed., Continental Drift: Royal Soc. Canada Spec. Pub. 9, p. 3-17.

———— 1966b, Did the Atlantic close and then re-open?: Nature, v. 211, p. 676-681.

Wilson, R. C. L., 1975, Atlantic opening and Mesozoic continental margin basins of Iberia: Earth Planetary Sci. Letters, v. 25, p. 33-43.

Ziegler, P. A., 1975, North Sea Basin history in the tectonic framework of N-W Europe, *in* Woodland, A. W., ed., Petroleum and the Continental Shelf of N-W Europe: London, Halsted-Wiley, v. 1, p. 131-150.

Ziegler, W. H., 1975, Outline of the geological history of the North Sea, *in* Woodland, A. W., ed., Petroleum and the Continental Shelf of N-W Europe: London, Halsted-Wiley, v. 1, p. 165-190.

Part II

AFRICA/ARABIA

The Afro-Arabia belt (chain) of rift valleys, zones, and systems has been described and discussed in *Rift Valleys: Afro-Arabian* (Benchmark Papers in Geology, Volume 60, 1982). The present volume reprints only papers on rifts outside that belt: the Benue valley in West Africa and the Central Arabian grabens.

RIFTING AND THE BREAKUP OF AFRICA

In 1965 Holmes stated the definitive short description of the crustal structure of Africa: "Africa, to which Arabia must be added from a structual point of view, is a vast continental shield, margined with mountainous folded belts only in the extreme north and south. . . . For hundreds of millions of years the movements of the shield have been persistently epeirogenic, giving a structural pattern of broad basins separated by irregular swells which rise towards the east to form a coalescing series of plateaus, the latter being traversed by a spectacular series of rift valleys" (Holmes, 1965, p.1053; see also Cahen et al., 1984; and Clifford, 1970). The active rift valleys of Africa, and indeed also the passive, are located on the swells, as described by Holmes, and not in the basins

The "breakup" of the original Afro-Arabian continent (Burke, 1977) appears to have followed the separation of South America and the formation of the South Atlantic by a three-rift system of which the Benue rift (aulacogen) is the failed arm. This lies between elevated regions. The separation of Arabia with the formation of the Red Sea and the Gulf of Aden was by a three-rift system on the Arabo-Ethiopian domal uplift. The third (failed) arm is the main Ethiopian rift (Gass, 1970).

Some believe that the East African rift systems (eastern and western) are the manifestations of future breakup of the continent. Crust-mantle structure, determined geophysically and geochemically, is advanced as evidence. There is conflict in interpretation, however.

The rifts of these systems are located on the East African swell and on the Kenya domal uplift in a manner similar to the location of the Red Sea, Gulf of Aden, and Ethiopian rifts on the Arabo-Ethiopian dome. This similarity could be cited as evidence of East African potential continental breakup. Wilson (in Jacobs, Russell, and Wilson, 1974) referred to the East African system as the prototype of the first stage of the Wilson cycle (see Introduction).

Recently it has been suggested that the breakup of Africa can be taking place along geophysically located "rifts" lying not within but between swells, that is, in the basins (Bermingham, Fairhead, and Stuart, 1983; Browne and Fairhead, 1983). Historical comparision and geological constraints as outlined above are against the location of true rifts in the basins, the configuration of which is, in any case, only partially known.

REFERENCES

This list of references includes articles on subjects within the Afro-Arabian belt, supplementing those that appeared in Benchmark Volume 60. The articles are drawn chiefly from symposia and reports that appeared after 1979, that is, after the Rio Grande symposium (publication 16 in the Bibliography), the two volumes of reports on the Advanced Study Institute on paleorift systems at Oslo in 1977 (publications 14 and 15 in the Bibliography), and the Dead Sea rift symposium in Jerusalem in 1979 (publication 18 in the Bibliography), all of which are dealt with in Volume 60.

Bahat, D., 1984, Fracture interaction in the Gregory rift, East Africa, *Tectonophysics* **104:**47–65.

Barberi, F., G. Capaldi, P. Sasperini, G. Marinelli, R. Santacroce, R. Scandone, M. Treuil, and J. Varet, 1980, Recent basaltic volcanism of Jordan and its implications on the geodynamic history of the Dead Sea shear zone, in *Geodynamic Evolution of the Afro-Arabian Rift System,* Accademia Nazionale dei Lincei, Rome, pp. 667–683.

Bartov, Y., G. Steinetz, M. Eyal, and Y. Eyal, 1980, Sinistral movement along the Gulf of Aqaba—its age and relation to the opening of the Dead Sea, *Nature* **285:**220–221.

Bender, F., 1983, On the evolution of Wadi Araba–Jordan Rift, in *Proceedings of the First Jordanian Geological Conference 1982,* Jordanian Geologists' Association, Amman, pp. 415–445.

Bermingham, P. M., J. D. Fairhead, and G. W. Stuart, 1983, Gravity study of the Central African Rift System: a model of continental disruption. 2. The Darfur domal uplift and associated Cainozoic volcanism, *Tectonophysics* **94:**205–222.

Boucarut, M., and M. Clin, 1980, Structural history of the Republic of Djibouti—a summary, in *Geodynamic Evolution of the Afro-Arabian Rift System,* Accademia Nazionale dei Lincei, Rome, pp. 541–544.

Browne, S. E., and J. D. Fairhead, 1983, Gravity study of the Central African Rift System: a model of continental disruption, 1. The Ngaoundere and Abu Gabra rifts, *Tectonophysics* **94:**187-204.

Burke, K., 1977, Aulacogens and continental breakup, *Ann. Rev. Earth and Planetary Sci.* **5:**371-396.

Cahen, L., N. J. Snelling, J. Delhal, and J. R. Vail, 1984, *The Geochronology and Evolution of Africa,* Clarendon Press, Oxford, 512p.

Clifford, T. N., 1970, The structural framework of Africa, in *African Magmatism and Tectonics,* T. N. Clifford and I. G. Gass, eds., Oliver & Boyd, Edinburgh, pp. 285-298.

Crane, K., and S. O'Connell, 1983, The distribution and implications of heat flow from the Gregory rift in Kenya, *Tectonophysics* **94:**253-275.

Crossley, R., 1980, Structure and volcanism in the S. Kenya rift, in *Geodynamic Evolution of the Afro-Arabian Rift System,* Accademia Nazionale dei Lincei, Rome, pp. 89-98.

Crossley, R., and M. J. Crow, 1980, The Malawi Rift, in *Geodynamic Evolution of the Afro-Arabian Rift System,* Accademia Nazionale dei Lincei, Rome, pp. 77-87.

Fairhead, J. D., and G. W. Stuart, 1982, The seismicity of the East African Rift System and comparisons with other continental rifts, in *Continental and Oceanic Rifts,* G. Palmason, ed., Report of Working Group, Geodynamic Series 8, American Geophysical Union/Geological Society of America, pp. 41-61.

Gass, I. G., 1970, Tectonic and magmatic evolution of the Afro-Arabian dome, in *African Magmatism and Tectonics,* T. N. Clifford and I. G. Gass, eds., Oliver and Boyd, Edinburgh, pp. 285-299.

Girdler, R. W., 1983, Processes of planetary rifting as seen in the rifting and break-up of Africa, *Tectonophysics* **94:**241-252.

Holmes, A., 1965, *Principles of Physical Geology,* Nelson, London, 1288p.

Jacobs, J., R. D. Russell, and J. T. Wilson, 1974, *Physics and Geology,* 2nd ed., McGraw-Hill, New York, 622p.

Kohn, B. P., and M. Eyal, 1981, History of uplift of the crystalline basement at its relation to opening of the Red Sea as revealed by fission track dating of apatites, *Earth Planetary Sci. Letters* **52:**129-141.

McConnell, R. B., 1980, A Precambrian origin of the Protorift dislocation belt of Eastern Africa, in *Geodynamic Evolution of the Afro-Arabian Rift System,* Accademia Nazionale dei Lincei, Rome, pp. 35-43.

Mikbel, S., and M. Atallah, 1983, Tectonics of Jordan: contribution to the structures of North Jordan, *Proceedings of the First Jordanian Geological Conference,* 1982, Jordan Geological Association, Amman, pp. 446-459.

Milanovsky, E. E., 1980, Neotectonics of the East African rift system, in *Geodynamic Evolution of the Afro-Arabian Rift System,* Accademia Nazionale dei Lincei, Rome, pp. 49-63.

Mohr, P., 1981, Musing on continental rifts, in *Continental and Oceanic Rifts,* G. Palmason, ed., Geodynamic Series, vol. 8, (Rep. of Working Group 4), American Geophysical Union/Geological Society of America, Washington, D.C., pp. 293-309.

Morgan P., and B. H. Baker, eds., 1983, Processes of Continental Rifting, *Tectonophysics* **94**, 680p.

Nolet, G., and St. Mueller, 1982, A model for the deep structure of the East African rift system from simultaneous inversion of teleseismic data, *Tectonophysics* **84:**151-178.

Quennell, A. M., 1983, Evolution of the Dead Sea rift—a review, *Proceedings of the First Jordanian Geological Conference, 1982,* Jordan Geologists Association, Amman, pp. 460–482.

Quennell, A. M., 1984, Western Arabia rift system, in *The Geological evolution of the Eastern Mediterranean,* Geol. Soc. London, Spec. Pub. 17, pp. 587–600.

Richard, O., and J. Varet, 1980, Study of transition from deep oceanic to emerged rift zone-Gulf of Tadjura, in *Geodynamic Evolution of the Afro-Arabian Rift System,* Accademia Nazionale dei Lincei, Rome, pp. 569–582.

Tamsett, D., 1984, Comments on the development of rifts and transform faults during continental break-up; examples from the Gulf of Aden and northern Red Sea, *Tectonophysics* **104:**37–46.

Williams, L. A. J., and P. H. Truckle, 1980, Volcanic sequences in the Kenya rift, in *Geodynamic Evolution of the Afro-American Rift System,* Accademia Nazionale dei Lincei, Rome, pp. 123–132.

Editor's Comments on Papers 6 and 7

6 CRATCHLEY and JONES
Excerpts from *An Interpretation of the Geology and Gravity Anomalies of the Benue Valley, Nigeria*

7 AL KADHI and HANCOCK
Excerpts from *Structure of the Ḍurmā-Nisāh Segment of the Central Arabian Grabèn System*

Paper 6 (by C. R. Cratchley of the British Geological Survey and G. P. Jones of University College, London) comprises extracts from a description of the geology of the Benue Valley, Nigeria, with a gravity survey and its interpretation. This paper does not speculate on the origin of the depression, but the recorded geology and gravity are all consistent with its being an aulacogen of the "penetrating" type (Milanovsky, Paper 4). It is the failed arm of a triple junction (Burke, Dessauvagie, and Whiteman, 1971).

Paper 7 (by A. Al Khadi and P. L. Hancock) describes the Central Arabian graben and trough system set in a sequence of Phanerazoic carbonates and clastic sedimentary rocks overlying the eastern margin of the Arabian shield. The segment described is 150 km long, in a belt concave to the northeast. A study was made of the related mesoscopic fracture patterns. The intraplate movements of blocks are related to those of the plate margins of Arabia, especially of its northeastern and eastern margins. A. Al Khadi is at the University of Riyadh, and P. L. Hancock is at the University of Bristol, England.

REFERENCES

Adighije, C. I., 1981, A gravity interpretation of the Benue Trough, Nigeria, *Tectonophysics* **7:**109–228.

Artsybashev, V. A., 1975, Crustal structure of the Benue Valley area (Nigeria), *Geol. Rundschau* **64:**324–329.

Brown, G. F., 1972, Tectonic map of the Arabian peninsula, *Dir. Gen. Min. Res., Saudi Arabia,* Geol. map AP-2, Scale 1:4m.

Burke, K., 1969, Seismic areas of the Guinea Coast where Atlantic fracture zones reach Africa, *Nature* **222:**655.

Burke, K., T. F. J. Dessauvagie, and A. J. Whiteman, 1971, Opening of the Gulf of Guinea and geological history of the Benue Depression and Niger Delta, *Nature Phys. Sci.* **233:**51-55.

Fitton, J. G., 1983, Active versus passive continental rifting: evidence from the West African system, *Tectonophysics* **94:**473-481.

Hancock, P. L., and A. Al Kadhi, 1978, Analysis of mesoscopic fractures in the Dhruma Nisah segment of the central Arabian graben system, *Geol. Soc. London Jour.* **135:**339-347.

Hancock, P. L., S. O. Alkhateib, and A. Al Kadhi, 1981, Structural and photogeological evidence for the boundaries to an East Arabian block, *Geol. Mag.* **118:**533-538.

Hoffmann, P., J. F. Dewey, and K. Burke, 1974, Aulacogens and their genetic relation to geosynclines, *Society Economic Paleontol. Mineral. Spec. Pub.* No. 19, pp. 38-55.

Peters, S. W., 1978, Stratigraphic evolution of the Benue trough and its implication for the Upper Cretaceous paleogeography of West Africa, *Jour. Geology* **86:**311-322.

Wood, C. A., 1983, Continental rift jumps, *Tectonophysics* **94:**529-540.

Wright, J. B., 1981, Review of the origin and evolution of the Benue trough in Nigeria, *Earth Evol. Sci.* **2:**98-103.

6

Reprinted from pages 1 and 20–25 of *Geophys. Paper No. 1,* 1965, 26p.

AN INTERPRETATION OF THE GEOLOGY AND GRAVITY ANOMALIES OF THE BENUE VALLEY, NIGERIA[1]

C. R. Cratchley, M.Sc.,[2] and G. P. Jones, M.Sc.[3]

[1]Published by permission of the Director, Geological Survey of Nigeria.

[2]Overseas Geological Surveys, formerly of Geological Survey of Nigeria.

[3]University College, London, formerly of Geological Survey of Nigeria.

C. R. Cratchley and G. P. Jones

ABSTRACT

ON THE AVAILABLE geological evidence there is little to choose between the two apparently contradictory hypotheses that the Benue 'trough' is a buried Cretaceous rift valley produced by tension, or a minor geosynclinal furrow formed by compression. The gravity survey was undertaken in an attempt to resolve this problem.

Stratigraphical evidence indicates that sinking along a linear depression must have commenced by mid-Albian time and continued until perhaps late Senonian, interrupted by some uplift and possibly folding in the late Albian. Compressional folding of the Cretaceous rocks began in the Senonian and resulted in the formation of a series of long, narrow folds whose axes are generally parallel to the boundaries of the linear basin. Minor intrusions and lead-zinc mineralisation probably followed immediately afterwards along the centre of the basin. A limited amount of volcanic activity took place in Tertiary and Recent times.

The Bouguer isogals are, in general, parallel or sub-parallel to the Cretaceous-crystalline basement boundaries. Over the Cretaceous basin a medial zone of high gravity is flanked to north and south by narrow elongated negative anomalies. Over the bounding areas of crystalline granites and gneisses, the Bouguer values decrease away from the basin. This pattern is interpreted as being due to (a) thinner crust beneath the basin than beneath the flanking basement, (b) a central zone of intruded rocks where basement may be at shallow depth and (c) narrow troughs filled with lighter Cretaceous rocks at either side. The belt of lead-zinc mineralisation shows a striking coincidence with the central zone of positive anomaly. The depth of sedimentary rocks in the northern trough may be 18,000 ft. while in the southern trough it probably does not exceed 13,000 ft. The contacts at both edges of the main Benue trough are probably faulted beneath Cretaceous cover. The gravitational interpretation leads to the conclusion that the Benue Valley originated as a rift valley under tension which continued while the sediments were being laid down. Intrusion and uplift occurred at a later stage.

The stratigraphical and structural evidence enables an approximate sequence of events to be worked out. The combined interpretation leads to a tentative structural history of the Benue Valley which supports the hypothesis of Bucher (1957) that geosynclinal 'welts' are formed under tension, the later folding being due to compression. The tensional origin of the Benue trough in Albian time is consistent with L. King's hypothesis on the separation of South America from Africa.

[*Editor's Note:* Material has been omitted at this point.]

Geological and Gravitational Interpretation

Geological Interpretation

The stratigraphic details presented previously are summarised in diagrammatic fashion in Fig. 14, and the lack of direct correlation between the sequences and major unconformities of the 'South-West' and 'North-East' is further emphasised (*see* Fig. 1 for localities).

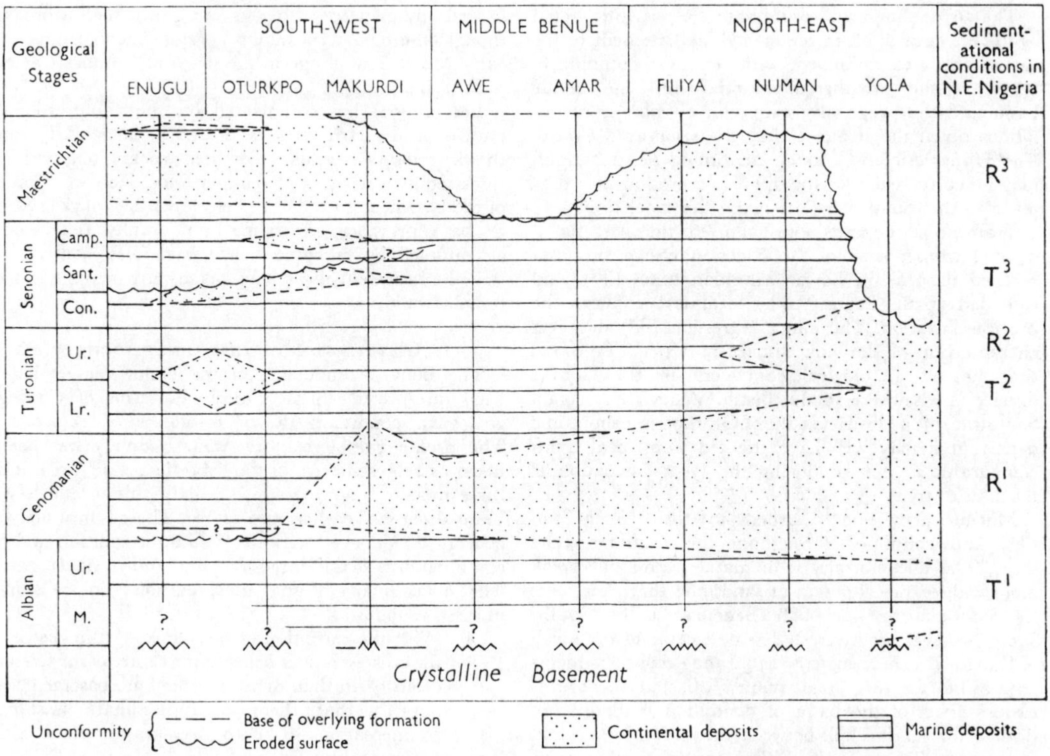

FIG. 14. Distribution of marine and continental deposits in the Cretaceous sequence of the Benue Valley.

This affords much ground for conjecture, and it is all the more unfortunate therefore that the succession in the critical area of the 'Middle Benue' is neither particularly complete nor systematically known. Occasionally abundant fossil material has enabled a zonal scheme to be worked out for the Cretaceous of the respective areas, and to some extent this substantiates the field evidence for the major unconformities. The period of uplift that occurred in the south towards the end of the Albian and the beginning of the Cenomanian had no comparable erosional break in the 'North-East'. Similarly, the absence of the Santonian time stage from southern Nigeria and 'South-West' has been correlated with a period of folding which gave rise to an angular unconformity. The recognition of the Santonian type fossil in 'North-East' Nigeria within an apparently unbroken Cretaceous sequence from Albian to Maestrichtian would indicate that the area was not affected by the Santonian earth movements of the south and 'South-West'. Conversely, the passage from Cretaceous to Tertiary in southern Nigeria is essentially conformable whereas in the 'North-East' and 'Middle Benue' an unconformity marks the end of the Cretaceous and is roughly coincident with large-scale folding, which led to locally strong erosion and subsequent outpourings of basaltic lava. There was no interruption of sedimentation in southern Nigeria at this time, and, in fact , a gradual change was in progress from paludal to deeper water marine conditions—a feature that is not compatible with large-scale disturbance in the vicinity.

The stratigraphic interpretation presented here accepts the framework provided by the faunal evidence but fills in the details by reference to more detailed work recently carried out in the 'North-East' by one of the authors (G.P.J.). It was there possible to build up a sequence of events based upon movement of the shoreline throughout the Cretaceous, with identification of three transgressive and three regressive phases (Fig. 14).

A feature of some significance is the identification of at least 4,000 ft. Albian marine and continental strata in the upper reaches of the Benue Valley. The only explanation that meets all the evidence requires the formation of a linear basin during or prior to the initiation of sedimentation in Middle Albian time (T^1).

The succeeding regressive phase (R^1) is interpreted as the effect of uplift in the crystalline basement to the east of the area examined, with resultant dumping of coarse, freshly weathered material into the Benue basin and the progressive extension of shore-line conditions down the Benue Valley to as far as Makurdi. The Upper Albian-Cenomanian timing of such uplift may be correlated with the uplift and erosion that took place in the 'South-West' at much the same time.

Transitional deposits foreshadowed the best documented transgression of the Nigerian Cretaceous that reached its peak in Lower Turonian times (T^2), and extended up the Benue Valley to almost as far as the previous invasion. The minor regression (R^2) that was characteristic of the Lau district in Upper Turonian time has no lithological equivalent in the Gombe district or elsewhere in the Benue Valley. The Agala Sandstone of the district south of Oturkpo is a shore-line deposit that may well have resulted from local uplift comparable to that producing the Jessu Formation of the Lau district.

Marine deposition returned to the 'North-East' (T^3) and continued throughout the Senonian time stage. The unconformity with marked erosional break and localised development of sandstone that represents the Santonian division of the Senonian in the 'South-West' becomes progressively less noticeable to the north of Enugu. It is here suggested that the earth movements responsible for this break mainly affected the south-eastern flank of the basin of deposition, and did not disturb the marine link between the 'Atlantic' sea and the upper Benue Valley. The extension of marine deposition to the south was subsequently followed by progressive shallowing of the sea over the whole country and the eventual expansion of swamp-deltaic conditions was a feature of both southern and north-eastern Nigeria (R^3). The severe earth disturbances that brought the Cretaceous period to a close in north-eastern Nigeria had as little effect on sedimentation in the 'South-West' as the Santonian movements there had on north-eastern deposition.

The structural pattern of the Benue Valley has the appearance of a single composite unit contained within a relatively narrow belt. The sub-parallel arrangement of the folds, their form, and the manner in which many continue on to the crystalline basement contact all suggest one contemporaneous period of folding, yet the ages of the angular unconformities in the respective areas demonstrate that such is not the case. The similarity of the folds suggests a repetition of compressional activity in different parts of the Benue Valley at differing times with initiation in the 'South-West' in Santonian time and completion in the 'North-East' towards the close of the Maestrichtian. Conclusive proof of such a 'pincers' movement would require evidence of a late Campanian unconformity in the intervening Amar-Muri district. Subsequent erosion has, however, removed any evidence either for or against such a break, though there is a possibility of an unconformity of early Maestrichtian age to the south of Gombe (Carter *et al.*, 1963, p. 56).

Lees (1952) has commented on the compressional nature of the structures, and, more particularly, on the flexuring of crystalline basement rocks along with a thick sediment cover in the larger folds. However, some of the basement cores have been shown by Jones (1960) to be expressions of original topography that were accentuated by subsequent flexuring. Lees' contention that the basement rocks have taken part in the folding remains valid.

Gravitational Interpretation

The Bouguer anomalies of the Benue Valley have been interpreted almost entirely in terms of varying geological conditions down to depths of 18,000 ft. The profile E-F across the Amar anomaly has been taken as representative of the 'Middle Benue' area and three models (Fig. 10) which will fit this profile have been discussed earlier (pp.14-16). There is insufficient geological evidence to allow of a unique solution and a combination of the three fits the known facts best. The main points of the interpretation can be summarised as follows:

(a) With the exception of the Gboko-Arufu district, the crust is thinner beneath the centre of the Cretaceous basin than under the flanking basement. It appears that there is approximate isostatic compensation of both topography and the sedimentary basin.

(b) The maximum depths of Cretaceous are confined to two relatively narrow troughs between which lies a zone of intermediate intrusives where basement is probably at shallower depth than in either of the flanking troughs. The main centres of this postulated igneous activity are at Gboko, Awe and Amar and the intrusive material is thought to be contained in both the basement and the overlying Cretaceous rocks (shales of Albian age). Some uplift is thought to have taken place beneath this central zone and this probably occurred in pre-Cenomanian time. This is suggested by geological evidence and by the way in which the Bima Sandstone palaeocurrent directions run parallel to the eastern end of the Amar positive anomaly (Fig. 13).

(c) The contact between the Cretaceous and basement rocks on the southern side of the Benue trough is fairly steeply dipping and probably faulted along a line which runs NE from Donga to Mutum Biyu (Fig. 5). On the northern side of the Benue Valley from Lafia to Wase and north-eastwards to the Gombe-Bauchi road the contact could be step faulted beneath Cretaceous cover.

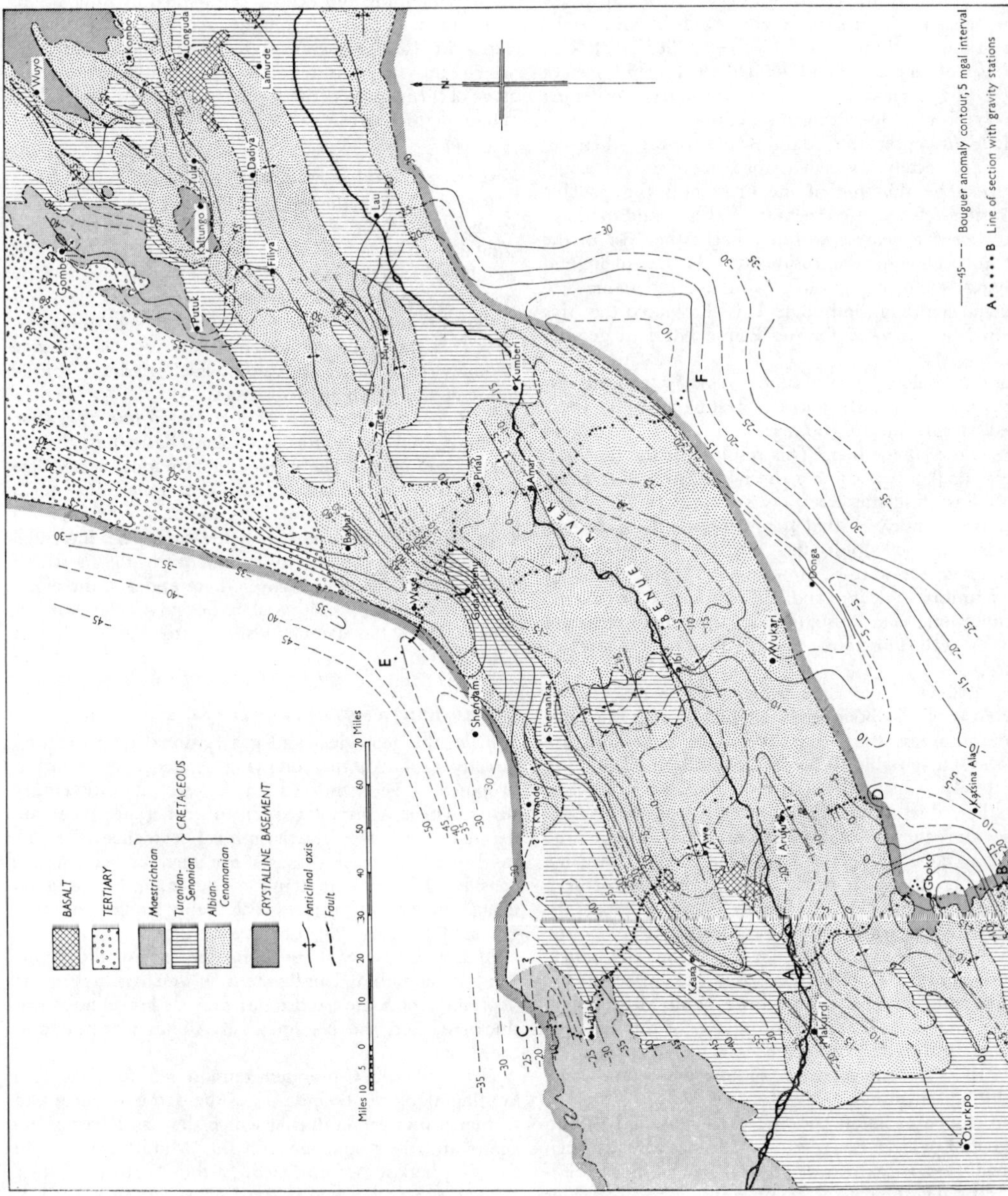

FIG. 15. The geology and Bouguer isogals of the Benue Valley.

This structural pattern is not consistent with a minor geosyncline in which sediments accumulated in a down buckle produced by compression. The thinning of the crust beneath the centre of the trough and the fairly steep contacts at the edges support the hypothesis that the Benue basin originated as a rift valley under tension. A process is envisaged in which the crust gradually thinned due to stress relief by flow at the base of the crust and 'creep' nearer the surface caused the formation of a continually deepening linear basin in which sediments simultaneously accumulated. Faulting at the edges would result whenever the stress locally exceeded that required to cause fracture. The thinning of the crust and the possible blanketing effect of sedimentary cover could perhaps cause excessive heating within or near the base of the crust and subsequent intrusion into the basement and overlying sediments, possibly with accompanying uplift in the centre of the trough. It is considered that this interpretation is valid for the Benue basin in general terms.

The anomalous area around Gboko has been discussed in detail (p. 16) and the postulated uplift there caused closure of the southern basin, most probably at the end of the Albian. This resulted in the northern trough (*see* p. 18) becoming the main link between the 'North East' and the area now occupied by the Niger delta. It is considered that this trough with its accompanying zone of uplift and intrusion to the south continues to the south-west where it is represented by the Anambra syncline and the associated Abakaliki anticlinorium. The outcrop of the coal measures (Fig. 1) well illustrates the positions of these two features.

The folding of the Cretaceous strata is reflected in two ways by the Bouguer anomalies. First, the local variations across the Amar and Awe anomalies are probably attributable to the folding, anticlines forming local positive anomalies and synclines local negative anomalies, due to the greater density of the Albian shales and the possibility of intrusions within the anticlinal cores. Second, the major Keana anticline occurs where the steep gravity gradient on the northern side of the Awe anomaly is found (Fig. 15). This anticline is asymmetrical, with the steeper limb to the north. A possible explanation for this is that the underlying northward sloping basement has affected the location and form in the way shown by de Sitter (1956, pp. 239 and 317). Both the asymmetry of the anticline and the trend of its axis are influenced by the wedging out of the sedimentary strata against the central zone of uplift and intrusion.

The close association between the lead-zinc belt and the zone of positive Bouguer anomaly has already been outlined. Geological evidence shows that the mineralisation post dated the emplacement of the minor intrusions in the south-west and that both followed the compressional period which caused the folding of the strata. The timing of the events is based on geological evidence. The emplacement of the Gboko intrusion is tentatively put at the end of the Albian. Similar intrusions below the Amar and Awe anomalies could have taken place at the same time and been accompanied by some uplift. The main period of folding followed after further deposition of sediments and opened the fissures by which the minor intrusives were emplaced. A late stage of the igneous activity was the lead-zinc mineralisation along fractures.

Comparison With Adjacent Areas

Recent work in the extreme south of the Chad Basin, to the east of the area discussed here, has revealed a similar pattern of Bouguer anomalies with an ENE trend. A central zone of positive anomaly flanked by two negative anomalies has been interpreted by Louis (1962) in an essentially similar manner to that described in this paper although he ascribes the positive anomaly to basaltic intrusion which perhaps caused uplift of the overlying sediments.

The continuation of the north-eastern negative anomaly into the south-west Chad Basin has already been mentioned. There is a continuous trough-like feature from Damaturu with an ENE trend to the north of Maiduguri (Fig. 1) at least as far as the shore of Lake Chad. Seismic refraction measurements indicated a depth of 10,000 ft. to basement to the north of Maiduguri (Cratchley, 1960). There appears, therefore, to be a strong possibility of a buried Cretaceous rift system from the Benue Valley into the south-west Chad Basin.

Tentative Structural History

From the geological and gravitational interpretation already outlined, the following sequence of events is proposed. The estimated thicknesses of sedimentary strata which accumulated in the different stages are derived from the stratigraphical evidence. In the south-west, these thicknesses are seen across what is considered to be the flank of the main depositional trough so that thicknesses within this trough could be greater than those given here.

Middle Albian: Sinking along the line of the Benue trough was initiated under tension and resulted in the deposition of some sandstones and shales of unknown thickness, and the beginning of Albian marine transgression.

Upper Albian: Continued tension led to some step faulting along the boundaries of the depressed strip and further transgression during which the Asu River group shales and their equivalent in the 'Middle Benue' and the 'North-East' accumulated. In the 'North-East' their thickness exceeds 4,000 ft. and in the 'South-West' it is of the order of 6,000 ft.

Post-Albian: Geological evidence requires that uplift of the crystalline basement took place on the south side of the Benue trough. Uplift, accompanied by the emplacement of igneous intrusions of intermediate composition, also possibly occurred along the centre of the Benue trough resulting in the formation of a northern and a southern linear basin and along an axis from Arufu to Abakaliki and beyond. The uplift at Gboko and Arufu caused almost complete closure of the southern trough.

Cenomanian: The post-Albian earth movements resulted in intense erosion and deposition of continental sandstones both within and on the flanks of the Benue basin. Sinking under tension must have continued within the basin by downwarping and further movement along step faults beneath the already accumulated sediments, which deformed mainly by flexuring. The main areas of accumulation were now within the closed southern trough and the continuous northern trough. Farther south-west marine conditions still prevailed and the sandstones would grade laterally into shales. The Bima Sandstone formation of the 'North-East' is 6,000 ft. thick while the Makurdi Sandstone is of the order of 4,000 ft. in thickness.

Turonian: Continuation but lessening of the tension was accompanied by marine transgression through the Benue Valley. This transgression was a world-wide event and is more probably associated with a general rise in sea-level than excessive sinking along the basin. This is borne out by the relatively thin sequence of 3,000 ft. in the 'South-West' and 1,500 ft. in the 'North-East' where a minor regressive phase took place in the Upper Turonian.

Senonian: Folding in which compression was directed NW-SE, possibly with sinistral movement along NNE transcurrent faults, began in the Santonian in the 'South-West'. Marine deposition continued in the 'North-East' until the beginning of the Maestrichtian and resulted in the accumulation of 1,700 ft. of sediment; the link with the Atlantic sea was almost certainly along the northern trough. The period of folding possibly reached the 'Middle Benue' by the end of the Senonian and closed the northern marine channel. The compression could have caused 'flow' of the basement as envisaged by Lees, particularly if large post-Albian intermediate intrusions within the basement were still in a fluid state. The folding of the Cretaceous rocks, influenced by the shape of the underlying basement, allowed emplacement of the minor intermediate and basic intrusions, particularly into anticlinal axes. Lead-zinc mineralisation resulted from a final stage of this igneous activity, the mineral-bearing solutions taking advantage of fractures created by the folding.

Maestrichtian: The return to continental and paludal conditions in both the 'North-East' and the 'South-West' could well have been a result of folding in the 'Middle Benue' at the end of the Senonian. During this stage 3,000 ft. of sediments accumulated in the 'South-West' where tension must have been renewed while 1,000 ft. accumulated in the 'North-East'.

Post-Maestrichtian: Folding in the 'North-East' was accompanied by faulting probably along older lines and possibly because of the great thicknesses of Bima Sandstone present. Marine deposition ensued in the 'South-West'. From Tertiary to Recent times minor outpourings of basalt have occurred in the 'Middle Benue' and the 'North-East'; apart from this the Benue Valley seems to have been relatively inactive since the Maestrichtian. Broad Tertiary warping which possibly caused elevation of the Jos Plateau also perhaps caused the River Benue to shift southwards from the northern trough to its present-day course.

The maximum thickness of Cretaceous rocks could well exceed 16,000 ft. in the deepest part of the northern trough, according to the stratigraphical evidence, and could exceed 18,000 ft. on the gravitational evidence. The major part of this (possibly 12,000 ft.) accumulated in the early Albian-Cenomanian stages.

The sequence of events supports the theory of Bucher (1957) that the linear geosynclinal 'welts' originated under tension which caused sinking along a narrow strip and deposition of sediments. This stage was succeeded by a reversal of stress which caused folding and uplift. In the major geosynclines, subsequent folding and thrusting has obliterated the original depositional conditions; the Benue Valley is thus possibly unique in that it has retained characteristics of both the depositional tension stage and of the subsequent compressional stage. The latter had not developed sufficiently to allow thrusting to become important.

The interpretation which postulates the initiation of a depressed fault-bounded strip under tension in the Albian together with the Albian marine invasion, lends some support to L. King's hypothetical rift valley system produced during a drift of South America from Africa (King, 1950). The main faults are hidden beneath the cover of Cretaceous sedimentary strata, however, and not found in the scarps which now stand well back from the present extent of the Cretaceous trough as he envisaged.

Summary

The stratigraphical evidence shows that marine transgression and the deposition of several thousands of feet of shales in the Albian was succeeded by the deposition of comparable thicknesses of continental sandstones in the Cenomanian. Further transgression of the sea in the Turonian coincided with world-wide marine deposition and was locally less important. Folding of the strata must have begun in the Santonian in the 'South-West' but did not affect the 'North-East' until post-Maestrichtian time. Emplacement of minor intermediate and basic intrusions is thought to have followed immediately after the folding, the lead-zinc mineralisation probably being a later phase of this igneous activity.

It is shown that the Bouguer anomaly map indicates that overall the region is approximately isostatically compensated and that the crust beneath the Benue trough is probably thinner than beneath its flanks. Within the trough, a central zone of positive anomalies is flanked over most of the 'Middle Benue' by two elongated negative anomalies. From the evidence of density values obtained on rock specimens, and on geological grounds, the positive anomalies are most reasonably ascribed to a zone of intrusion in which the oldest Cretaceous rocks of Albian age and the basement which underlies them are close to the surface and contain igneous material of intermediate to basic composition. The maximum depth of Cretaceous rocks is found under the relatively narrow northern negative anomaly and could be as much as 18,000 ft. Near the southern boundary the total thickness of Cretaceous rocks is unlikely to exceed about 13,000 ft. The gravity gradients across the northern and southern edges of the trough suggest that the contacts are probably faulted along most of their lengths. Basement which outcrops around Gboko and Arufu is thought to be part of the original floor of the trough which has been uplifted during subsequent earth movements. The general pattern of the gravity anomalies and the width of the Benue trough lead to the conclusion that this is a rift valley of Cretaceous age. The subsequent folding of the Cretaceous strata within the trough has locally modified the gravity field.

The tentative structural history of the Benue Valley, which forms the final section of this paper, is deduced from the geological and gravitational evidence and interpretation. The initial tension and infilling of the trough with sediments were followed by intrusion and a later period of folding due to compression. This structural history supports the hypothesis of Bucher (1957) that geosynclinal 'welts' are initiated under tension which results in sinking and accumulation of sediments, the later folding of the sedimentary strata being due to a reversal of the stress. The tensional origin of the Benue trough in Albian time is consistent with King's hypothesis on the separation of South America from Africa (King, 1950) in the break up of the Gondwana land mass.

[*Editor's Note:* Only the references cited in the preceding excerpts are reprinted here.]

REFERENCES

Bucher, W. H. 1957. 'The Deformation of the Earth's Crust.' 518 pp., figs., photos., refs., map. (New York: Hafner Publishing Co.)

Carter, J. D., Barber, W., Tait, E. A. and Jones, G. P. 1963. 'The geology of parts of Adamawa, Bauchi and Bornu Provinces in north-eastern Nigeria.' *Bull. No.* 30, *geol. Surv. Nigeria.* 99 pp., figs., photos., refs., col. geol. maps.

Cratchley, C. R. 1960. 'Geophysical survey of the south-western part of the Chad Basin.' Unpublished paper presented at the C.C.T.A. Conference on 'Geology', Kaduna, N. Nigeria, Feb., 1960.

De Sitter, L. U. 1956. 'Structural Geology.' 552 pp., figs., photos., refs. (London: McGraw Hill Book Co.)

Jones, G. P. 1960. 'Depositional history of the Bima Sandstone of north-eastern Nigeria.' Unpublished paper presented at C.C.T.A. Conference on 'Geology', Kaduna, N. Nigeria, Feb., 1960.

King, L. 1950. 'Speculations upon the outline and the mode of disruption of Gondwanaland.' *Geol. Mag.*, Vol. 87, No. 5, pp. 353-359, figs., refs.

Lees, G. M. 1952. 'Foreland folding.' *Quart. J. geol. Soc. Lond.*, Vol. 108, Pt. 1, pp. 1-34, figs., refs.

Louis, P. 1962. 'Interprétation géologique d'une partie de la carte gravimétrique du bassin du Logone (République du Tchad).' *C. R. Acad. Sci., Paris*, Vol. 254, No. 21, pp. 3732-3736, figs. [In French.]

7

Reprinted from pages 1-3 and 37-39 of *Mineral Resour. Bull. Jiddah No. 16,* 1980, 40p.

STRUCTURE OF THE ḌURMĀ-NISĀḤ SEGMENT OF THE CENTRAL ARABIAN GRABEN SYSTEM

By Abdullah Al Kadhi[1] and Paul L. Hancock[2]

Abstract.—A new 1:250,000-scale geologic map of the 150-km Ḍurmā-Nisāḥ segment of the central Arabian graben and trough system is presented. The structure of the region is dominated by the en-echelon Qarādān, Ḍurmā, Awsaṭ, and Nisāḥ grabens; the offset Bu'ayjā', Mughrah, and Sahbā' troughs; and the Marāghah and Wubayr monoclines. The grabens and troughs lie within an arcuate zone of structures that disturbs a gently tilted Jurassic-Cretaceous sequence of carbonate rocks, plus subordinate clastic rocks, within the Interior Homocline of the Arabian craton. The system was initiated in the Cretaceous and established by the late Paleogene. The Kharj Formation, presumed to be of Neogene Age, rests unconformably on a buried paleorelief in the flexed and faulted Mesozoic rocks. The formation is undeformed except around an inlier of Biyadh Sandstone (Lower Cretaceous) near Ashqar Marāghah, where the Neogene rocks are involved in a pericline. Grabens are bounded by high-angle normal faults and single or multiple inward-facing monoclinal flexures. Subsidiary synthetic or antithetic normal or extension faults also are common along graben margins. The Aswaṭ and Nisāḥ grabens each comprise two segments overlapping in a sense opposite to that of the four principal grabens. The two segments of the Awsaṭ graben overlap by 25 km rather than by 7 km, as investigators had thought earlier; within the region of overlap, there are previously unrecorded outcrops of the Arab Formation (Upper Jurassic). Trough margins are single or multiple inward-facing monoclines accompanied locally by subsidiary faults. Three systems comprising 10 sets of joints, veins, and small-scale faults are arranged symmetrically with reference to the trend of the nearest graben or trough. Paleostress trajectories inferred from these mesoscopic fractures indicate that the principal horizontal stresses were either perpendicular or parallel to grabens and that, although fluid pressures were probably high, differential stresses were small. The grabens and troughs of the Ḍurmā-Nisāḥ segment probably evolved in a tensile stress field generated by north-south stretching of the sedimentary cover above an arcuate major fault zone cutting the underlying crystalline rocks. The separation and relative displacement to the north of the block to the northeast of the graben system may have occurred during the overall northward motion of a semirigid Arabian subplate in the Late Cretaceous and early Paleogene.

INTRODUCTION

The central Arabian graben and trough system is an arcuate 560-km intraplate fault zone. It cuts the gently inclined Mesozoic-Cenozoic sedimentary succession of the Interior Homocline within the Arabian Shelf (Powers and others, 1966) (fig. 1). The significance of the system in the tectonic evolution of Arabia is unclear and disputed. Wolfart (1961) proposed that the system traced a sedimentary basin that started to fill in pre-Triassic times, whereas Powers and others (1966, p. D109) thought that it might represent "... a hinge line between a relatively stable land mass to the southwest and a periodically subsiding shelf to the northeast." Both Wolfart and Powers and others interpreted the faults of the system as normal faults related to regional tension. According to Brown (1972), subsurface structure contours on the base of the Mesozoic and on the top of the crystalline basement are offset to the west on the northern side of the eastern end of the system. Brown thought that this offset indicates either left-lateral displacement or net downthrow to the north. He noted that left-lateral movement would be sympathetic with that along the northwest-trending strike-slip faults of the Najd orogeny within the crystalline rocks of the Precambrian-Cambrian Shield.

The Ḍurmā-Nisāḥ segment of the system is about 150 km long and lies between lat 24°05′ and 24°45′ N. and long 46°00′ and 47°25′ E. (fig. 2). It was selected for the initial phase of a structural investigation of the whole system because (1) it includes the region of maximum curvature of the arc, (2) there is an unconformable cover of Neogene rocks, (3) its components are relatively well exposed, and (4) by the standards of the region, it is accessible. During the period 1970–1977, two aspects of the structure of the Ḍurmā-Nisāḥ segment were investigated—the distribution and arrangement of the major folds and faults and the geometry of the accompanying mesoscopic fractures. A brief outline of the field characteristics of the mesofracture pattern, together with a comprehensive interpretation of the paleostress conditions at the time of their development, is given by Hancock and Kadhi (1978). The pre-

[1]Department of Geology, University of Riyadh, Riyadh, Saudi Arabia.
[2]Department of Geology, University of Bristol, Bristol, England.

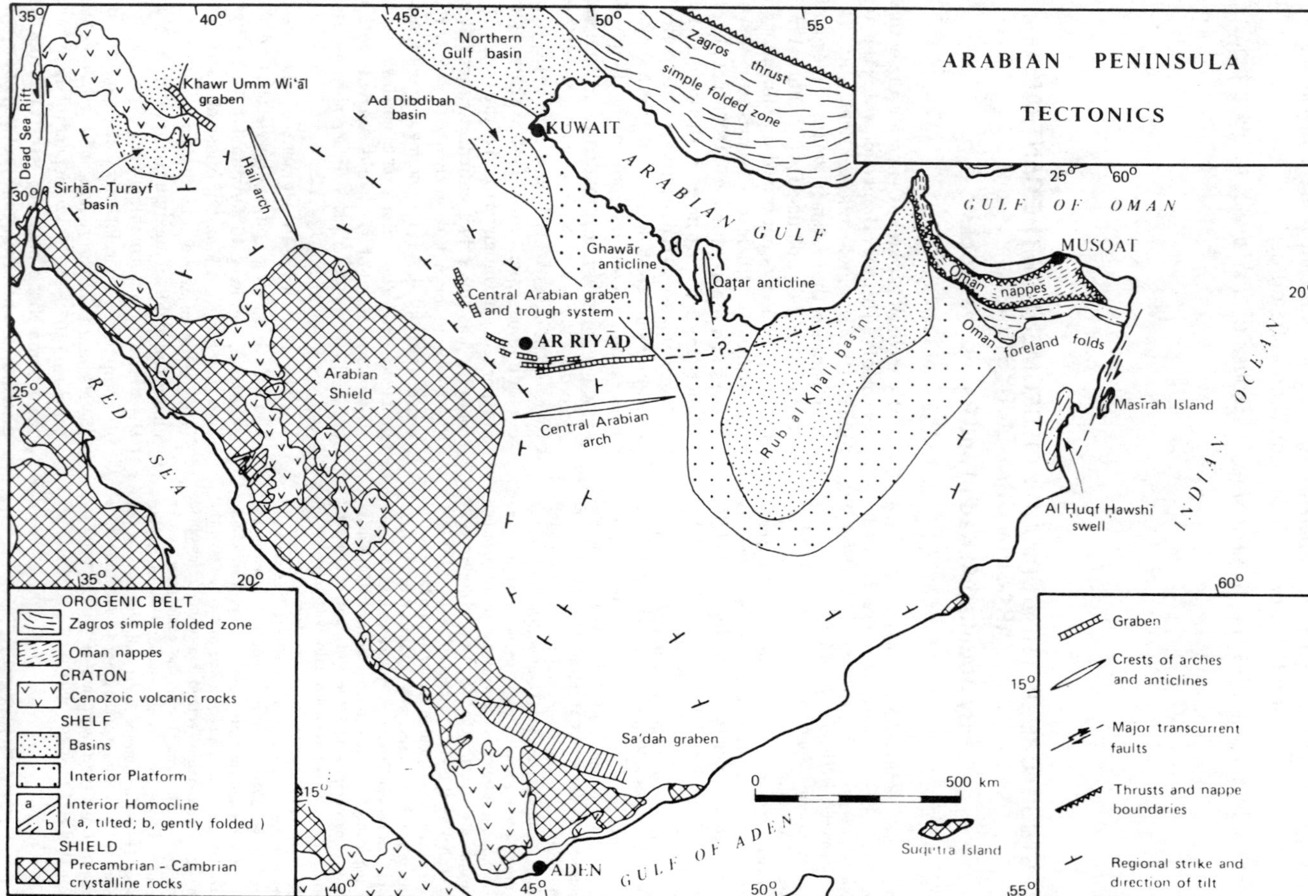

FIGURE 1.—Map showing the location of the central Arabian graben and trough system in relation to other major tectonic features and provinces. Compiled from Powers and others (1966, pl. 2), Brown (1972), and Hancock and Kadhi (1978, figs. 1, 2).

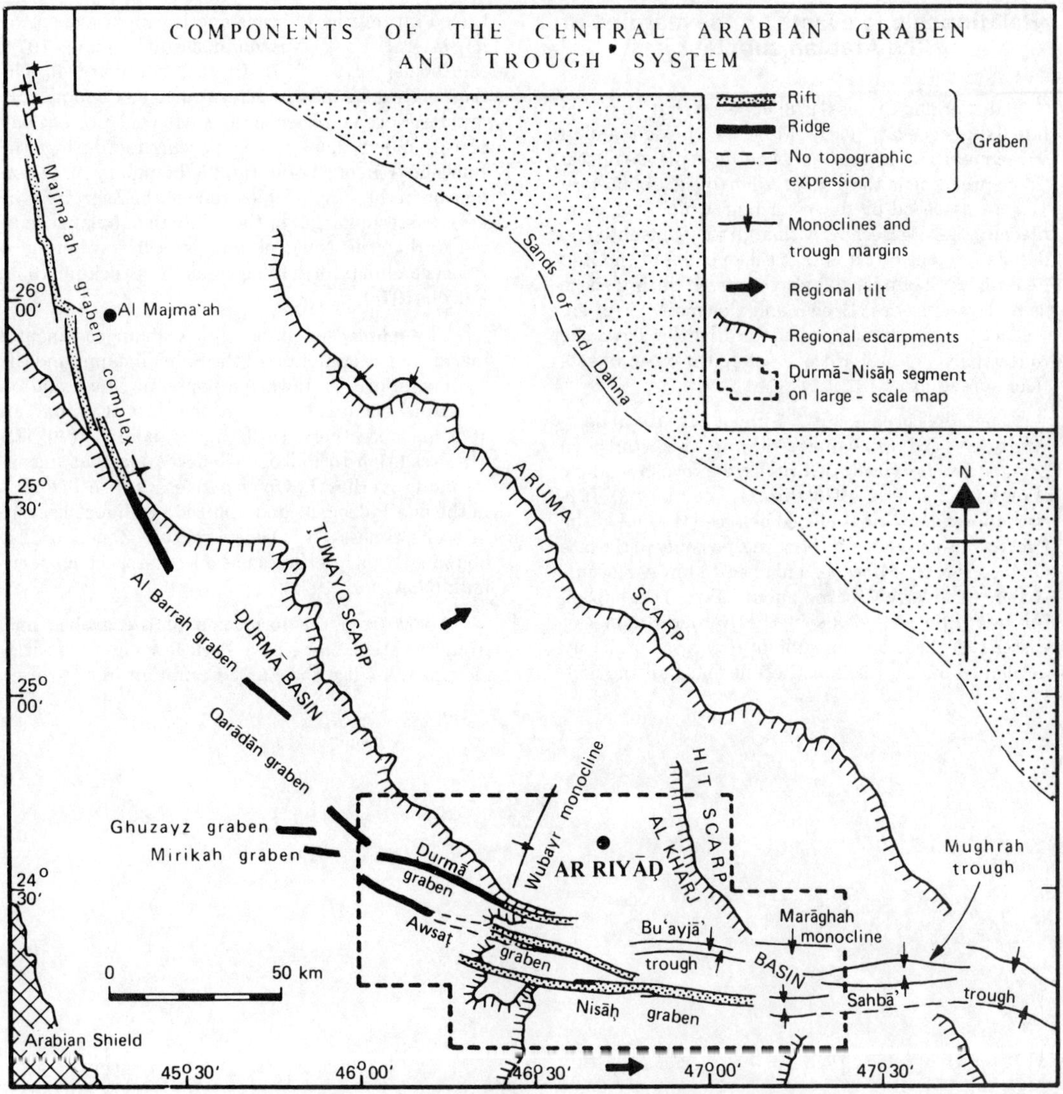

FIGURE 2.—Map of the principal structures and topographic features in the central Arabian graben and trough system. Based on Bramkamp and Ramirez (1958) and data collected by the authors.

[*Editor's Note:* Material has been omitted at this point.]

Relationship to events on the margins of the Arabian subplate

Girdler and Styles (1978) believed that an initial late Oligocene-late early Miocene episode and a subsequent late Miocene-present-day episode of seafloor spreading in the Gulf of Aden (fig. 21, boundary 1) were preceded by downwarping, which was complete by the late Eocene. Although they presented no details, they also stated that the opening of the Red Sea (fig. 21, boundary 2) also occurred during these same two episodes. Brown and Coleman (1972) envisaged tensional collapse of the Red Sea from the Maestrichtian to the Eocene before the onset of seafloor spreading.

There has been about 105 km of left-lateral post-Cretaceous displacement along the southern segments of the Dead Sea Rift, a transform boundary (Freund and others, 1970) (fig. 21, boundary 3). It is probable that 60 km occurred between the end of the Cretaceous and the early Miocene, possibly in the late Oligocene-early Miocene, and that 45 km has occurred since the early Miocene, about 10 km of the slip being post-Pliocene. Folds in the Border Ranges of eastern Turkey, on the subplate margin, probably developed during the late Miocene, but faulting continued during the Pliocene and is still active on the left-lateral East Anatolian Fault (Ilhan, 1974; McKenzie, 1976, 1977) (fig. 21, boundary 4). The Zagros thrust zone was active during a Cenomanian-middle Eocene phase and a Miocene-present-day phase, much displacement occurring during the Pliocene (Falcon, 1974) (fig. 21, boundary 5). Deformation in the simple folded zone of the Zagros on the northeastern margin of the Arabian subplate occurred during the Neogene and is still continuing; a Pliocene climax of folding is likely (Stocklin, 1968; Falcon, 1974).

The emplacement on the Arabian continental margin of the ophiolites of the Samā'il nappe and the sediments of the Ḥawāsina nappe in Oman and on Maṣirah occurred during the late Campanian (Glennie and others, 1973) (fig. 21, boundary 6). The Maestrichtian to Paleogene succession that unconformably overlies the Oman nappes was gently folded in the late Paleogene and uplifted in Oligocene-early Miocene times. The Owen Fracture Zone (fig. 21, boundary 7) is interpreted as a left-lateral transform fault (Nōwroozi, 1972).

Apart from the formation of the graben and trough system, there have been few vigorous earth movements since the Early Cambrian in central or eastern Arabia. The north-south folds of Ghawâr and Qaṭar, for example (fig. 1), started to develop as synsedimentary structures in the Late Jurassic or Early Cretaceous, amplified during the Cretaceous, and experienced accelerated amplification in the Paleogene (Brown and Coleman, 1972). Powers and others (1966) believed that the central Arabian arch was initiated in the Late Cretaceous and developed during the Paleogene and that there was little exaggeration of the structure from the Miocene onward. Brown and Coleman (1972) proposed that Pliocene epeirogeny elevated the sedimentary rocks fringing the Arabian Shield.

When the timing of tectonic events on the subplate margins is compared with the timing of those in central Arabia, it is clear that the initiation of the graben and trough system and the arch coincided approximately with the emplacement of the Oman nappes and the earlier movements on the Zagros thrust zone. The probable early Paleogene climax of movements on the graben and trough system occurred at about the same time as downwarping and the first episodes of opening in the Gulf of Aden and the Red Sea and while displacements on the Zagros thrust zone were continuing. The Dead Sea Rift may also have been active at this time. The important Neogene and Quaternary displacements that continued on all the subplate margins (except that of Oman) are not reflected by significant structures in central Arabia. The deformation of the Kharj Formation and the possible Quaternary faulting (Wolfart, 1961) are either localized or minor tectonic events. The authors conclude that the Arabian subplate, or at least its central part, was rigid by the late Paleogene. During the Late Cretaceous and early Paleogene, the subplate may have been semirigid, relative motion

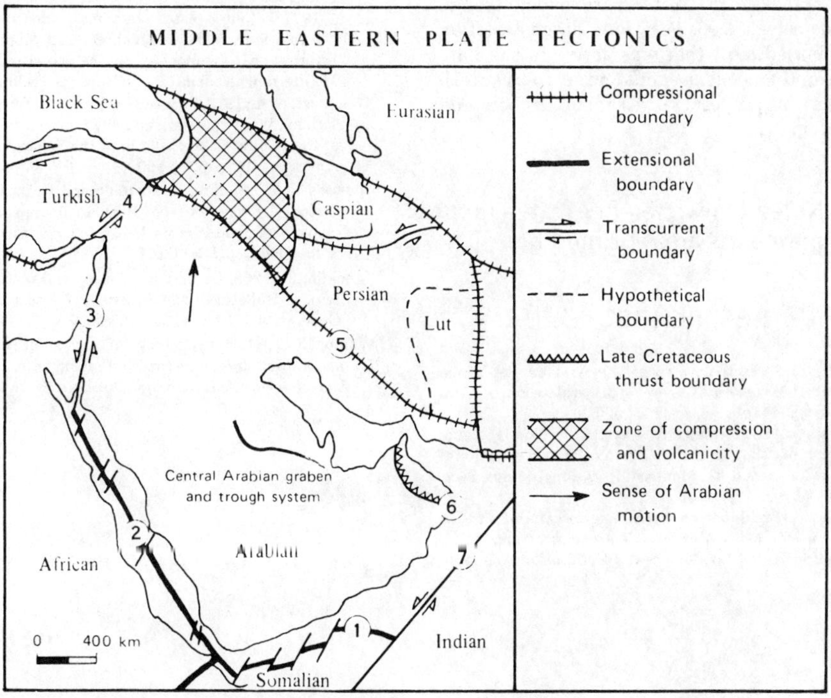

FIGURE 21.—Map showing plates and subplates of the Middle Eastern region. Locations and types of boundaries based on reports by McKenzie (1972, fig. 2*B*; 1976, fig. 1; 1977, fig. 2) and Nowroozi (1972, fig. 8). Transform faults of the Red Sea and the Gulf of Aden are shown schematically. Boundary 1, Gulf of Aden; boundary 2, Red Sea; boundary 3, Dead Sea Rift; boundary 4, East Anatolian Fault; boundary 5, Zagros thrust zone; boundary 6, Oman nappes; boundary 7, Owen Fracture Zone.

occurring across a major fault zone in the crystalline basement beneath the central Arabian graben and trough system.

If the arguments presented in the section on mechanistic aspects are reasonable, it is also necessary to account for the relative northward displacement of the block to the northeast of the graben and trough system. The precise present direction of motion of the Arabian subplate is debatable; it is about 004°, according to McKenzie (1972, fig. 2*B*; 1977, fig. 2), but about 020°, according to Nowroozi (1972, fig. 8) (fig. 21). If the overall motion of a semirigid Arabian subplate during the Late Cretaceous and Paleogene also was approximately northward, the northeastern block may have separated slightly from the southwestern block across an arcuate fault zone. The resulting slight north-south extension across the basement fault zone was propagated upward through the Mesozoic sedimentary cover to be expressed by the en-echelon structures of the central Arabian graben and trough system. Geometrical considerations suggest that an additional buried fault that was active in the Late Cretaceous and Paleogene should occur to the north or northeast of the central Arabian graben and trough system.

[*Editor's Note:* Only references cited in the preceding excerpts are reprinted here.]

REFERENCES

Brown, G. F., 1972, Tectonic map of the Arabian Peninsula: Saudi Arabian Directorate General of Mineral Resources Arabian Peninsula Map AP-2, scale 1:4,000,000.

Brown, G. F., and Coleman, R. G., 1972, The tectonic framework of the Arabian Peninsula: International Geological Congress, 24th, Montreal 1972, Proceedings, Sect. 3, p. 300–305.

Falcon, N. L., 1974, Southern Iran—Zagros Mountains, *in* Spencer, A. M. (ed.), Mesozoic-Cenozoic orogenic belts: Geological Society of London Special Publications, no. 4, p. 199–211.

Freund, R., Garfunkel, Z., Zak, I., Goldberg, M., Weissbrod, T., and Derin, B., 1970, The shear along the Dead Sea Rift: Philosophical Transactions of the Royal Society of London, v. 267A, p. 107–130.

Girdler, R. W., and Styles, P., 1978, Sea-floor spreading in the western Gulf of Aden: Nature, v. 271, p. 615–617.

Glennie, K. W., Boeuf, M. G. A., Hughes-Clark, M. W., Moody-Stuart, M., Pilaar, W. F. H., and Reinhardt, B. M., 1973, Late Cretaceous nappes in Oman Mountains and their geologic evolution: Bulletin of the American Association of Petroleum Geologists, v. 57, p. 5–27.

Hancock, P. L., and Kadhi, A., 1978, Analysis of mesoscopic fractures in the Dhruma-Nisah segment of the central Arabian graben system: Journal of the Geological Society of London, v. 135, p. 339–347.

Ilhan, E., 1974 Eastern Turkey, *in* Spencer, A. M. (ed.), Mesozoic-Cenozoic orogenic belts: Geological Society of London Special Publications, no. 4, p. 187–197.

McKenzie, D., 1972, Active tectonics of the Mediterranean region: Geophysical Journal of the Royal Astronomical Society, v. 30, p. 109–185.

_____1976, The East Anatolian Fault—A major structure in eastern Turkey: Earth and Planetary Science Letters, v. 29, p. 189–193.

_____1977, Can plate tectonics describe continental deformation?, *in* Biju-Duval, B., and Montadert, L. (eds.), International symposium on the structural history of the Mediterranean basins: Paris, Editions Technip, p. 186–196.

Nowroozi, A. A., 1972, Focal mechanism of earthquakes in Persia, Turkey, West Pakistan, and Afghanistan and plate tectonics of the Middle East: Bulletin of the Seismological Society of America, v. 62, p. 823–850.

Powers, R. W., Ramirez, L. F., Redmond, C. D., and Elberg, E. L., 1966, Geology of the Arabian Peninsula—Sedimentary geology of Saudi Arabia: U S. Geological Survey Professional Paper 560-D, p. D1–D147.

Stocklin, J., 1968, Structural history and tectonics of Iran—A review: Bulletin of the American Association of Petroleum Geologists, v. 52, p. 1229–1258.

Wolfart, R., 1961, Hydrogeology of the central Tuwaiq Mountains and adjoining regions (Saudi Arabia): International Association of Scientific Hydrology, Publ. No. 56, p. 98–112.

Part III

EUROPE

RIFTS: EUROPE AND THE NORTH SEA

This map was prepared expressly for this Benchmark volume by A. M. Quennell

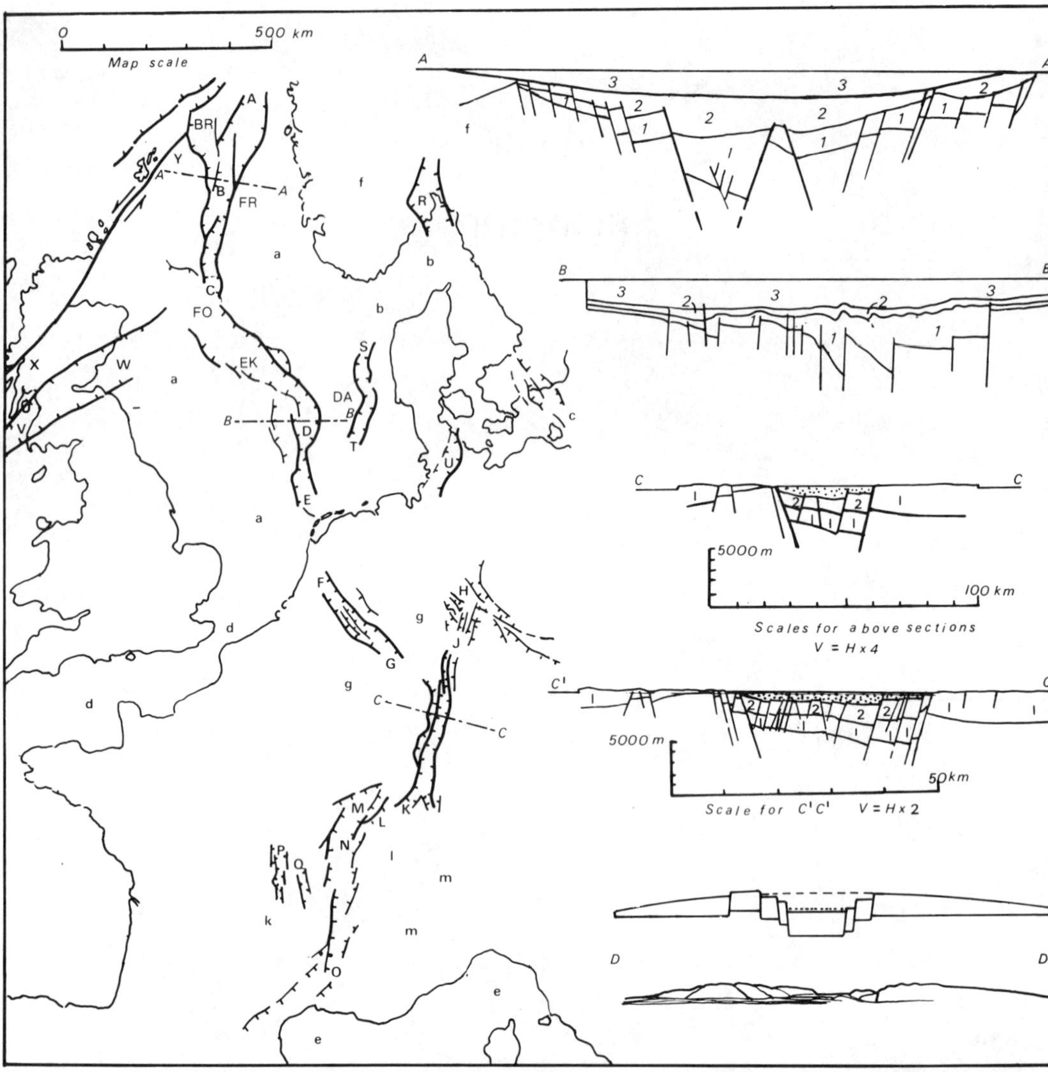

Fields: BR = Brent; FR = Frigg; FO = Forties; EK = Ekofisk; DA = Dan.

Rift valleys: AC = Northern North Sea Trough; AB = Viking Basin; BC = Viking Graben; CD = Central Graben (Ekofisk Trough); DE = Dutch North Sea Graben (North Netherlands trough); FG = Lower Rhine Embayment (graben); HJ = Hessische Graben; JK = Upper Rhine Graben; KL = Belfort Channel (trough); M = Saône Trough (graben); NO = Rhône Trough; P = Limagne Trough; Q = Loire Trough; R = Oslo Graben; ST = Horn Graben; U = Gluckstadt Trough; VW = Midland Valley Graben; XY = Great Glen Fault.

Geographic features: a = North Sea; b = Skagerrak; c = Baltic Sea; d = English Channel; e = Mediterranean Sea; f = Fenno-scandia; g = Rhenish Massif; h = Vosges Mts; j = Black Forest; k = Central Massif; l = Jura; m = Alps.

Sections: *A-A* = Northern North Sea Trough (Viking Graben); *B-B* = Central Graben (Ekofisk Trough); *C-C* = Upper Rhine Graben; *D-D* (upper) = Les Vosges, la plaine du Rhin et le Forêt-Noire, rue de la cime du Röthi-Fluhe; *D-D* (lower) = Soulèvement et écroulement combinés qui ont produit la plaine du Rhin.

Sediments: In *A-A* and *B-B*: *1* Syntectonic, Permian to Jurassic; *2* = Late tectonic, Cretaceous; *3* = Post tectonic, Cenozoic. In *C-C*: 1 = Pre-faulting, infaulted cover beds, Triassic and older; *2* = Post-faulting or infilling sediments, Cenozoic.

Editor's Comments on Papers 8 and 9

8 DUFRENOY and ELIE DE BEAUMONT
Excerpt from *Les Vosges*

9 SITTLER
The Sedimentary Trough of the Rhine Graben

Paper 8 is an extract from Volume 1 of the *Explanation of the Geological Map of France* (1841) and sketch (Fig. 27) and model (Fig. 28) of the Rhine graben, by Dufrenoy and Elie de Beaumont. The latter had previously (1827) published his views on the formation of the graben. A disciple of von Buch, Elie de Beaumont had his own views on the nature of lifting forces, regarding them as the result of horizontal stress from the shrinking of a cooling globe. Quennell (1982) reprints a paper by Willis (1936), who gave extracts from accounts by early workers. De Martonne (1897), followed by de Lapparent (1898), made comparison between the Rhine graben and the rift valleys of Africa. This was about the time of Gregory's first visit.

In Paper 9, Sittler states in summary: "Accurate knowledge of the sediments in the Rhine graben is not only essential for proper corrections of geophysical data but it provides the main tool to reconstruct the development of the graben in time and space." This is in fact the theme of this present volume, set out in the Introduction; Crough's axiom (1983, 35) bears repeating once again: "the ultimate test of any tectonic model must be its consistency with the known geological record."

The sections—longitudinal and cross-section—are of special value. In a written communication Sittler informs us that some small modifications have been introduced, particularly in the sections, but have not been incorporated in this reprint (Sittler, 1974). This rift valley is the most thoroughly researched in the world and could be said to have come to be thought of as the prototype. However, as Paper 5 shows, in its origin and evolution it should not be too closely equated with the Kenya or the Baikal rifts. They belong to different stages of the Wilson cycle and have differing crustal settings.

Much of the published literature on the Rhine graben deals with geophysical surveys. However, many important geological and structural descriptions are to be found only in the German and French literature. The structural evolution as it is now accepted is best described in Papers 3 and 5. Knowledge derived from boreholes and seismic surveys has made possible a better understanding of the complex faulting in the downthrown floor.

In addition to the references, the Bibliography contains symposia and reports containing papers on the Rhine graben: see especially publications 3, 6, 8, and 12. Of special interest is the recorded knowledge of Illies, and its application.

[*Editor's Note:* The reader is referred to the list following the comments on Papers 12 and 13 (pages **148-150**) for the references cited in these comments.]

8

Reprinted from pages 433–437 of *Explication de la Carte Géologique de la France,* Service de la Carte Géologique de la France, Paris, 1841

LES VOSGES

A. Dufrenoy and L. Elie de Beaumont

[*Editor's Note:* In the original, material preceded this excerpt.]

Les noms de Forêt-Noire et d'Odenwald s'appliquent, en effet, aux proéminences qui s'élèvent au milieu de la plaine ondulée de la Souabe, dans une acception précisément analogue à celle que nous avons reconnue au nom de Vosges. Toutes ces montagnes forment deux groupes en quelque sorte symétriques, qui se terminent, l'un vis-à-vis de l'autre, par deux longues falaises légèrement sinueuses, dont les directions générales sont parallèles l'une à l'autre et au cours du Rhin, qui coule entre elles depuis Bâle jusqu'à Mayence. Ces deux falaises sont principalement composées d'éléments rectilignes, orientés presque exactement du N. N. E. au S. S. O., et les montagnes dont elles sont, pour ainsi dire, les façades, présentent, les unes comme les autres, dans beaucoup de points de leur pourtour ou de leur intérieur, d'autres lignes d'escarpements parallèles aux précédents. Analogie des montagnes des deux bords du Rhin.

Ces lignes sont les traits caractéristiques du groupe naturel ou du système de montagnes dont nous parlons, système que M. Léopold de Buch a nommé *système du Rhin.* Elles se dessinent très-nettement sur une carte géologique de ces contrées, aussitôt qu'on y distingue par des couleurs différentes les deux formations, si souvent confondues ensemble, du grès des Vosges et du grès bigarré. Dans la Forêt-Noire et dans l'Odenwald, aussi bien que dans les Vosges, les escarpements ci-dessus mentionnés sont habituellement composés, en tout ou en partie, de grès des Vosges. Ils forment, en général, la tranche des plateaux plus ou moins étendus dont les couches de cette formation constituent la surface. Dans la Forêt-Noire et dans l'Odenwald, ils paraissent dus, comme dans les Vosges, à de grandes fractures, à une série de failles parallèles, qui ont rompu et diversement élevé ou abaissé les différents compartiments dans lesquels elles ont divisé la formation du grès des Vosges, à une époque où cette formation n'était encore recouverte par aucune autre. Le bouleversement dans lequel elles se sont produites est, par conséquent, antérieur au dépôt du système du grès bigarré, du muschelkalk et des marnes irisées, qui, tout autour des Système du Rhin; ses traits caractéristiques.

montagnes des deux bords du Rhin, s'étend jusqu'au pied des falaises dirigées du N. N. E. au S. S. O., mais qui, malgré les traces de dislocation très-nombreuses et souvent fort étendues qu'on y observe, ne s'élève jamais, comme le grès des Vosges, en véritables montagnes. Ce groupe de couches s'arrête toujours au pied des montagnes que constituent les formations, ses aînées, dans une sorte d'attitude respectueuse, qui est un des caractères géologiques les plus remarquables de la contrée : cela seul donne aux montagnes du système du Rhin un cachet d'ancienneté qui les distingue éminemment du Jura, des Pyrénées, des Alpes, et, en général, de toutes les chaînes plus modernes et plus hautes sur les flancs desquelles des formations récentes se montrent à de grandes hauteurs.

Petits accidents des contrées rhénanes qui appartiennent à d'autres systèmes.

Les accidents que présente la plaine secondaire et tertiaire qui environne la Forêt-Noire, les Vosges et l'Odenwald, sont le prolongement des accidents du Jura et de la Côte-d'Or, qui ne se terminent pas subitement. La convulsion à laquelle les Alpes occidentales doivent les principaux linéaments de leur relief a aussi déterminé l'apparition, à la surface du globe, des collines phonolithiques de Hohentwiel et du groupe du Kaiserstuhl; elle doit même s'être prolongée au delà dans l'intérieur de la plaine du Rhin, mais en n'y produisant que de très-faibles aspérités. Ces accidents, d'un ordre tout à fait secondaire pour la contrée qui nous occupe, ne forment aussi qu'un trait accessoire de peu d'importance dans le tableau des grandes révolutions, dont les traits principaux sont ailleurs.

Étendue probable du système du Rhin.

Les événements plus remarquables arrivés auparavant sur le même sol ne sont, sans doute, eux-mêmes, que des fractions de faits beaucoup plus généraux : ainsi les grandes failles, dirigées vers le N. N. E., auxquelles est dû l'isolement dans lequel se trouvent les montagnes du système du Rhin au milieu des plaines ondulées qui les entourent, ne sont qu'une petite partie d'un système de dislocations beaucoup plus étendu qui traverse le sol d'une partie considérable de l'Europe.

La ligne presque droite suivant laquelle se terminent à l'E. les grauwackes du Westerwald près de Homburg, de Giessen, de Marburg, est dans le prolongement presque exact de la faille qui limite les basses Vosges de Wissembourg à Wachenheim.

La bande de terrains houillers, en lambeaux intercalés, pour la plupart, dans les replis des roches cristallines, qui traverse le centre de la France en ligne droite, de Decise à Pleaux, marque une dislocation parallèle aux précédentes, et qui en est probablement contemporaine.

Les montagnes du département de l'Aveyron, celles entre la Saône et la Loire, celles même de l'Écosse et de l'Irlande, depuis les îles Orcades et Shetland jusqu'aux granites de Wicklow et de Carlow, en présentent aussi qui sont exactement dans le même cas. Mais leur étude nous écarterait de notre objet actuel.

Nous pouvons d'autant mieux nous borner ici à considérer les dislocations des montagnes des deux rives du Rhin, que leurs résultats offrent une simplicité qui permet de les embrasser dans leur ensemble comme si elles formaient un tout complet, représenté dans le relief extérieur par la disposition symétrique qu'affectent les montagnes des deux rives du Rhin. *La symétrie des montagnes des deux bords du Rhin en fait un ensemble.*

Cette symétrie ne se manifeste jamais si bien que lorsqu'on peut apercevoir à la fois l'un et l'autre groupe, en totalité, d'un point un peu éloigné vers le midi. J'ai déjà indiqué comment, des collines de la Haute-Saône, et particulièrement de la colline de la Motte, près de Vesoul, on voit le profil des Vosges, qui est très-bas et très-plat vers le Val-d'Ajol, se relever et se bosseler assez fortement plus à l'E. dans la région des ballons. Les montagnes de la Forêt-Noire présentent une disposition correspondante dans un sens diamétralement contraire : on peut en juger en choisissant, pour les examiner, un point situé par rapport à elles comme l'est la Motte de Vesoul par rapport aux Vosges. *Points favorables pour juger cette symétrie. La Motte, près de Vesoul.*

De la cime de l'Uetliberg, au midi de Zurich, on distingue à l'horizon la ligne monotone de la Forêt-Noire. Elle paraît bombée, mais très-peu festonnée; moins que les Vosges ne semblent l'être lorsqu'on les voit de la Franche-Comté. Cette ligne de la Forêt-Noire s'élève vers l'O. avec une extrême uniformité, à partir des plaines du Wurtemberg, auxquelles elle fait parfaitement continuité, ce qui rappelle le raccordement des Vosges avec les plateaux qui bordent le Val-d'Ajol, tel qu'on l'observe de la Motte de Vesoul. L'œil ne rencontre d'autre point d'arrêt, depuis le milieu de la Forêt-Noire jusque bien loin dans les plaines du Wurtemberg, que le rebord des Randen qu'on aperçoit de l'Uetliberg au N. 1° E. *L'Uetliberg.*

Mais, pour voir à la fois, avec un égal développement, les Vosges et la Forêt-Noire, il faut monter, par un temps serein, sur une des hautes cimes du Jura, placées dans le prolongement méridional de la plaine du Rhin. Me *Röthi-Fluhe.*

trouvant, le 28 juillet 1836, au lever du soleil, par un ciel sans nuages, sur la cime du Röthi-Fluhe, au-dessus de Soleure, je détournai un instant mes regards du spectacle si attachant que m'offraient les Alpes et leurs magnifiques glaciers, pour considérer les lignes moins hardies de la partie septentrionale de l'horizon. Les Vosges présentaient alors les pentes abruptes de leur flanc S. E., par-dessus les crêtes successives du Jura et la plaine de Beffort, et je remarquais en même temps la terminaison escarpée qu'elles offrent en se prolongeant vers le N. le long de la plaine du Rhin. Je suivais de l'œil leur bord oriental jusqu'à la montagne de Sainte-Odile. Je distinguais aussi très-nettement le profil de la Forêt-Noire. L'horizon de la Souabe s'élevait doucement vers ce large massif, qui ne se découpait un tant soit peu que vers le Belchen, presque sur le bord de la plaine du Rhin. Le Feldberg se détachait à peine de la ligne générale. La chute rapide du Blauen vers la vallée du Rhin était très-sensible. Mes regards s'étendaient sur cette plaine unie, du milieu de laquelle je voyais surgir le petit groupe isolé du Kaisertuhl, semblable à une taupinière dans le fond d'un large fossé.

Fig. 27.

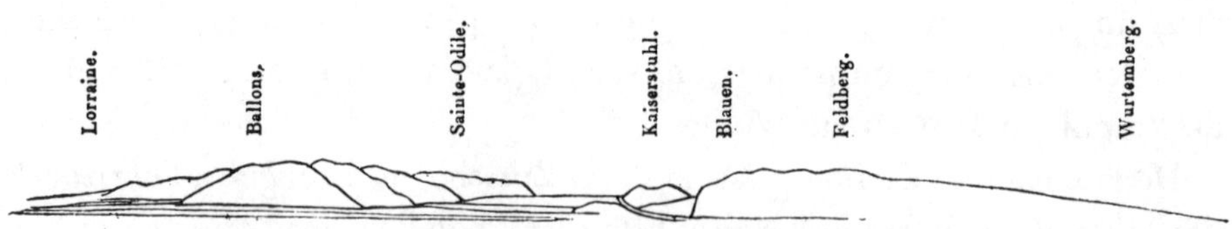

Les Vosges, la plaine du Rhin et la Forêt-Noire, vues de la cime du Röthi-Fluhe.

Origine des deux groupes montagneux et de la plaine qui les sépare.

L'imagination se représentait aisément cette plaine, remplacée par des masses aussi élevées que les Vosges et la Forêt-Noire entre lesquelles elle s'étend, formant de ces deux groupes une seule proéminence légèrement bombée, dont la voûte extrêmement surbaissée s'inclinait légèrement, d'un côté vers la Lorraine, et de l'autre vers le Wurtemberg. Il semblait qu'il ne manquait que la clef de cette voûte, qui se serait un jour abîmée pour donner naissance à la plaine du Rhin, flanquée de part et d'autre par ses culées restées en place, de manière à former sur ses flancs deux escarpements ruineux en regard l'un de l'autre. C'est ce qu'exprime le diagramme ci-dessous, qui, en figurant un terrain bombé, fissuré, puis écroulé, me paraît indiquer l'origine la plus probable des failles, qui forment le caractère essentiel des montagnes du système du Rhin.

Fig. 28.

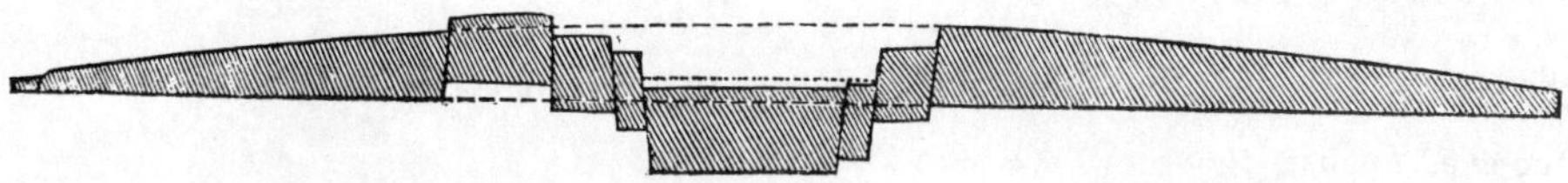

Soulèvement et écroulement combinés qui ont produit la plaine du Rhin.

Ce mode de formation aurait beaucoup de rapports avec celui que M. de Buch assigne, d'après les observations de M. Krug de Nidda, à la grande vallée, bordée de masses trachytiques, qui traverse l'Islande du N. E. au S. O.[1].

[1] Léopold de Buch, *Revue générale des phénomènes volcaniques, placée à la suite de la description physique des îles Canaries;* traduction française, page 352. (Paris, Levrault, 1836.)

9

Reprinted from *Tectonophysics* **8**:543-560 (1969)

THE SEDIMENTARY TROUGH OF THE RHINE GRABEN

C. SITTLER[1]

Laboratory of Geology and Paleontology, University of Strasbourg, Strasbourg (France)

(Received August, 1968)
(Resubmitted January, 1969)

SUMMARY

Accurate knowledge of the sediments in the Rhine graben is not only essential for proper corrections of geophysical data but it provides the main tool to reconstruct the development of the graben in time and space. A large number of borehole data and reflexion seismic results have been made available by mining and oil prospecting companies. Depth charts of the basis of Tertiary sediments of different Mesozoic layers and sometimes of the surface of the crystalline basement have been designed. Longitudinal and cross sections of the trough show tectonic features in detail, e.g. the north-south tilting, the basins and sills within the rift, the existence of deeper depressions at the edges of the rift. Aeromagnetic and gravity data are used for additional evidence.

INTRODUCTION

To reconstruct the development of the Rhine graben through time and space, knowledge of the sedimentary trough is essential, but it also provides the main tool for the correction of geophysical data. For detailed study of the Rhine rift I refer to the International Rhine Graben Research Group (1967).

The outline of the Rhine graben drawn by the masterly tectonician Cloos (1955) shows the principal features (Fig.1): the rift crossing the Hercynian structures, the Jura mountains in the south and the Taunus in the north, the Vosges and the Black Forest mountains with their steep faults, the big field of faults of Saverne and its correspondence with the Kraichgau depression, the narrowing structure of the graben between the Odenwald Massif and the Mainz Basin, then finally, the northern extension to the Hessen depression with the big Vogelsberg volcano in contrast to the little Kaiserstuhl within the Rhine rift.

THE MESOZOIC SUBSTRATUM OR THE PRE-GRABEN "BASEMENT"

We have to distinguish the underlying Mesozoic beds which bear no relation to the graben, from the Tertiary filling of this graben. The Permian and the Mesozoic deposits, especially those of the Triassic were affected

[1]Present address: Institut de Géologie, 1, rue Blessig, Strasbourg 67 (France).

Fig.1. Structural sketch of the Rhine graben. (After Cloos, 1955.)

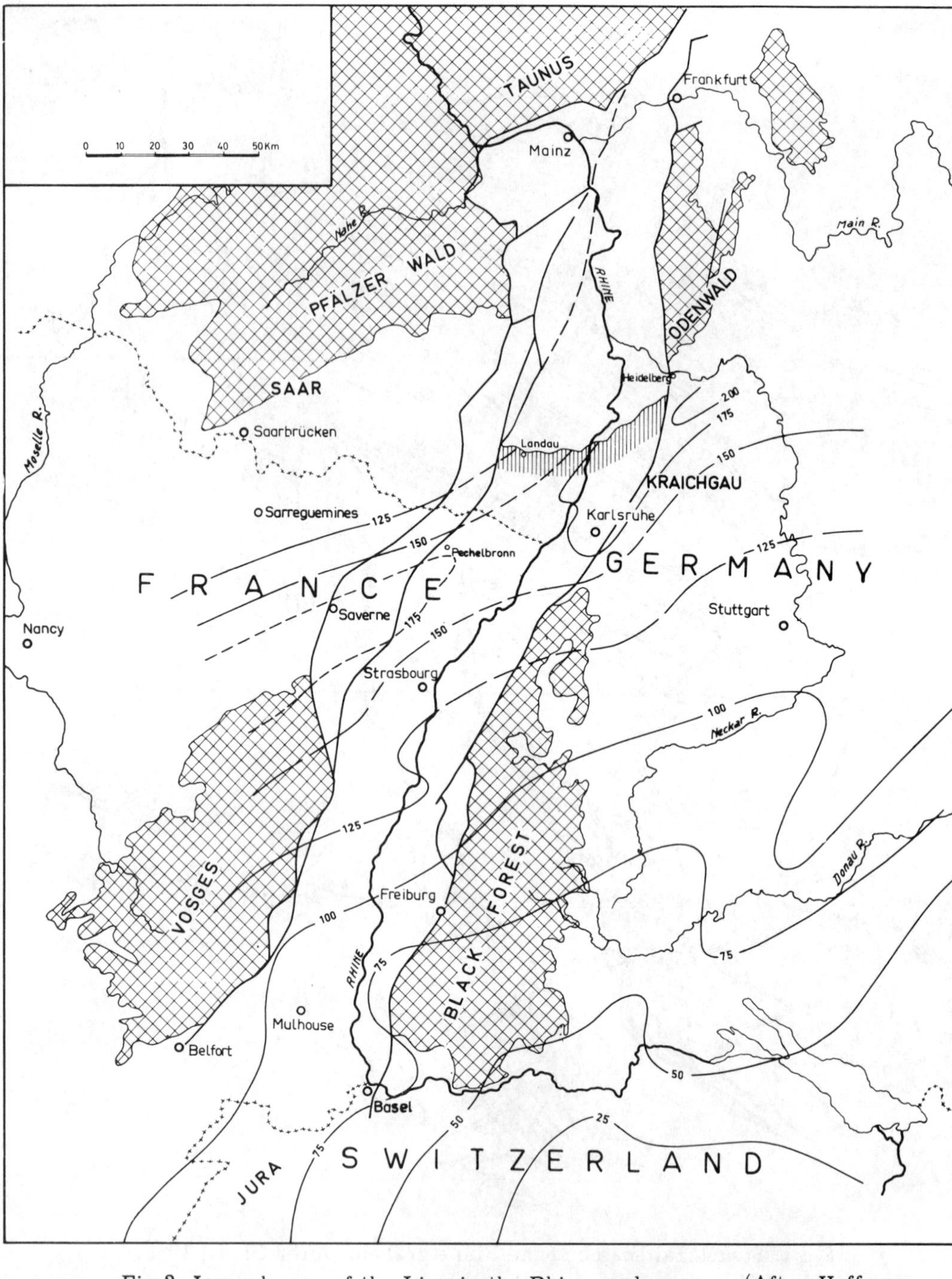

Fig.2. Isopach map of the Lias in the Rhine graben area. (After Hoffmann, 1967.)

by the uplift and trough zones of the old Hercynian surface. As they cross the Rhine graben area, and from north to south we note the following Variscan orientated structures (topography on Fig.2): the Permian Saar–Nahe trough, the Odenwald uplift zone passing westward to the Triassic depression of Sarreguemines and southward to the general Mesozoic depression of Saverne–Kraichgau, the Vosges and Black Forest mountains general uplift with smaller depressions, a foredeep or a trough at the foot of the Vindelician land in southeastern direction. The northern and the southern trough zones show obviously the larger extension of the rift toward the Hessen depression in the north and the Saône–Rhône depression in the south.

Generally speaking, the thicknesses of the Mesozoic deposits (more than 1,000 m in all) decrease in a southern direction due to the influence of the Vindelician mainland. An isopach map points out these facts for the Lias (Fig.2). Erosion took place during Late Mesozoic so that we find, north of Landau for example (Fig.2), neither Lias in the Mesozoic substratum, nor reworked Liassic material in the infilling Tertiary beds. Just before the Tertiary, the Rhenish shield emerged more - and therefore was more eroded - in the north than in the south.

THE TERTIARY INFILLING OF THE RHINE GRABEN

A longitudinal section of the whole rift (Fig.3) shows the differential erosion and the tectonic features of the Mesozoic bottom moulded onto the Variscan basement. The Tertiary infilling also reflects these structures, and in the southern part the sediments accumulated in variable thicknesses and in different facies in a regular succession of basins and sills (Fig.3). Gravity and aeromagnetic charts confirm these structures (Fig.4). From south to north we find (compare Fig.3, 4): the Mulhouse Basin, the Colmar Sill, the Sélestat Basin, the Erstein Sill, the Strasbourg Basin, the Haguenau Sill and so on.

In the Middle Eocene, some lacustrine depressions foreshadowed the accentuated deepening of the trough during the Late Eocene and Oligocene. The sea then invaded the rift from the south, probably from the perialpine area (Sittler, 1965), and not from the Paris Basin as previously assumed, whilst the subsidence maintained the different basins and sills of the eroded pre-Tertiary surface (Fig.3).

During the Early Oligocene (Lattorfian) the subsidence was greatest in the south: 1,500–2,000 m of saliferous marls deposited in the Alsatian potassium salt basin of Mulhouse, while in the north up to 500 m of oil bearing sandy marls were laid down (petroliferous series of Pechelbronn). During the Middle Oligocene (Rupelian) the transgression from the north - certainly from the North Sea Basin - filled up the trough with a constant thickness of grey marls. In Late Oligocene, Miocene and Pliocene times (Fig.3) the trough subsided predominantly in the north, to which the sea drew back: here 2,000 m of marine, brackish and progressively freshwater formations are covered by 300 m of Quaternary alluvium. On the contrary, in the south an uplift brought about the erosion of previous deposits.

An isopach map of the Tertiary and Quaternary filling of the trough is not shown here, but it indicates that the axis of greatest subsidence follows the eastern side of the Rhine stream, with a maximum depression between

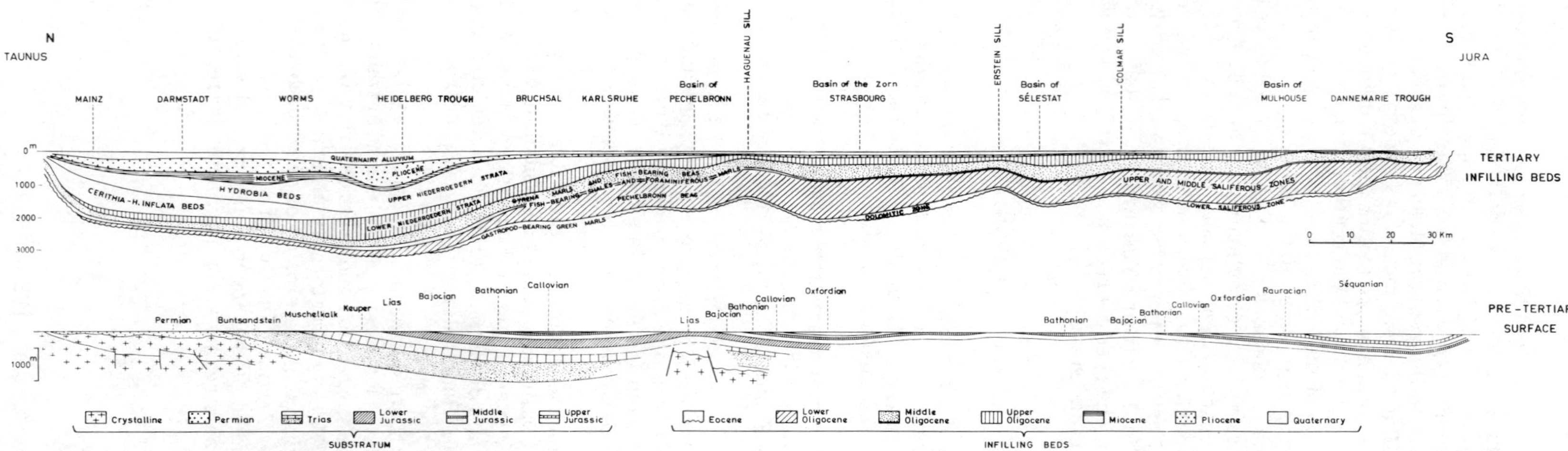

Fig.3. Longitudinal section of the Rhine graben. (After Sittler, 1965a.)

pp.547–549

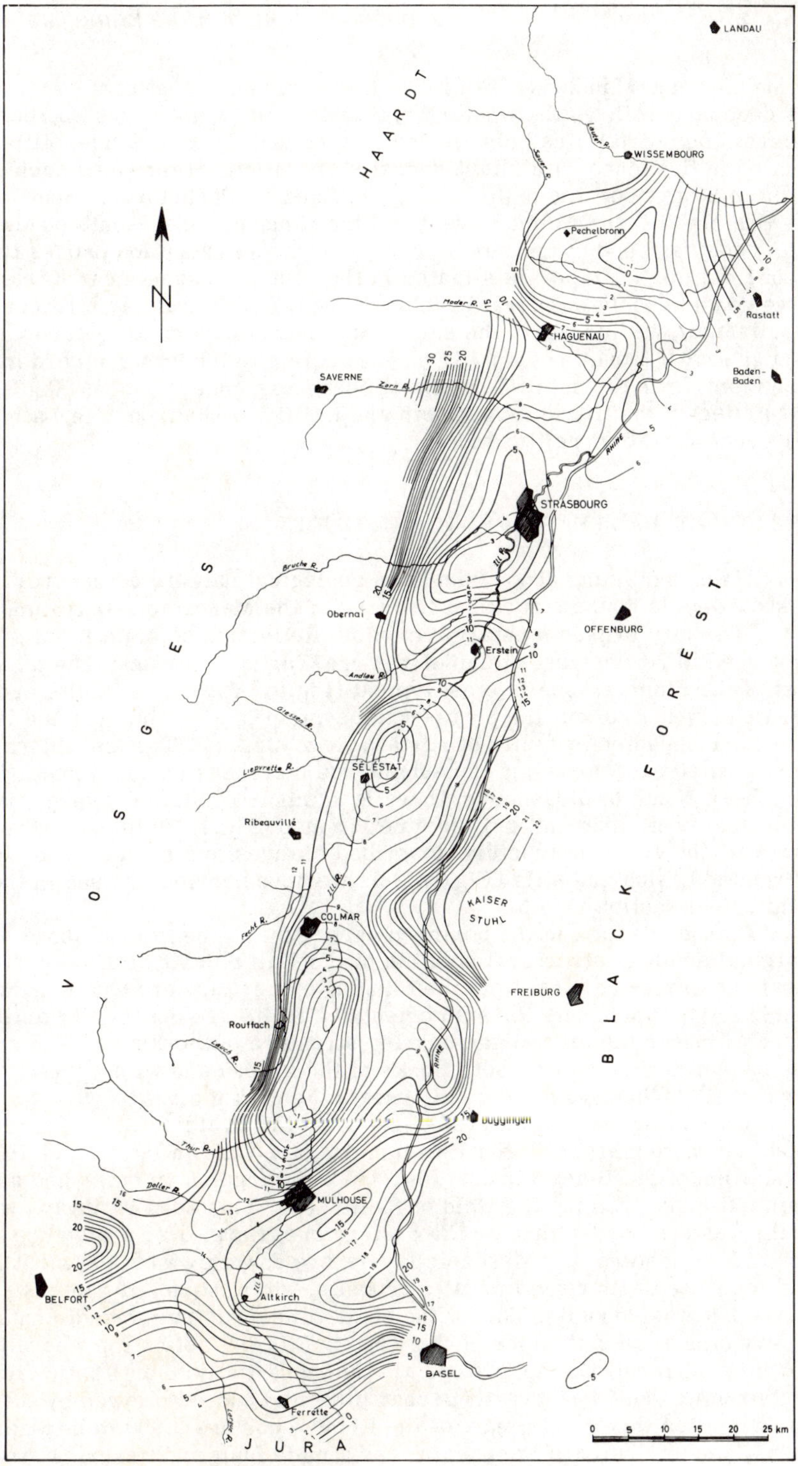

Fig.4. Experimental gravimetric map of the southern part of the Rhine rift (After Maikovsky, 1952). Equidistance of the isogals = 1 mgal. For private and local purposes, the base of the isogals are related to an arbitrary value of +0.550 mgal. (Bouguer anomaly). The isogal 0 of this map corresponds with −34 of the gravimetric map of France (density 2.7) since the gravity values have been calculated on the base of the Paris Observatory (980,943.00).

pp.550–552

Heidelberg and Mannheim (Fig.5–7). In the northernmost part of the graben, the deep zone follows the western main rift fault, while in the southern third the maximum depth lies between the Vosges and the Rhine. The oilfields are located in the zones with slight deposits or structural terraces such as: the horst of Stockstadt, the highs of Speyer, Landau, Pechelbronn–Soufflenheim, the Erstein Sill, the Colmar swell and the Sundgau horst (south of Mulhouse) at the southern end of the rift. But, as always "an exception proves the rule", an oilfield, for example, is situated in the central channel near Karlsruhe. If we consider only one stratigraphical level, e.g., the Early Miocene (Aquitanian), we find again the same outstanding structural features of this asymmetrical trough not only from the eastern to the western border, but also from north to south where the Miocene sequence thins out (Fig.3). This is an effect of the tilting to the north and infilling of the trough in Late Oligocene and Miocene times.

THE STRUCTURAL FEATURES WITHIN THE RIFT

Maps combining geophysical and geological data have been worked out to show details of the structural features of the Mesozoic substratum and of the Tertiary synsedimentary tectonics. Reflection seismic results have been used to reconstruct a map of the pre-Tertiary surface. The middle part of the Rhine graben is most typical (Fig.5). The major faults are Rhenish orientated, but the north–south system is also common and is sometimes accompanied by faults of northwestern striking. The fault blocks are mostly tilted eastwards and the subsidence increases in this direction (Fig.5–8). North of Strasbourg, there is no clear relation between this geological depth map and the gravimetric map (Fig.4). In the southern part, however, the general structure described changes to a series of northeast orientated basins and sills (Fig.4, 5) as we have previously seen in the longitudinal section (Fig.3).

A cross section in the northern part of the Rhine graben shows the marginal and high structural position of the Mainz Basin with respect to the subsiding graben (Fig.6). Note the thinning of Tertiary and Quaternary in Mainz Basin. Numerous Rhenish orientated faults are essentially antithetic, especially near the oil and gas bearing structure of Stockstadt. The crystalline basement outcrops on both flanks of the graben and within the rift it was found at its highest point 2,250 m beneath the drift plain.

South of Strasbourg, a cross section in the Erstein Sill area has been established, running from the Permian basin of Villé in the Vosges to the crystalline of the Black Forest (Fig.7). We distinguish an outer and an inner main rift fault, framing the field of faults zone. This zone is always wider on the western border than on the eastern one. Here, over a distance of some hundreds of meters, the Mesozoic layers are downthrown 2,400 m: it is a real fore-deep with conglomeratic deposits, whilst some kilometers eastwards salt was deposited during the same time. This saline sedimentation is very typical on both sides of the Erstein Sill, but not on top where the Tertiary thins out. Collapse steps are due essentially to antithetic faults. The structure rises eastward (Eschau oilfields) and is followed by the central gutter whose axis, east of the Rhine, subsides 2,000 m beneath the plain. Symmetrically to Eschau, beyond a main fault and flexuring structure, lie the Kork and Legelshurst oilfields which also have very thin Tertiary

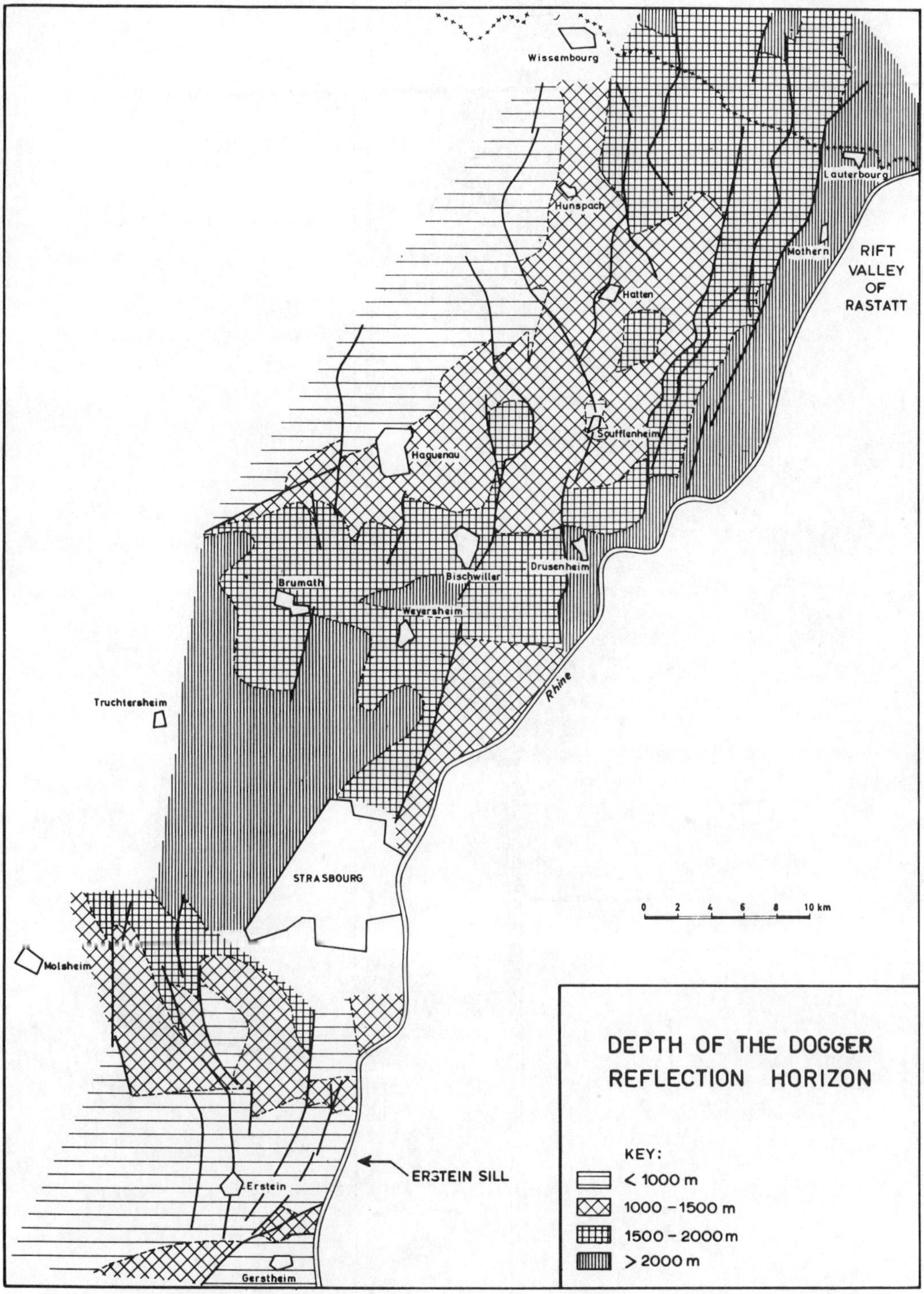

Fig. 5. Depth of the Mesozoic substratum in the middle part of the Rhine graben: geophysical and geological data. (After Simler and Masclaux, 1967.)

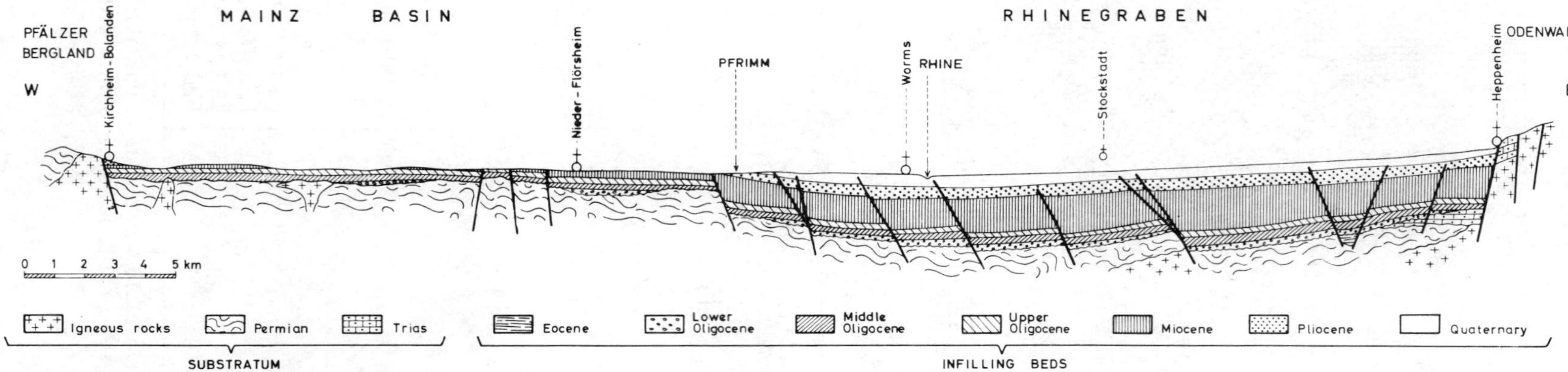

Fig.6. Cross section of the Rhine trough in its northern segment, the Mainz Basin area. (After Wagner, 1938, and Straub, 1962.)

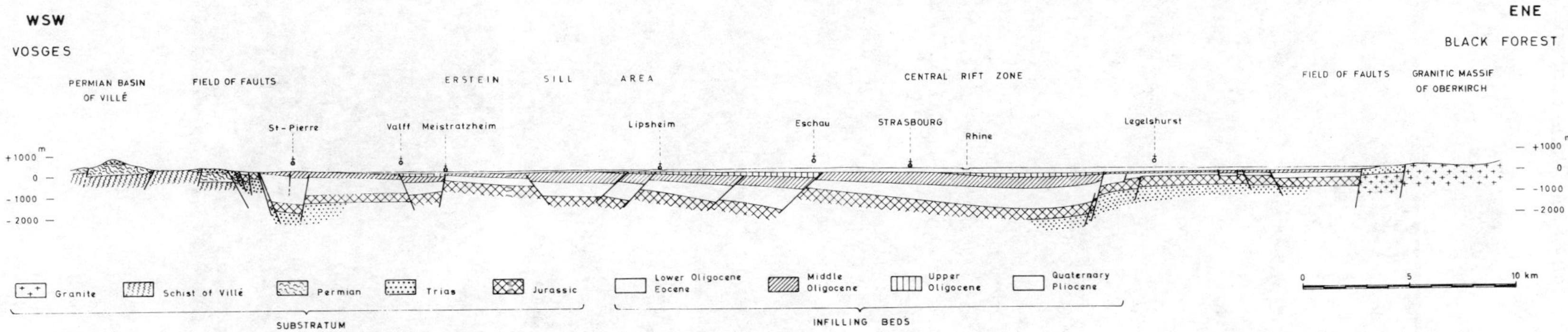

Fig.7. Cross section of the Rhine trough in its middle segment, the Erstein Sill area.

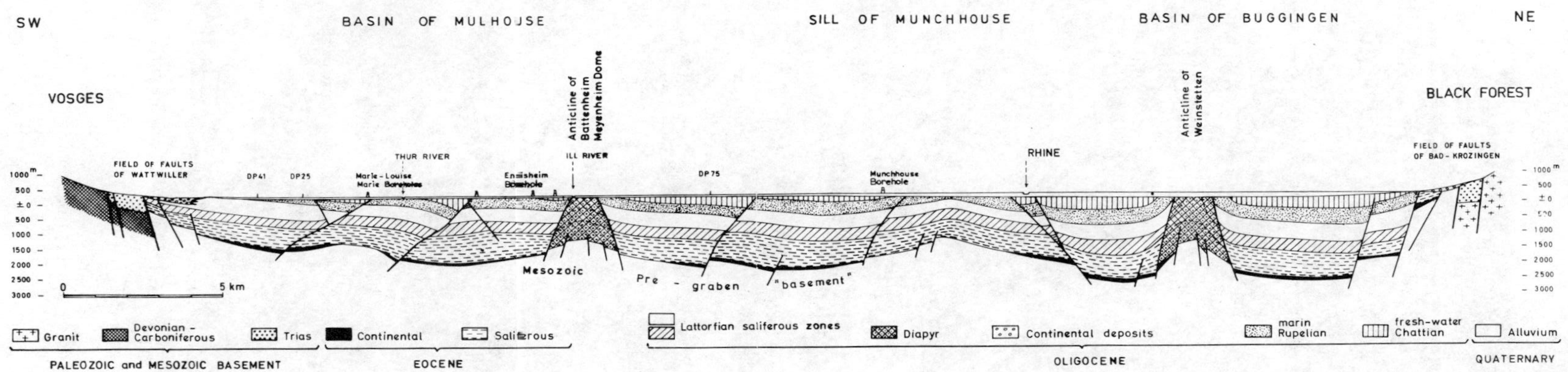

Fig.8. Cross section of the Rhine trough in its southern segment, the potassium salt basins of Mulhouse and Buggingen.

pp.557–558

Fig.9. Relief view of the Rhine graben.

deposits, representative of active high zones. On the eastern border, the rift rises quickly, beyond a fore-deep and two main faults, to the granites of the Black Forest.

In the southern Rhine graben the beds infilling the rift have a more plastic structure (Fig.8); the saliferous Early Oligocene suffers deformation in a plastic way with flat and sometimes curved faults. Diapyric injections are frequent.

Thus, from the border to the centre of the graben, the rift may be subdivided in five zones: (1) the mountain border zone; (2) the field of faults zone, with Mesozoic outcrops; (3) the marginal rift zone, more or less reduced, beyond the main rift fault, sometimes with a fore-deep, or with Early Oligocene "outcrops" under the Pliocene–Quaternary sediments; (4) the middle rift zone with Middle Oligocene "outcrops"; and (5) the central rift zone where Late Oligocene and Miocene "outcrops" and where the underlying Mesozoic beds had their maximum subsidence. The extension of these zones is not symmetrical on both sides of the central gutter; they are much wider on the western side of the rift.

The stratigraphical subdivisions and the correlations of the graben filling are now known very precisely. Also we now have a good idea of the seismic velocities, the electrical conductivities and the rock densities in the area. Aerogeological documents are also of great help to the tectonician and the geophysicist trying to explain, from the surface (Fig.9), the deep structure of our Rhine graben.

REFERENCES

Cloos, H., 1955. Geologische Strukturkarte der Mittelgebirge. Geol. Rundschau, 44: 480.

Cogné, J., Géze, B., Goguel, J., Grolier, J., Letourneur, J., Pellet, J., Rothé, J. et Sittler, C., 1966. Les "rifts" et les failles de décrochement en France. Rev. Géograph. Phys. Géol. Dyn., Sér. 2, 8(2): 123–131.

Hoffmann, K., 1967. Die stratigraphischen, faziellen und paläogeographischen Verhältnisse des Lias im Oberrheintal zwischen Heidelberg und Offenburg, unpublished. Also: Der Oberrhein-Graben zwischen Heidelberg und Offenburg in der Lias-Zeit (abstract). Z. Deut. Geol. Ges., 118(2): 236.

International Rhine Graben Research Group, 1967. The Rhine graben Progress Report, 1967. Abhandl. Geol. Landesamtes Baden-Württemberg, 6: 1–146; Publ. Serv. Carte Géol. Alsace Lorraine, 1–146.

Maikovsky, V., 1952. Le pétrole dans le Haut-Rhin: le sondage Staffelfelden IV-DP 25. Bull. Soc. Ind. Mulhouse, 1: 25–34.

Simler, L. et Masclaux, R., 1967. Cartes des profondeurs de deux horizons sédimentaires dans la plaine d'Alsace. Publ. Serv. Carte Géol. Alsace Lorraine, 26: 68.

Sittler, C., 1965a. Le Paléogène des fossés rhénan et rhodanien. Études sédimentologiques et paléoclimatiques. Mém. Serv. Carte Géol. Alsace Lorraine, 24: 392 pp.

Sittler, C., 1965b. Monographie structurale du fossé rhénan. Synthèse géologique réalisée dans le cadre du Projet international du Manteau supérieur. Inst. Géol., Strasbourg, 41 pp., unpublished.

Straub, E.W., 1962. Die Erdöl- und Erdgaslagerstätten in Hessen und Rheinhessen. Abhandl. Geol. Landesamtes Baden-Württemberg, 4: 123–136.

Wagner, W., 1938. Das Mainzerbecken. Jahresber. Mitt. Oberrhein. Geol. Ver., 27: 25–62.

Editor's Comments on Papers 10 and 11

10 RAMBERG AND SPJELDNAES
Excerpts from *The Tectonic History of the Oslo Region*

11 ZIEGLER
Excerpts from *Geology and Hydrocarbon Provinces of the North Sea*

Paper 10 is by Ivar B. Ramberg of the University of Oslo (Norway), and Nils Spjeldnaes of the University of Aarhus (Denmark). The Oslo rift zone is an intracratonic feature penetrating the Fennoscandian or Baltic shield. The rifting in the Oslo region began in the mid-Carboniferous and continued chiefly after Permian times. Basaltic eruption began early and filled the graben, accompanied by the subsidence of calderas lying along a north-south trending zone. In the final paragraph, the stages of formation are briefly listed. The most important conclusions are of general application. This is in the relationship between zones of lithosphere weakness and disturbance in the asthenosphere and mantle. The model presented emphasizes mantle control, the details of graben features being determined by crustal control. (Relevant papers are found in publications 13 and 14 of the Bibliography.)

Paper 11 is by P. A. Ziegler of Shell International Petroleum Co., in the Hague (Netherlands). It is clear from the Paleogeographic maps of the North Sea region that the important rift faulting began there in the Triassic and was completed by the end of the Cretaceous, when Tertiary sediments masked all that had gone before. The scale of the faulting far exceeded that of the Rhine rifts (compare section cc with *AA* and *BB*). Nothing would have been known of this structural pattern were it not for the drilling and seismic surveys carried out in the search for oil and gas.

10

Reprinted from pages 167-169, 174-177, and 188-193 of *Tectonics and Geophysics of Continental Rifts*, I. B. Ramberg and E.-R. Neumann, eds., D. Reidel, Dordrecht, 1978, 444p.

THE TECTONIC HISTORY OF THE OSLO REGION

Ivar B. Ramberg[1] and Nils Spjeldnæs[2]

[1] Department of Geology, University of Oslo, Box 1047, Blindern, Oslo 3, Norway
[2] Palaeocology Dept., Århus University, DK-8000 Århus C, Denmark

1. INTRODUCTION

The classical studies of the tectonics of the Oslo Region were by Brögger who in a series of papers towards the end of the last century [1-4] described the structural features in great detail. Later important contributions include those by Cloos [5], Störmer [6,7], Holtedahl [8-11] and Gleditsch [12]. In connection with the extensive studies of the alkaline igneous rock series of the region, numerous observations were made on the tectonics also, and various petrogenetic hypotheses were offered, especially by Barth [13-14] and Oftedahl [15-17]. In recent years special emphasis has been given to geophysics, gravity in particular. These data, and their bearing on our understanding of the geodynamic evolution of the rift, have been analysed by Ramberg [18]. The latter paper also includes references to much of the older literature, as well as appropriate descriptions and maps of the region. Recently, Ramberg and Larsen [19] reviewed the tectonomagnetic evolution of the rift, particularly discussing the lavas and the main, Permo-Carboniferous taphrogenetic event. The present paper is an extension of the latter two papers, and emphasizes the chronology of the development of the region, and succession and significance of tectonic events from Precambrian to Recent.

2. TERMINOLOGY AND TECTONIC SETTING

The Oslo Region is a non-genetic geographical term loosely referring to an area of about 11000 km^2 in Southern Norway (Fig. 1) in which the rocks are younger than the surrounding Precambrian terrain [18-20]. The Oslo Graben specifically refers to the relatively

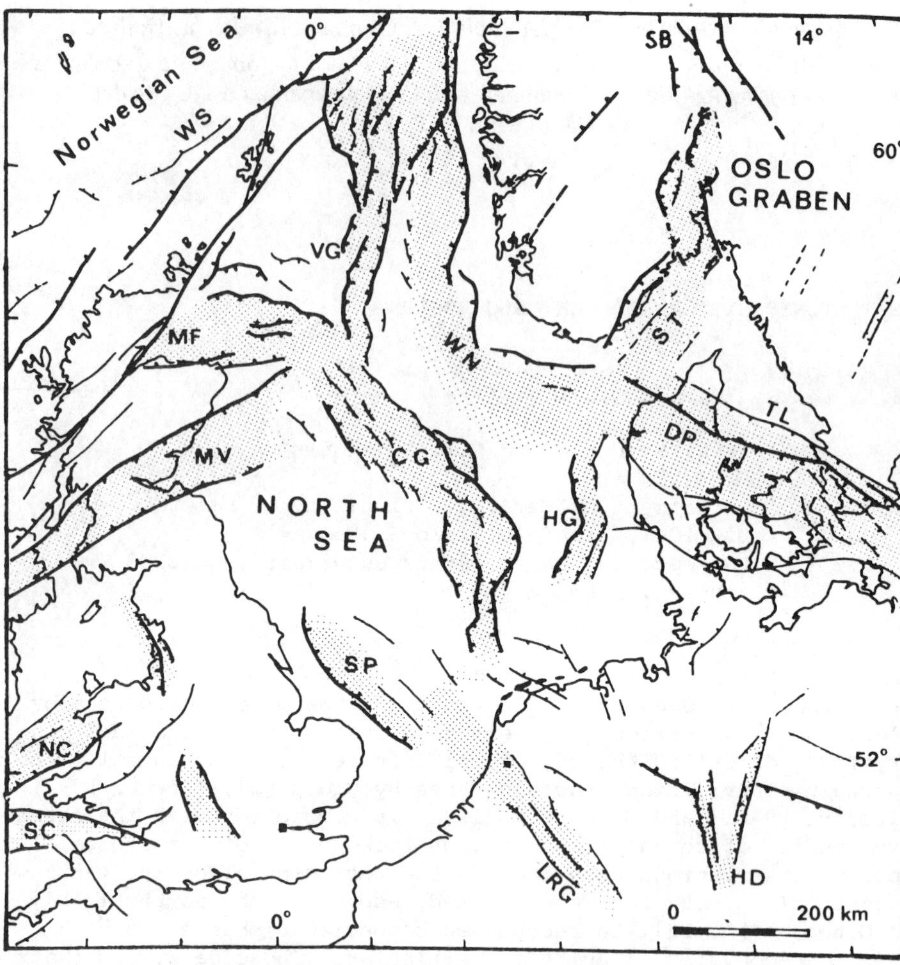

Fig. 1. Tectonic setting of the Oslo Graben. CG-Central Graben, DP-Danish Polish Depression, HD-Hessen Depression, HG-Horn Graben, LRG-Lower Rhine Graben, MF-Moray Firth Trough, MV-Midland Valley, NC-North Celtic Sea Trough, SB-Sparagmite Basin, SC-South Celtic Sea Trough, SP-Sole Pit Basin, ST-Skagerrak Trough, TL-Tornquist Line, VG-Viking Graben, WN-West Norway Trough, WS-West Shetland Basin.

depressed crustal block of about 220 km length on land. The graben itself represents only a part of the Oslo Rift, an extensive fracture zone connecting the Oslo Graben SSW-ward with the North Sea system of grabens, the development of which was also initiated in Late Paleozoic time [21].

The graben consists of two individual, north-trending segments arranged in an en echelon manner relative to each other. These two segments define the overall NNE direction of the rift [19]. Both graben segments are asymmetrically tilted towards the central N-trending Oslofjord-Randsfjord faults, which together have the characteristics of a scissor fault. To the SSE there is evidence of one or more graben segments below the Skagerrak Sea.

In detail, the graben and surrounding areas are criss-crossed by a number of major and minor faults and joints, forming a complex fracture pattern [22]. Whereas north trending fractures predominate in the central graben zone, the Precambrian terrain on both sides of the graben is characterized by NW and NE trends. The diagonal trends have been interpreted as an old regmatic shear pattern. Many of the Permian faults and dikes exhibit a strong influence by older structural trends.

[*Editor's Note:* Material has been omitted at this point.]

It is observed, therefore, that the Oslo Region was a sedimentary basin, with up to 2 km sediments which were eroded down to the Precambrian on parts of the east side in the Late Silurian. A possible explanation is to assume a dome-like uplift of about 2 km over the basins. Such domes are well known from active rifts, and its supposed absence was regarded as one of the peculiarities of the Oslo graben. Subsequent to rifting, the thermal uplift decreased and the dome seems to have been further removed by peneplanation, down to a level approaching that of the previous Precambrian peneplain (Fig. 3D).

4. THE PERMIAN EVENT

The formation of the Oslo Graben (Fig. 4) represents a major event which involved the development of extensional tectonics, volcanism and subsidence. The activity has generally been regarded as Early Permian, but a more extensive period of (repeated?) tectonic and volcanic activity seems now more likely in view of more recent dating results [34], tectonomagmatic [19] and general geological considerations.

4.1 Dating of activity

The first dating of the volcanic activity was based on paleontological evidence. The rather rich fossil beds found at Semsvik in 1931 [35], just below the lowest basalt beds, indicated the volcanism commenced in the Late Paleozoic, probably in the Early Permian. Analysis of the original descriptions of fossils (including fishes [36], plants [37-39] and non-marine lamellibranchs [40]), reveals that the evidence from fishes and molluscs give an age between Upper Carboniferous and Permian. The plant evidence

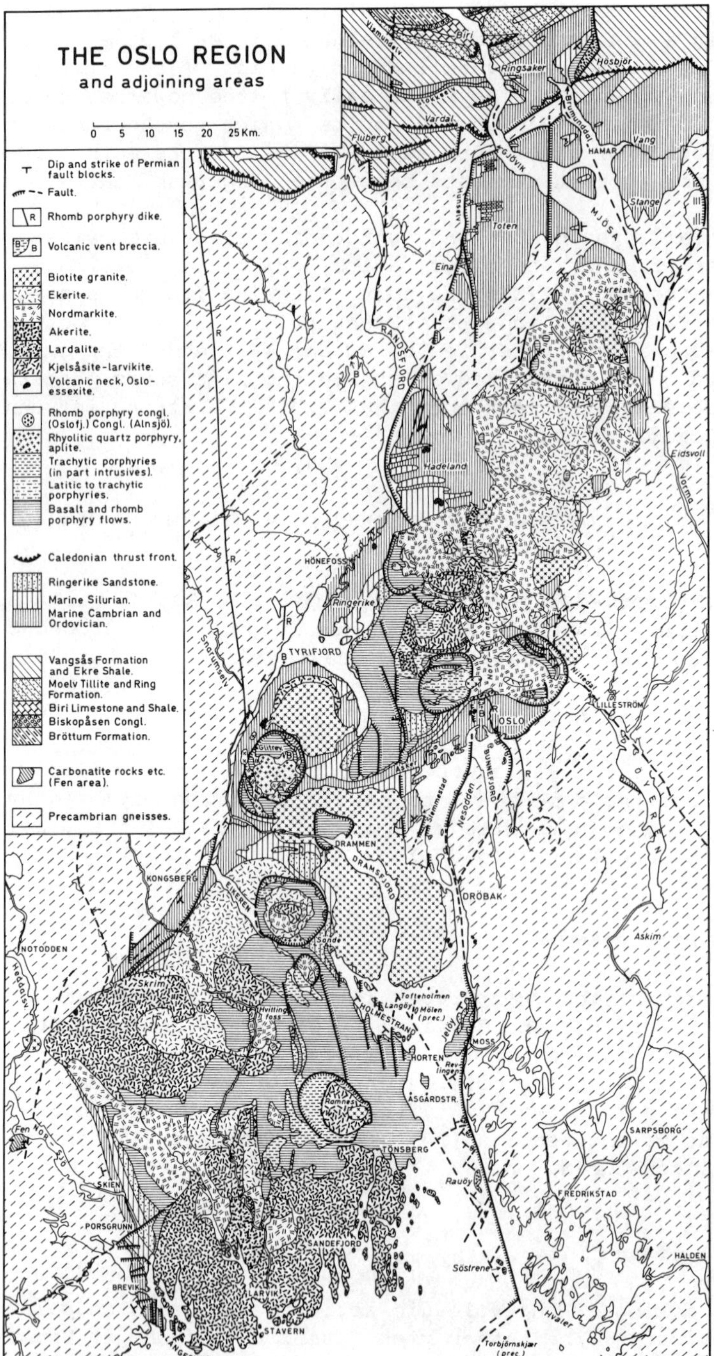

Fig. 4. Geological map of the Oslo Region (after [18]).

narrows the age to Rotliegendes (probably lowest part) and the uppermost Carboniferous, but a more precise age determination for the initial volcanism in the central part of the graben seems extremely difficult on the basis of the Semsvik macro-fossils. If they are of Permian age, they belong to the oldest part of the period.

Since the fossiliferous (and underlying) beds are rich in volcanic material, and since the lowest (B_1) basalt in the southern part of the graben is considered to be older than those at Semsvik, the definite possibility remains that the eruptive activity in the Oslo Region started already in the Upper Carboniferous.

The end of the activity is more difficult to define. It has been suggested [41] that the porous Brumunddal sandstone resting on the youngest rhomb porphyry lavas in the northern part of the graben (Fig. 4) is of Triassic age. This is based on a small number of eolian cross-beddings, which conform to the prevailing wind-directions of the British Bunter, and differ from that of the generally E to W Permian direction. The paleomagnetic directions [42] are normal, both in the sandstone and in the underlying, deeply weathered lavas, indicating a post-Permian (or Post-Kiaman) age for the remanent magnetism.

No ash which could have come from the Oslo eruptions has been reported from the Zechstein salt deposits in the North Jutland basin (Fig. 1). Since igneous rocks of the Oslo type are found not only in the Oslo Graben to the north, but also on the Fyn-Ringkøbing Ridge [43] and possibly in the boring C-1 and D-1 west of Jutland [44], the evaporite basin would have been a likely recipient for ash from the copious eruptions. The salt basin also shows a quiet development, indicating that there were no large-scale tectonic disturbances during the deposition of the salt. This seems to imply that the whole eruptive activity, and the rifting, took place during the Rotliegendes ($\sim$ 280 to 250 m.y.) and was finished before salt deposition began during Zechstein time. Alternatively, and more likely, there was a quiet interval in the activity during the Zechstein, with active periods both before and after. Pre B_1 faulting, for instance at the Krokskogen lava plateau [45], and downfaulting of the supposedly Triassic Brumunddal sandstone add to the impression of an extended period of repeated tectonic activity.

Radiometric age determinations, mainly by the Rb/Sr and the K/Ar-methods [34,46], generally support the latter hypothesis. The main magmatic activity apparently lasted from about 290 to 245 m.y., while especially K/Ar clay alteration ages on contact and dike-related deposits [46], extend this period to about 220 m.y. The same age (219 $\pm$ 6 m.y.) is also found for a diabase dike near Oslo [47]. A possible Triassic pulse of activity may have surfaced mainly as extensional faulting and emplacement of dikes.

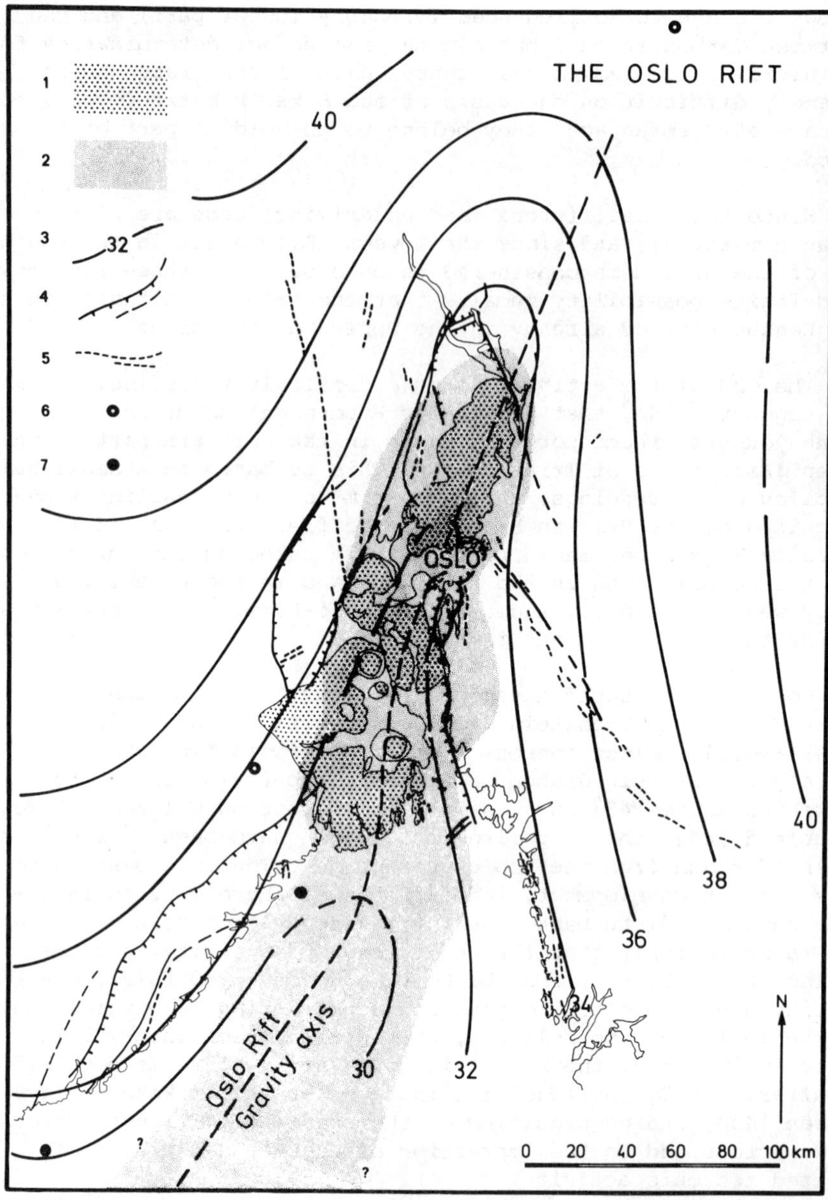

Fig. 5. The Oslo Rift with exposed intrusive rock complexes, outline of deep-seated mafic body, and crustal thickness indicated. 1 - Exposed plutonic rocks, 2 - Outline of deep-seated intrusion (at about 20 km depth), 3 - Moho contours, 4 - Faults, observed and inferred, 5 - Dikes, dike swarms, 6 - Peralkaline complexes (Fen, Särna), 7 - Submarin, mafic plugs or complexes.

[*Editor's Note:* Material has been omitted at this point.]

4.4 Structural trends and directions

The fault pattern in the Oslo Region largely consists of a predominant N-S direction combined with two 'diagonal' trends (NN-SE and NE-SW) [22]. All three directions are partly inherited from the Precambrian basement; the faults seem to have been active intermittently during the Late Precambrian, and probably also during the post-Permian.

These features are especially well demonstrated for the great fault SW of the graben [66] (see Fig. 5). On a smaller scale, the parallelism between Permian faults and older Precambrian features can also be observed for instance at Nærsnes (near Slemmestad, Fig. 4) some 20 km SSW of Oslo. Here the fault pattern, with directions parallel to the Permian directions, form characteristic three-sided depressions in the fine-grained Precambrian rocks. One of the three-sided cavities has been found filled with Middle Cambrian fossiliferous sediments [67], indicating the fault pattern was in existence in the Early Paleozoic and that it was reactivated during the subsequent rifting.

Because of the predominance of N-S trending faults, it has generally been supposed that the axis of tension (or least compression) generating the rift was in an E-W direction. An analysis of the intricate fault pattern in the vicinity of the city of Oslo, and especially the trace of the Tyveholmen rhomb porphyry dike (the "type rock" of RPs), suggests that the tensional (σ_1) axis was probably directed somewhat to the north of west, more in harmony with the overall NNE trend of the graben. The trace of the Tyveholmen dike (Fig. 9) reveals that it has been repeatedly displaced in a direction parallel to the Caledonian fold axes.

It is, however, not unlikely that the dislocation pattern (in Fig. 9) may be derived from different episodes of crustal movement. That such events actually took place and also involved contrasting stress fields, follows from the fact that several zones of Permian (?) thrusting and folding occur with axial planes

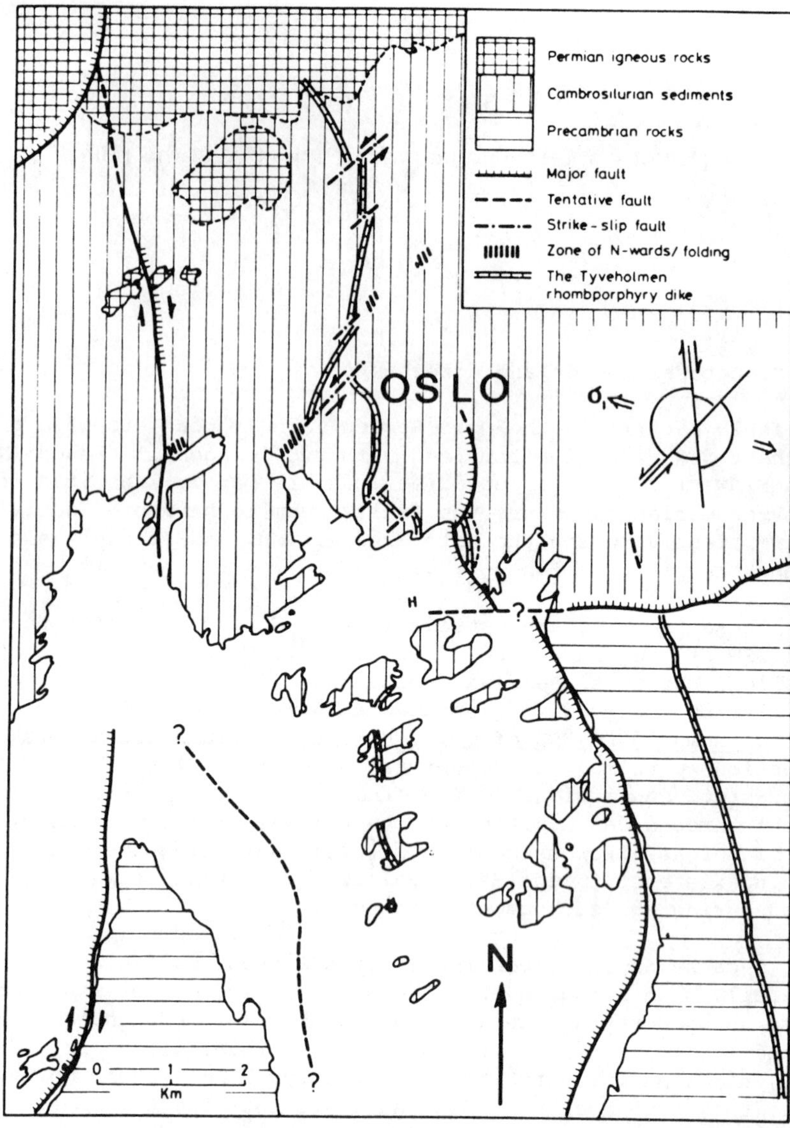

Fig. 9. Tectonic setting of the Tyveholmen rhomb porphyry dike (modified from the Geological Map of the Oslo area, in scale 1:50 000, by Holtedahl & Dons, 1953). Several oblique faults cut the dike; relative movement along major N-S trending faults is also indicated. A tentative stress field solution is shown (insert). Also note that the Tyveholmen dike, which is cut by several faults, was itself injected into the intrusive (nordmarkitic) massif to the north. Its continuation S-ward on the islands is complicated by largescale block-faulting.

of folds generally striking in the NE-SW direction (Fig. 9), i.e. parallel to the oblique faults that offset the dikes of the area. In some cases Permian sills have been folded or thrust indicating a syn- or post-Permian age for the compressional movements [e.g. 10]. Such folds seem to have a NW-ward vergence in contrast to the Caledonian folds which commonly face SE-ward.

5. POST-PERMIAN ACTIVITY

As discussed previously, the main rift-forming phase culminated in Permian time. Except for a possible extension of the activity into the Triassic, there are few direct signs of later tectonic and igneous activity within the graben area proper. Along the Norwegian Skagerrak coast, however, i.e. the SSW-ward extension of the rift zone, there are geological and geophysical indications of Cretaceous and Tertiary activity, including Tertiary dolerite dikes [68], post-Jurassic basalt [69] and possible Tertiary volcanic centres in the Skagerrak [70]. Within the graben area, Sæther [71] showed that in contrast to most of the dike rocks in this region, there exists a "late" group of diabase dikes which are conspiceously less altered, and which may be related to Mesozoic and Tertiary events in the North Sea and Skagerrak areas.

The western part of the Fennoscandian Shield was uplifted in connection with the opening of the Norwegian Sea. This uplift can be recognized not only in the topography, but also in the altitude of the Pre-Cambrian peneplain [10]. The exact age of the uplift has been debated, but may be as late as late Tertiary [72] or as early as early Tertiary. The peneplain is almost perfectly flat in the east, but west of the Graben rises with a constant gradient towards the coast. The Oslo Region apparently acted as a type of hinge-line for the uplift in this area, but the geological and geophysical consequences have not been studied properly.

Repeatedly during the Pleistocene and probably late Pliocene, the Oslo Region as part of the Fennoscandian area, was depressed under the load of ice, the area rebounded isostatically when the ice-sheets melted during interglacial times. These movements may have reactivated the old fractures, but direct evidence for this is only fragmentary. Some discrepancies in the recent uplift rates may reflect such movements, or alternatively errors in the measurements. Some of the fault lines are unusually well marked geomorphologically, something which again might indicate late tectonic activity. Finally, Brögger [2, pp. 398-99, text-fig. 29] published a second-hand report about a kettle-hole in the apatite-mine in Bamble (just to the SW of the Oslo Graben) being cut by a fault. The locality is no longer available, and although it is quite probable that small scale movements took place along the fractures even in the Pleistocene, hard evidence is lacking. The historic seismic activity [55] in the Oslo Region has been relatively modest but higher than that of the surrounding Pre-Cambrian terrain; the epicentre patterns seem to follow old fault zones both to the south and to the north of the graben.

SUMMARY AND CONCLUSION

The main features of spatial/temporal evolution of the Oslo Region are now reasonably well known. Recent geophysical and geological studies have provided a) an estimate of mass and geometry of the various plutonic bodies and b) an improved picture of the gross crustal structure along the rift zone. These results, combined with new petrologic data, have served as a basis for the tectonic model outlined above. This model emphasizes the mantle control on the overall location and evolution of the rift, and the crustal control on the detailed setting and trends of individual graben features. The main taphrogenetic event in Carbo-Triassic time has been separated into several stages including 1) a Pre-Permian stage of fracturing and intrusion, and subsidence of sedimentary basins; 2) an initial (basalt) volcanic stage, possibly associated with crustal doming; 3) normal faulting, continued fissure eruptions and rift valley formation, accompanied by granitic intrusion; 4) a central volcano and cauldron stage accompanied by explosion vents and diatremes; 5) the emplacement of major composite batholiths; and 6) a terminal stage of faulting and dike intrusion.

A number of basic problems, however, remain unsolved: Is it purely accidental that the Late Paleozoic rifting followed the same axial trend as the Lower Paleozoic sedimentary basins? Or does this coincidence violate the idea of rifts being formed by an asthenospheric disturbance whose origin is independent of the overlying lithosphere? In other words, do rift formation and magmatism simply represent the response of pre-existing anisotropy to stresses acting upon the entire plate? Or does intra-continental rifting <u>require</u> the coincidence of a zone of lithosphere weakness with an underlying disturbance? What types of processes cause igneous activity to continue over tens of millions of years? Whether the rifting is caused by processes below or within the lithospheric plates might be analysed by studying surface features such as volcano spacing and frequency, trends and spacing of faults, and age gradients along the rift. Other problems of importance in the Oslo Region include present seismicity, and its possible tectonic implications, and the relation between the Oslo rift and the North Sea rift system.

ACKNOWLEDGEMENT

Peter Vogt is gratefully acknowledged for critically reviewing the paper and for correcting the style.

[*Editor's Note:* Only the references cited in the preceding excerpts are reprinted here.]

REFERENCES

1. W.C. Brögger, Universitetsprogram, 1882.
2. W.C. Brögger, Nyt Mag. Naturvid., 28, 253, 1884.
3. W.C. Brögger, Nyt Mag. Naturvid., 30, 95, 1886.
4. W.C. Brögger, Norg. geol. Unders., 139, 1, 1933.
5. H. Cloos, Fortschr. Geol. Pal., 21, 241, 1928.
6. L. Störmer, Nor. geol. Tidsskr., 15, 43, 1935.
7. L. Störmer, Geol. Soc. London Quart. J., 123, 183, 1967.
8. O. Holtedahl, Am. J. Sci., 29, 504, 1935.
9. O. Holtedahl, Nor. Vidensk.-Akad. Oslo, I. Mat. Natur. Kl., 1943, No. 1, 89 pp., 1943.
10. O. Holtedahl, Norges Geologi, Norg. geol. Unders., 164, 1118 pp., 1953.
11. O. Holtedahl, (ed.) Geology of Norway, Norg. geol. Unders., 208, 540 pp., 1960.
12. Chr. Gleditsch, Nor. geol. Unders., 181 & 182, 1952.
13. T.F.W. Barth, Nor. Vidensk.-Akad. Oslo, I. Mat.-Natur. Kl., 1944, No. 9, 104 pp., 1945.
14. T.F.W. Barth, Nor. Vidensk.-Akad. Oslo, I. Mat.-Natur. Kl., 1954, No. 4, 20 pp., 1954.
15. C. Oftedahl, Nor. Vidensk.-Akad. Oslo, I. Mat.-Natur. Kl., 1952, No. 3, 64 pp., 1952.
16. C. Oftedahl, Permian rocks and structures in the Oslo Region, in: Geology of Norway, ed. by O. Holtedahl, Norg. geol. Unders., 208, 298, 1960.
17. C. Oftedahl, Geol. Rundsch., 57, 203, 1967.
18. I.B. Ramberg, Norg. geol. Unders., 325, 1, 1976.
19. I.B. Ramberg and B.T. Larsen, Tectonomagmatic evolution, in: The Oslo Paleorift, ed. by J.A. Dons and B.T. Larsen, Norg. geol. Unders., 337, 55, 1978.
20. J.A. Dons, Terminology and history of investigations, in: The Oslo Paleorift, ed. by J.A. Dons and B.T. Larsen, Norg. geol. Unders., 337, 3, 1978.
21. P.A. Ziegler, this volume.
22. I.B. Ramberg, R.H. Gabrielsen, B.T. Larsen and A. Solli, Geol. Mijnbouw, 56, 295, 1977.
34. B. Sundvoll, Isotope and trace-element chemistry, geochronology, in: The Oslo Paleorift, ed. by J.A. Dons and B.T. Larsen, Norg. geol. Unders., 337, 35, 1978. Also: B. Sundvoll, Proceedings of the NATO ASI in Oslo, Vol. I, 1978.
35. O. Holtedahl, Nor. geol. Tidsskr., 12, 323, 1931.
36. A. Heintz, Nor. geol. Tidsskr., 14, 176, 1934.
37. O.A. Höeg, Nor. geol. Tidsskr., 16, 1, 1936.
38. O.A. Höeg, Nor. geol. Tidsskr., 16, 279, 1937.
39. O.A. Höeg, Nor. geol. Tidsskr., 16, 280, 1937.
40. E. Dix and A.E. Trueman, Nor. geol. Tidsskr., 15, 25, 1935.

41. N. Spjeldnæs, Er Brumundsandstenen Triassisk? X Nordic Geol. Meeting, Oslo, Jan. 6, 1972.
42. K.A. Storetvedt and N. Petersen, Zeitsch. Geophys., 36, 569, 1970.
43. P. Holmsen, Medd. Dansk Geol. Foren., 14, 61, 1958.
44. L.B. Rasmussen, Geol. Surv. Denm., III Ser., 42, 1, 1974.
45. B.T. Larsen, Krokskogen lava area, in: The Oslo Paleorift, ed. by J.A. Dons and B.T. Larsen, Norg. geol. Unders., 337, 143, 1978.
46. P.M. Ihlen, P.R. Ineson and J.G. Mitchell, K/Ar dating of clay-mineral alteration associated with ore deposits in the northern part of the Oslo Region, in: Petrology and Geochemistry of Continental Rifts, ed. by E.-R. Neumann and I.B. Ramberg, Reidel Publ. Comp., 1978.
47. J.A. Dons, News from the Oslo Region Res. Group, 4, 29, 1972.
55. E.S. Husebye and I.B. Ramberg, Geophysical investigations, in: The Oslo Paleorift, ed. by J.A. Dons and B.T. Larsen, Norg. geol. Unders., 337, 41, 1978.
66. R. Selmer-Olsen, Nor. geol. Tidsskr., 28, 171, 1950.
67. N. Spjeldnæs, Nor. geol. Tidsskr., 34, 105, 1955.
68. K.M. Storetvedt, Nor. geol. Tidsskr., 48, 153, 1968.
69. A. Noe-Nygaard, Medd. Dansk Geol. Foren., 17, 285, 1967.
70. K. Åm, Norg. geol. Unders., 287, 1, 1973.
71. E. Sæther, Nor. Vidensk.-Akad. Oslo, I. Mat.-Natur. Kl. 1947, No. 3, 60 pp., 1947.
72. N. Spjeldnæs, Norg. geol. Unders., 316, 116, 1975.

11

Reprinted from pages 7-8, 10-18, 19, 21, 26, 28, and 31-32 of *Geojournal* **1**:7-32 (1977)

GEOLOGY AND HYDROCARBON PROVINCES OF THE NORTH SEA

Manuscript received August 1976

P. A. ZIEGLER, Den Haag *

Summary: The technically recoverable hydrocarbon reserves of the North Sea are currently estimated at 23 x 10^9 bbls of oil, condensate and NGL and 90 x 10^{12} scf of gas.

The North Sea underwent a complex geological evolution. The oil and gas occurrences in the central and northern North Sea are tied to Mesozoic and Tertiary sediments; the gas fields of the southern North Sea are mainly contained in Permian series.

The proven hydrocarbon provinces cover at best one third of the total area of the North Sea.

It is estimated that in the early 1980's the oil production from the North Sea will meet 22 % of the oil demand of Western Europe.

Zusammenfassung: Die technisch ausbeutbaren Kohlenwasserstoffreserven der Nordsee werden gegenwärtig auf 23 x 10^9 bbls Öl, Kondensat und NGL 90 x 10^{12} scf Gas geschätzt.

Das Nordseegebiet durchlief eine komplexe geologische Entwicklungsgeschichte. Die Öl- und Gasfelder der zentralen und nördlichen Nordsee sind hauptsächlich an mesozoische und tertiäre Sedimente gebunden. Die Gasfelder der südlichen Nordsee sind hauptsächlich in permischen Serien enthalten.

Die nachgewiesenen Kohlenwasserstoffprovinzen bedecken höchstens ein Drittel der Gesamtoberfläche der Nordsee.

Es wird angenommen, daß die Ölproduktion der Nordsee während der ersten Hälfte der 1980er Jahre 22 % des Ölbedarfs von West-Europa decken wird.

Résumé: Les réserves exploitables de la mer du Nord sont actuellement estimées 23 x 10^9 bbl de pétrole, condensat et NGL et 90 x 10^{12} scf de gaz.

L'évolution géologique du bassin de la mer du Nord est très complèxe. Les champs de pétrole et de gaz des parties centrales et septentrionales de la mer du Nord se trouvent surtout dans des séries mésozoiques et tertiaires, tandis que la majorité des champs de gaz des régions méridionales sont situés dans des réservoirs permiens.

Les provinces petrolfières connues couvrent moins d'un tiers de la surface totale de la mer du Nord.

La production de pétrole de la mer du Nord est estimée correspondre pour les premièrès années de 1980 à 22 % de la demande en pétrole de l'Europe occidentale.

Abbreviations

scf = standard cubic feet (atmospheric conditions at 15° C)
TCF = 10^{12} scf
bbls = barrels
NGL = Natural Gas Liquids

Conversion Factors:

1 Nm^3 gas (normal cubic meter) = 37.225 scf (atmosphere conditions at 0° C)
1 m^3 oil = 6.28 bbls
1 ton of 35° API oil = 7.39 bbls
6000 scf = 1 bbls oil assuming calorific value of gas 9.400 kcal/Nm^3

* Dr. P.A. ZIEGLER, Shell Internationale Petroleum Mij. B.V. Carel van Bylandtlaan 30, Den Haag, Netherlands

Introduction

The North Sea is the most important hydrocarbon province of Western Europe. Present technically recoverable reserves (i.e. proven and probable reserves) are estimated to amount, in established accumulations, to some 23 x 10^9 bbls of oil, condensate and NGL and about 90 x 10^{12} scf of gas. About one third of this gas is directly associated with the oil; for this reason only a part thereof may ultimately become available as sales gas. Expressed in energy equivalents the reserves proven up in the North Sea correspond to about 3.2 times those contained in the giant Slochteren gas field in the Netherlands. Remaining oil reserves in onshore Western Europe amount to about 1 x 10^9 bbls with remaining gas reserves including the Slochteren field being in the order of 100 x 10^{12} scf; furthermore these gas reserves have to be discounted somewhat for their generally lower calorific value as compared with the North Sea gas. The hydrocarbon reserves of the North Sea are, however, dwarfed by the reserves of the Middle East, which are estimated to be in the order of 340 x 10^9 bbls of oil and 520 x 10^{12} scf of gas.

The exploration history of the North Sea spans about 15 years. After the discovery of the giant Dutch onshore gas-field Slochteren in 1959 the interest of the industry began to focus on the North Sea. First concessions were allocated in 1963 by Denmark and in 1964 by Germany and the U.K., by which time exploration started in earnest. Initially, exploration efforts were mainly concentrated on the gas prospects of the southern North Sea, for which the spectacular Slochteren field stood as a model. However, with the gradual recognition of the oil potential of the central North Sea the exploration emphasis began to shift northward. Whereas gas exploration in the southern North Sea was rewarded with success shortly after the allocation of the first concessions, exploration for oil in the central and northern North Sea had to labour through a long series of dry holes before the first two giant oil fields were discovered in 1969 and 1970, Ekofisk in Norwegian and Forties in British waters.

By now the industry has drilled over 800 exploration and appraisal wells in the North Sea. These have resulted in the discovery of some 90 oil and/or gas accumulations; however, not all of these discoveries are backed up by sufficient reserves to support development. Currently some 30 mobile drilling units are engaged in exploration drilling and in the evaluation of new discoveries.

Whereas development of the gas fields in the southern North Sea is well advanced, only seven of the oil fields of the central North Sea are currently on stream. An additional four are anticipated to be on stream by late 1976. A major network of oil and gas pipelines linking the producing offshore areas with the onshore is currently under construction.

The present offshore operations in the North Sea are certainly the largest of their kind. They take place in a climatic environment that is decidedly harsh. Development of the newly discovered oil fields in these northern waters with depths between 100 and 200 metres requires new technologies. Correspondingly, cash demands are very large.

Although these newly discovered North Sea energy sources are of the greatest importance for the economy of Europe, it should be realised that only large – by onshore standards very large – fields can be economically developed. Whether smaller fields can be economically produced, depends not only on the size of the reserves but also on the development costs, which increase rapidly with increasing water depth, on the distance from shore and nearby outlet facilities, on the crude realisation price, on the concession conditions and last but not least on the type and degree of government take, such as royalties and taxation.

In the North Sea the industry has not only to cope with huge technological, financial and concessionary problems but has also to absorb the risk of an expensive exploration venture in a geologically complex basin. Industry-wide exploration expenditures are estimated to have passed by late 1975 the US $ 2 x 10^9 mark. Capital investment for offshore field development is currently running at a level of some US $ 4 to 5 x 10^9 per year.

Although the spectacular string of discoveries scored by the industry during the last years in the North Sea appears to continue, a creaming-off effect will sooner or later result in a gradual dropping off of the rate at which new reserves are added to those already established.

The prospectivity of the North Sea is by no means uniform. Even a quick glance at the distribution pattern of the currently known oil and gas fields reveals that these are concentrated along the north-south axis of the central and northern North Sea (Fig 1). An exception are the gas fields of the southern North Sea that form part of a broad belt, stretching east-west from Germany to southern England.

A summary of the geological evolution of the North Sea and its bearings on the distribution of these hydrocarbon accumulations is presented in the next chapter, which is followed by a discussion of the stratigraphic and structural characteristics of the individual oil and gas provinces of the North Sea.

North Sea Basin History

The North Sea occupies large parts of the North West European intracratonic sedimentary basin which extends from Poland through Germany, Denmark and the Netherlands to eastern England and through the North Sea to the Atlantic Shelf (Fig 2). This large basin is flanked to the north east by the Fennoscandian Precambrian shield, to the east by the stable Russian platform, to the south by the late Paleozoic Variscan massifs and to the west by the Caledonides of Scotland. The North West European basin is open to the north, where it grades into the Atlantic Shelf basins of western Norway and the Shetland Islands.

The North Sea as such came into existence during the Tertiary Age. The sedimentary sequence underlaying the North Sea reaches maximum thickness in excess of 9 kilo-

[*Editor's Note:* Material has been omitted at this point.]

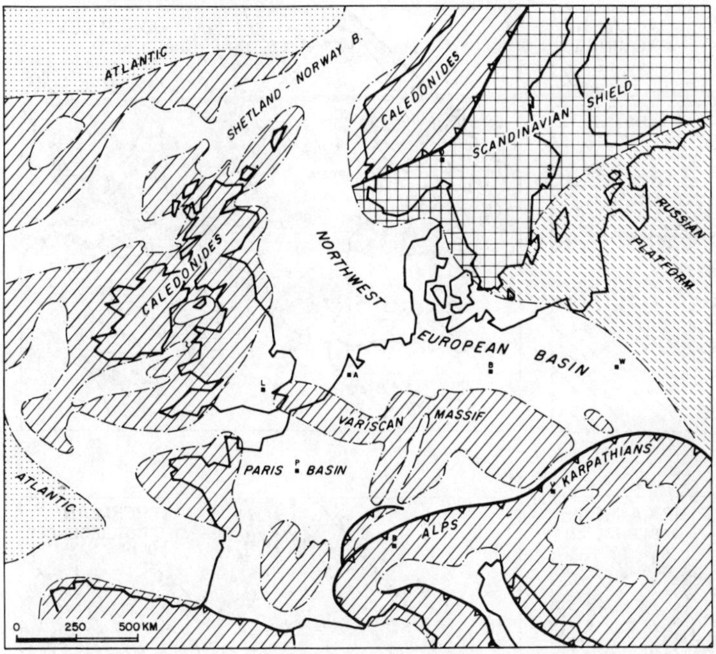

MEGATECTONIC FRAME OF NW-EUROPEAN BASIN

Sedimentary Basins — Highs — Shields — Platforms — Ocean Floors

Fig 2 Regional Tectonic Setting

Fig 3 Paleographic maps (Basement provinces, Devonian Upper Carboniferous, Rotliegend Zechstein, Triassic)

(for legend see p. 13)

metres with sediments ranging in age from lower Paleozoic to Quaternary. The North Sea area underwent a long and complex evolution (**Kent** 1975, **Walmsley** 1975, **P. A. Ziegler** 1975, **W. H. Ziegler** 1975) during which it was occupied by a number of genetically different sedimentary basins. The superposition of these basins in various combinations in different parts of the North Sea controls the hydrocarbon potential of the respective provinces.

1. The Basin Floor

The metamorphic and igneous rocks which form the floor or basement of the North Sea basin have been reached only by a small number of wells. In much of the North Sea the top of the basement is located beyond conventional drilling depths. In the deeper parts of the North Sea basin the basement cannot even be mapped by reflection seismic. Insufficient refraction data are available to construct a basinwide top basement structure map.

Based on projections from onshore areas and on radiometric age determination of a few of the critical offshore basement samples a reconstruction of the major age provinces of the basement in the general North Sea area (Fig 3a) can be drawn up.

Offshore data confirm the longstanding hypothesis of a link between the Caledonides (400 million years and older) of Scotland and those of Norway across the northern North Sea. More novel, however, are the findings of Caledonian basement rocks in the central North Sea. These give evidence for the existence of a late Caledonian fold belt extending from Norway southward into the northern parts of the Dutch and German offshore. It can only be speculated that this branch of the Caledonides may have extended through northern Germany to northern Poland where subsurface data again give evidence for Caledonian folding. Furthermore it cannot be ruled out that the Caledonides of the Ardennes and those of the central North Sea link up through the southern North Sea, in which case the London-Brabant massif with its Precambrian basement (700 million years plus) and lower Paleozoic undeformed sedimentary cover may represent an intramontane stable platform. The boundary between the central North Sea Caledonides and their north eastern undeformed foreland is, however, still ill-defined. Precambrian basement rocks (age 690 to 870 million years) form the Ringkøbing-Fyn high (**Larsen** 1971); seismic and limited drilling data indicate that unmetamorphosed lower Paleozoic series overlay a Precambrian basement in the Skagerrak, in the Kattegat and in Jutland.

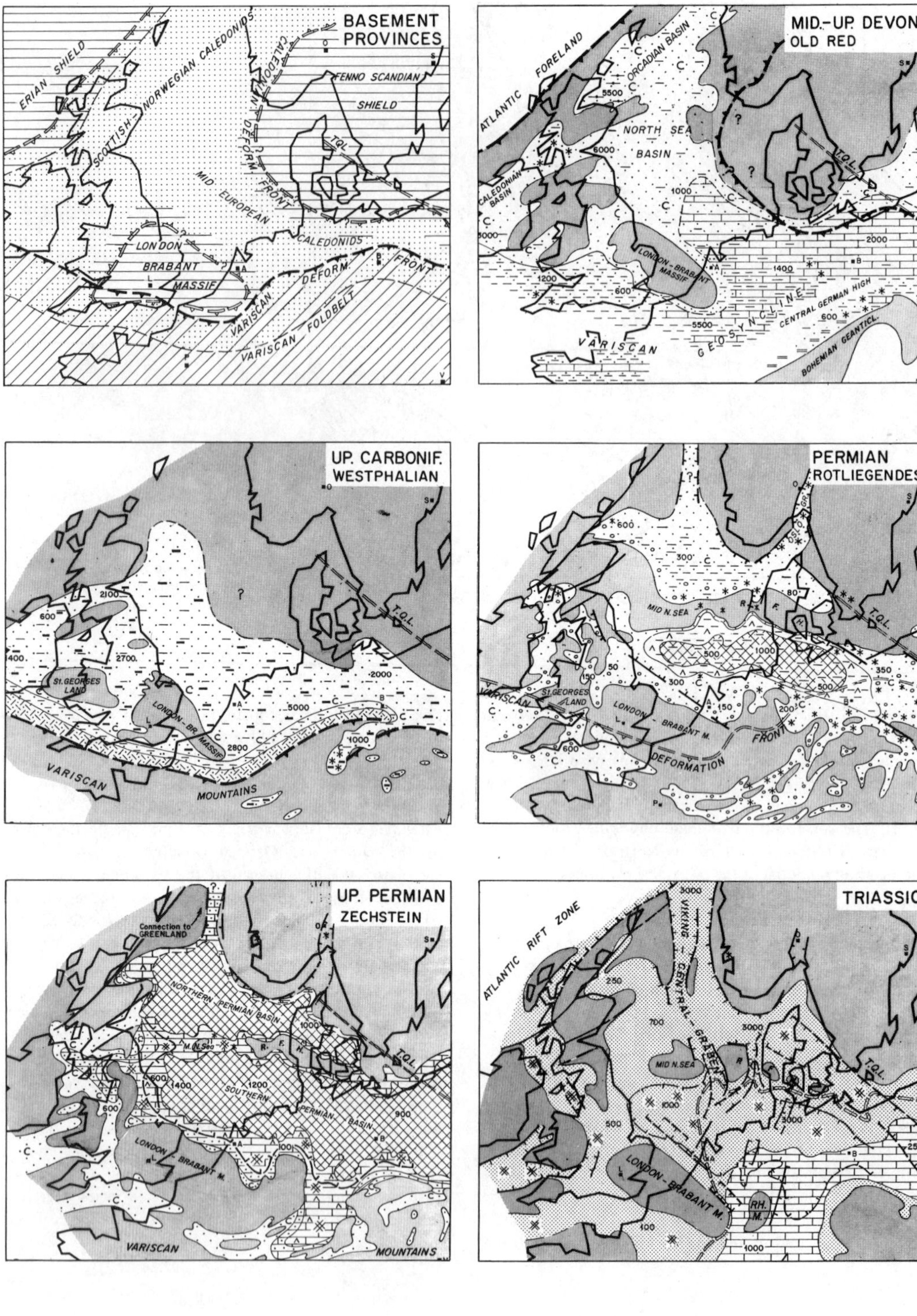
BASEMENT PROVINCES
ERIAN SHIELD
SCOTTISH-NORWEGIAN CALEDONIDS
CALEDON. DEFORM. FRONT
FENNO SCANDIAN SHIELD
T.Q.L.
MID EUROPEAN CALEDONIDS
LONDON BRABANT MASSIF
VARISCAN DEFORM. FRONT
VARISCAN FOLDBELT
MID.-UP. DEVON. OLD RED
ATLANTIC FORELAND
ORCADIAN BASIN
NORTH SEA BASIN
CALEDONIAN BASIN
LONDON-BRABANT MASSIF
VARISCAN GEOSYNCLINE
CENTRAL GERMAN HIGH
BOHEMIAN GEANTICL.
UP. CARBONIF. WESTPHALIAN
St. GEORGES LAND
LONDON-BR. MASSIF
VARISCAN MOUNTAINS
PERMIAN ROTLIEGENDES
OSLO GRABEN
MID N. SEA
LONDON-BRABANT M.
DEFORMATION FRONT
UP. PERMIAN ZECHSTEIN
Connection to GREENLAND
NORTHERN PERMIAN BASIN
SOUTHERN PERMIAN BASIN
VARISCAN MOUNTAINS
TRIASSIC
ATLANTIC RIFT ZONE
VIKING-CENTRAL GRABEN
MID N. SEA
LONDON-BRABANT M.
RH. M.

The Tornquist line, also referred to as the Fennoscandian border zone, is thought to mark the boundary between the generally older Precambrian basement rocks of both the Fennoscandian shield and the Russian platform (900 to 3000 million years) and the younger Precambrian basement provinces of central and western Europe.
In the Hebrides and in the Scottish Northern Highlands Precambrian basement rocks (Laxfordian 1600–1400 m.y.) are overlain by unmetamorphosed late Precambrian and lower Paleozoic sediments. This remnant of the western foreland of the Scottish-Norwegian Caledonides extends, based on well data, north westward under the West Shetland shelf.

Thus recent drilling has yielded important new information on an as yet ill defined but complex puzzle of the various basement provinces making up the basin floor of the greater North Sea area. The subsequent development of the North Sea basin was at least in part influenced by the differential behaviour of the various basement blocks and their internal structural grain.

2. The Basin Fill

The sediments overlaying the crystalline basement of the North Sea were deposited through time in different types of sedimentary basins. Each of these basins developed in response to a different tectonic setting. Thus the development history of the North Sea area can be subdivided into a number of more or less distinct tectonic as well as depositional cycles.

a) Caledonian Cycle

This cycle spanned the Cambrian to Silurian times and closes with the Caledonian orogeny which, as seen above, has affected much of the North Sea area. Essentially undeformed Lower Paleozoic sediments containing hydrocarbon source rocks occur in the Baltic where they were deposited under platform conditions in the Caledonian miogeosyncline. Remnants of similar sediments are present in the Danish onshore, in the Kattegat, in Scania and possibly in the Skagerrak.

Although productive in the Baltic, the hydrocarbon potential of the Lower Paleozoic series in the North Sea is highly speculative in view of their generally high degree of organic metamorphism.

b) Variscan Cycle

The Variscan tectonic cycle spans the Devonian and Carboniferous periods and terminates with the final phases of the Variscan orogeny during the late Carboniferous.

During the Caledonian diastrophism the Variscan geosyncline which extended from Ireland to Eastern Europe became accentuated. The southern North Sea area formed during the Devonian and Carboniferous part of the Variscan foredeep in which a thick essentially clastic wedge was deposited on the deeply eroded Caledonides. Deposition started in the Devonian with the continental Old Red sandstones. During the Middle Devonian (Fig 3b) marine incursions originating from the Variscan geosyncline reached northwards into the central North Sea where locally carbonates were deposited.

The postorogenic collapse of the North Atlantic Caledonian foldbelt led during the Devonian to the development of large intermontane basins such as the Orcadian Old Red basin of Scotland. In these basins great thicknesses of continental clastics were accumulated. Well data point towards a possible link-up between the Orcadian Basin and the scattered Old Red outcrops on the Norwegian coast. Furthermore a link may have been established between the Old Red basin of the southern and central North Sea and the Orcadian Basin.

During the Carboniferous (Fig 3c) occasional marine ingressions pushed again northward from the Variscan geosyncline into the central North Sea. Paralic conditions prevailed during the late stages of the Variscan cycle leading to the deposition of thick upper Carboniferous coal-measures mainly in the southern and in part also in the west-central North Sea. These coals are of particular interest since they form the source rock for the gas occurrences in the southern North Sea. Coal measures are apparently missing in the northern North Sea, in the Norwegian and in much of the Danish offshore.

The late Carboniferous terminal phase of the Variscan orogeny led to a consolidation of the Variscan fold belt with the Craton of north western Europe. Postorogenic uplifting and erosion affected not only the Variscan foldbelt but also its foreland and gave rise to the regional pre-Permian "Saalian" unconformity.

c) The Pre-Permian Surface

The Pre-Permian sediments of onshore north western Europe contain by North Sea standards only minor hydrocarbon occurrences.

Lower Paleozoic oil accumulations in the Baltic area are tied to Cambrian and Silurian oil source rocks. Carboniferous series contain a number of small oil fields in Lincolnshire in eastern England and in the Midland Valley of Scotland. In Germany and in the Netherlands minor gas accumulations have been proven up in Upper Carboniferous sands. Only a small fraction of the total North Sea hydrocarbon reserves are contained in pre-Permian reservoirs.

As already indicated above the Westphalian coal-measures of the southern North Sea represent an outstanding gas

LEGEND TO PALEOGEOGRAPHIC MAPS FIG. 3 and 5

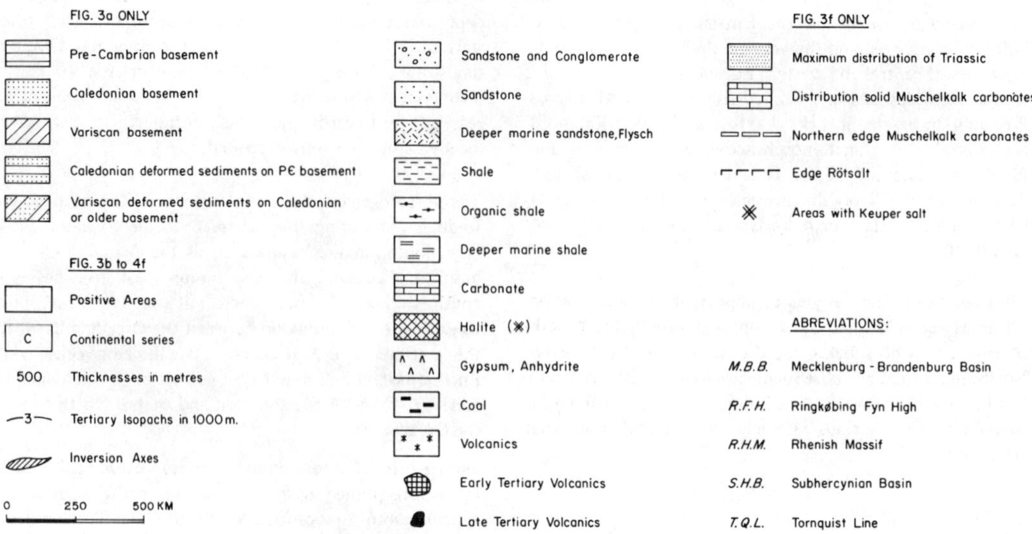

source rock. As such they have heavily contributed to the hydrocarbon reserves of the southern North Sea. Although pre-Westphalian coal measures occur locally in the central North Sea and Devonian oil source rocks have been reported from the onshore parts of the Orcadian Basin these have contributed little if any to the hydrocarbon reserves of the North Sea.

The present-day distribution of the pre-Permian series in the North Sea is summarised in Fig 4. Although locally modified by subsequent tectonism and Mesozoic erosional phases this map represents essentially the subcrop below the "Saalian" unconformity and depicts as such the main pre-Permian tectonic elements of the North Sea area.

In early Permian times the southern margin of the present-day North Sea was formed by the highlands of the Variscan foldbelt stretching from Ireland to Poland. Much of the North Sea area was occupied by deeply eroded highs exposing Precambrian and Caledonian basement rocks and intervening depressions underlain by Devonian and Carboniferous basins. The Lower Paleozoic Danish trough formed part of a large low-relief high which occupied the eastern North Sea.

Fault trends active during pre-Permian times are the Great Glen fault, the faults bounding the Midland Valley, as well as the Tornquist line. These faults were repeatedly re-activated during the Permian and Mesozoic.

Purely Permian and younger features shown in Fig 4 are the Oslo Graben, the Bamble Trough, the Horn and the Central Graben.

d) Permo-Triassic Variscan Foreland Collapse

In contrast to the compressive setting of much of the North Sea during the Variscan cycle its Permo-Triassic development was dominated by extensional stresses. These resulted initially from post-orogenic uplifting and collapse of the Variscan foldbelt and its foreland. However, during the course of the Triassic stresses related to early rifting movements in the northern North Atlantic began to dominate the North Sea area.

Differential subsidence of two essentially east west orientated intracratonic basins started mainly during the Lower Permian Age in the Variscan foreland (Fig 3d). Simultaneously a northwest southeast orientated fault system came into evidence.

Effusion of widespread volcanics, mainly along the margins of major stable blocks, preceded the subsidence of the basinal areas. The mid North Sea Ringkøbing-Fyn High came into evidence as a continuous barrier separating a northern basin extending from Scotland to Denmark from a southern one which reached from eastern England to Poland. Deposition started in both basins under desert conditions with the middle Permian Rotliegend clastics. Supplied by a system of intermittent streams originating in the rapidly degraded Variscan mountains a thick belt of wadi deposits and dune sands was accumulated along the southern margins of the southern Permian basin. These sands grade northwards into up to 1000 metres thick sabkah shales and salts (**Lutz** et al 1975, **Thomas** 1975). The Rotliegend sands are the main gas reservoir in the southern North Sea and the adjacent onshore areas.

In the northern Permian basin facies patterns are as yet ill-defined. However, being fringed by lower relief highs this basin received apparently less coarse material. An outstanding feature of the northern Permian basin is the volcanic Oslo Graben.

Continued subsidence of both the southern and the northern Permian basins, possibly below sea level, brought about in the upper Permian the apparently catastrophic transgression of the Zechstein seas (Fig 3e).

Faunal evidence indicates that this transgression originated from the marine Arctic Permian basins of Spitzbergen and Greenland, thus giving evidence for an early seaway through the northern North Atlantic and the Viking Graben in the northern North Sea. In both the southern and the northern Permian basins up to 1000 metre thick Zechstein halites were deposited; these are offset along the basin margins and on local highs by carbonate and sulphate banks. Zechstein carbonates contain significant gas reserves in the German and Netherlands onshore areas but play only a subordinate role as oil and gas reservoirs in the North Sea.

At the end of the Permian renewed rifting movements may have again closed off the early northern North Atlantic seaway. Thus the onset of the Triassic is marked in the North Sea by a regional regression and a return to a continental depositional regime.

The Permian basin edges were gradually overstepped during the Triassic. With the progressive wearing down of the fringing highlands the supply of coarse clastics to the northern and southern North Sea basins gradually diminished. Marine incursions during the Middle Triassic Muschelkalk originated from the Alpine Geosyncline but did not reach further north than the southern North Sea basin. During the Upper Triassic the Muschelkalk sea retreated again and much of North West Europe was covered by a wide lagoonal plain with local evaporite and salt pans receiving only a limited amount of sands.

During the Triassic the Permian tectonic framework of the North Sea was modified by the emplacement of an essentially north south trending Graben system. Earliest indications of rifting movements coupled with uplifting of intervening highs occurred during the Bunter, giving rise to the regional Hardegsen unconformity **W. H. Ziegler**, 1975).

Seismic and drilling data give clear evidence for the rapid subsidence of a major rift system in the northern North Sea, the Viking Graben. Further south the North Sea – Ringkøbing-Fyn High, was breached during the Triassic by the Central Graben and the Horn Graben. Similar grabens along the Atlantic seaboard are the West Shetland – Minch basin, the Cardigan Bay, Celtic Sea and the Western Approaches basins. Major subsidence areas in the Dutch and German onshore areas are the Emsland Graben, the Glückstadt trough and the Hessian depression. Rapid subsidence during the Triassic marked also the Danish Embayment, the Bamble trough and the Egersund basin where up to 3000 metres of Triassic red beds were deposited (Fig 8). Emplacement of these new Triassic tectonic elements as well as continued subsidence of the older Permian basins heralded the Jurassic-Cretaceous tectonic regime of the North Sea.

The great thickness of the Triassic sediments triggered in the Permian basins the diapirism of the Zechstein salts which continued through much of the Mesozoic and Tertiary Ages.

Although Triassic series make up a major part of the North Sea basin fill, they only play a very minor role in the North Sea hydrocarbon potential. Significantly Triassic sediments are devoid of oil and gas source rocks and offer only a limited reservoir potential.

e) Jurassic Lower Cretaceous Taphrogenic Stage

The over 1000 kilometres long rift system of the Viking and Central Graben which was in essence emplaced during the Triassic developed during the Jurassic into the dominant structural element of the North Sea. Subsidence of this graben system continued in several basinwide distinct phases that are correlating with similar rifting phases in the northern North Atlantic (**P. A. Ziegler**, 1975). The major oil provinces of the North Sea are tightly linked to the Viking and Central Graben. Sediments deposited during the Jurassic-Cretaceous taphrogenic stage of this rift system contain oil and gas source rocks as well as excellent reservoir horizons. The majority of the oil reserves of the central and northern North Sea are contained in Jurassic sediments.

At the Keuper Rhaetic boundrary the basinwide "early Kimmerian" tectonic pulse resulted in a mild accentuation of the main structural elements of the North Sea area.

This phase was followed by the regional Rhaetic-Hettangian transgression during which sands were deposited along the margins of the main positive elements. In this context the Statfjord sands of the Viking Graben deserve particular mention as a reservoir horizon. During the Pliensbachian fully marine conditions were established throughout most

Fig 4 Permian Subcrop Map

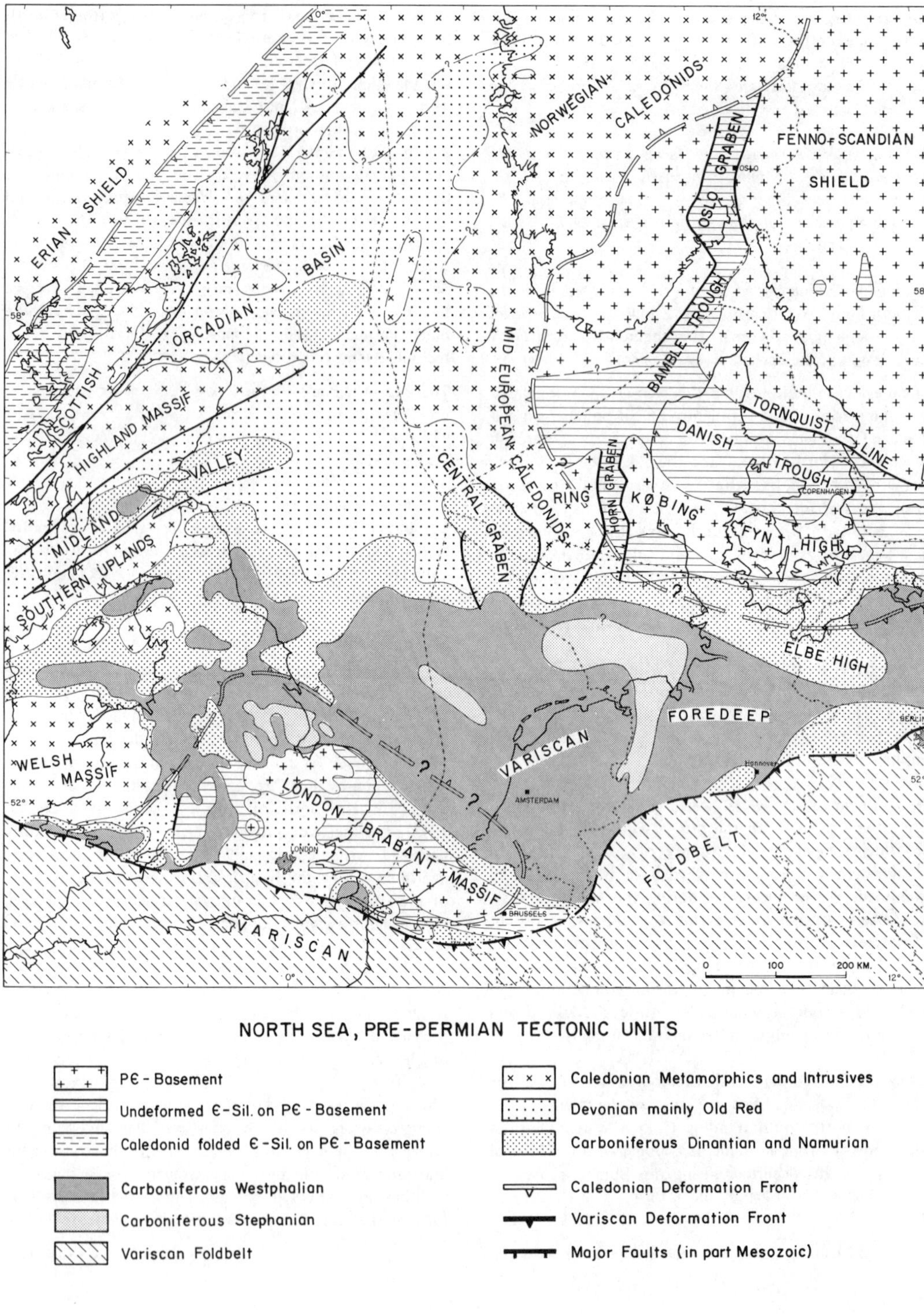

NORTH SEA, PRE-PERMIAN TECTONIC UNITS

of the North Sea with transgressions originating both from the Tethis as well as from the by now firmly established northern North Atlantic seaway. The Liassic deposits of the North Sea generally are characterised by open-marine relatively shallow-water shales (Fig 5a). Minor amounts of sands were derived from sources in Poland and East Germany but never reached the actual area of the North Sea. During the late Liassic the Toarcian organic Posidonomya shales were deposited in the southern North Sea and in the adjacent Dutch and German onshore. They form the main oil source rock in the Lower Saxony and West Netherlands basins. However, their equivalents in the northern North Sea and in the Danish Embayment are developed in a non-source rock facies.

During the early Dogger a major tectonic pulse, colloquially referred to as the "Mid Kimmerian" phase, affected much of the North Sea area and resulted in major paleogeographic changes (Fig 5b).

Differential subsidence of the Viking Graben continued with the deposition of the regressive-transgressive up to 300 metres thick Brent Sands (Bajocian-Bathonian), which form an outstanding hydrocarbon reservoir. In contrast to this much of the central North Sea was uplifted initially and subjected to erosion before deposition resumed in the Central Graben and in the Egersund-Danish Embayment. This uplift was parallelled by a distinct rift volcanism with a main effusion centre located in the area of the Moray Firth Basin, Central Graben and Viking Graben triple junction; subsidiary volcanic centres are recognised in the Egersund basin, the Viking Graben and on the Norwegian coast (**Al–Kassim** 1975). Uplifting of the Mid North Sea – Ringkøbing-Fyn High during the Mid-Kimmerian phase resulted in a renewed separation of the northern North Sea basin from the southern one. Clastics derived from this trend of highs were deposited under paralic conditions in the Egersund-Danish Embayment but were also shed southwards into the more marine Lower Saxony, West Netherlands and Solepit basins.

Minor rifting pulses preceded a regional transgression during the Callovian and Oxfordian which led ultimately, during the Kimmeridgian, to the development of deep water troughs in the Viking and Central Graben. Whereas the Moray Firth and the Egersund Basin as well as the Danish Embayment continued to subside during the Upper Jurassic, the essentially Triassic Horngraben became inactive.

Throughout the northern as well as parts of the central North Sea the Kimmeridgian (Fig 5c) is represented by organic rich shales that constitute the main oil source rock in these areas. However, Kimmeridgian shales lose their source rock potential southward from the Egersund basin into the Danish Embayment. Furthermore, Kimmeridgian source rocks grade laterally in the northern Netherlands offshore into sandy, paralic series.

Sedimentation during the Upper Jurassic gradually overstepped the edges of the Middle Jurassic basins. During this transgression upper Jurassic shallow-water sands were locally deposited on shelf areas flanking the North Sea graben system. Darived from these shallow-water sand accumulations and transported by density currents, the grabens received locally deep-water sands. Both shallow water and deeper water Upper Jurassic sands contain major oil accumulations in the central and northern North Sea.

Subsidence of the mid North Sea High during the upper Jurassic led to a renewed marine connection between the northern and the southern North Sea basins. In the latter upper Jurassic series display a wide variety of facies types, ranging from carbonates and evaporites in the Lower Saxony basin to paralic sand series in the West Netherlands basin and ultimately to shales containing only minor carbonates and clastics in the Sole Pit basin. These three basins offset the London-Brabant and Rhenish Massif and are referred to as "Marginal Troughs" (**Voight** 1962). They were accentuated during the middle and upper Jurassic whereby local warping during the Dogger and early upper Jurassic is thought to be related to minor transcurrent movements.

The rift volcanism of the central North Sea ceased with the subsidence of the Central Graben during the Upper Jurassic; however, in the Dutch Wadden Sea, a new, this time upper Jurassic volcanic centre developed. Upper Jurassic dates have also been obtained from dykes in the northernmost North Sea and on the West Norwegian coast.

At ethe Jurassic-Cretaceous boundary a further major rifting phase, the "Late Kimmerian Phase", affected the entire North Sea Graben system. In the Viking Graben this phase accentuated the already existing sea floor topography and gave rise to a strong, blockfaulted submarine relief of some 2000 to 3000 metres. In the Central Graben rift tectonics were less obvious since they were in part obscured by the intense diapirism of the Zechstein salts. Simultaneously the Mid North Sea – Ringkøbing-Fyn High and the London-Brabant High – as well as the Rhenish-Bohemian Massif were uplifted; this went parallel with a sharp accentuation of the Marginal Troughs. The base of the Cretaceous is marked throughout the entire North Sea by a regional unconformity which is referred to as the "late Kimmerian unconformity". This break in sedimentation was followed by a major transgression that inundated during the Lower Cretaceous most of the Jurassic highs, thus shutting off the supply of clastics from these highs to the Central and

Fig 5 Paleographic Maps
(Liassic, Dogger, Malm, Lower and Upper Cretaceous, Tertiary) (for legend see p. 13)

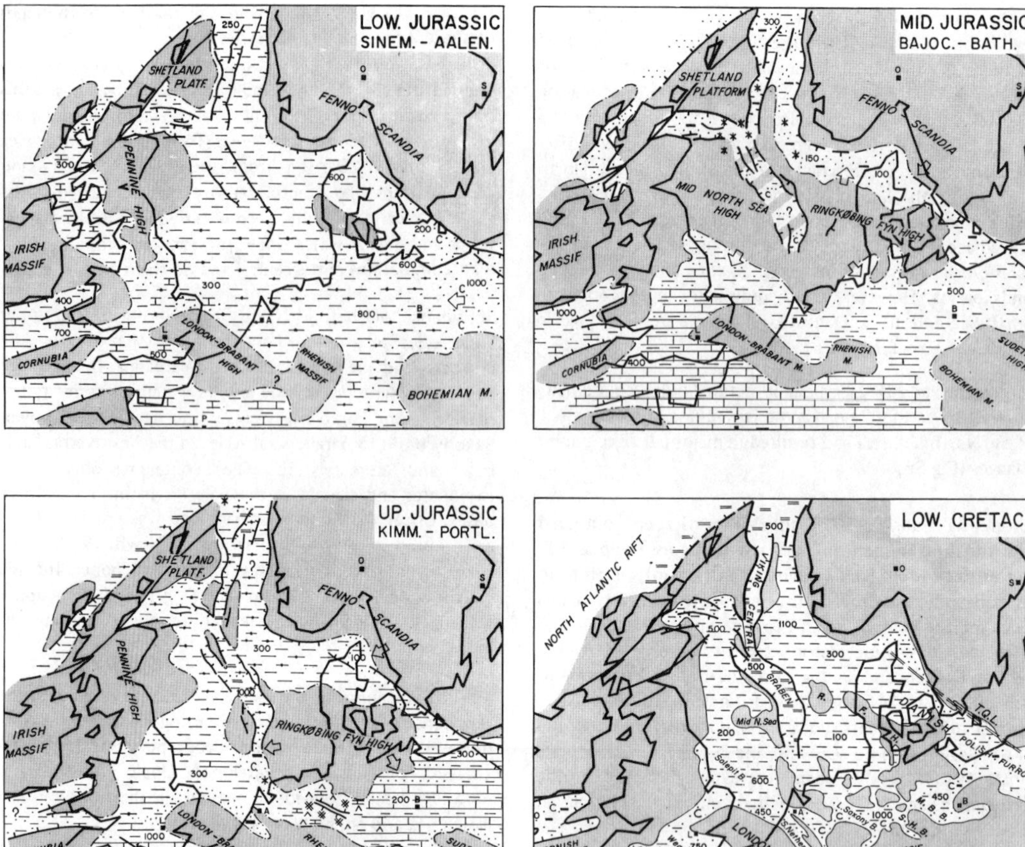
LOW. JURASSIC
SINEM. - AALEN.
SHETLAND PLATF.
PENNINE HIGH
IRISH MASSIF
CORNUBIA
LONDON-BRABANT HIGH
RHENISH MASSIF
BOHEMIAN M.
FENNO-SCANDIA
MID. JURASSIC
BAJOC.-BATH.
SHETLAND PLATFORM
MID NORTH SEA HIGH
RINGKØBING FYN HIGH
IRISH MASSIF
CORNUBIA
LONDON-BRABANT M.
RHENISH M.
SUDETIC HIGH
BOHEMIAN M.
FENNO-SCANDIA
UP. JURASSIC
KIMM. - PORTL.
SHETLAND PLATF.
PENNINE HIGH
IRISH MASSIF
CORNUBIA
LONDON-BRABANT M.
RINGKØBING FYN HIGH
RHENIS M.
SUDETIC HIGH
BOHEMIAN M.
FENNO-SCANDIA
LOW. CRETAC.
NORTH ATLANTIC RIFT
VIKING
CENTRAL GRABEN
Mid N. Sea
DANISH
T.Q.L.
POLISH FURROW
CORNISH PLATF.
Weald B.
LONDON-BRABANT M.
RHENISH MASSIF
BOHEMIAN MASSIF
LATE CRETAC.
VIKING
CENTRAL GRABEN
T.Q.L.
S. GERMAN HIGH
BOHEMIAN M.
TERTIARY

Viking Graben (Fig 5d). The occurrence of Lower Cretaceous sands is essentially restricted to the Moray Firth Basin, the Dutch parts of the Central Graben and the Marginal Troughs.

During the Lower Cretaceous the topography of the North Sea rift became gradually buried in a thick shale sequence. This process was interrupted during the early Albian by a minor rifting phase (Austrian phase) which in turn was followed by the basinwide transgression of the Albian and Aptian shales.

f) Late and Post-Taphrogenic Stage

Rifting movement in the North Sea petered out gradually during the Upper Cretaceous and ceased entirely after the early Tertiary Laramide phase. This last rifting phase coincides roughly with the onset of ocean floor spreading in the northern North Atlantic. With this the active extensional forces that dominated the Mesozoic setting of the North Sea apparently ceased to influence the area. A regime of regional subsidence that appears to continue to the present day characterised the Tertiary evolution of the North Sea basin. Following the Albian-Aptian, transgression sedimentation in the North Sea changes over from mainly shales to carbonates. The purity of the Upper Cretaceous chalk series in the North Sea is an indication to what extent the surrounding highlands were inundated by the Cretaceous seas which reached their maximum extent during the Senonian (Fig 5e).

In the central and northern North Sea Upper Cretaceous times corresponded to a period of general quiescence that was only interrupted by the mid-Senonian, Subhercynian phase which gave rise to a subregional disconformity. Both the Viking and the Central Graben continued to subside rapidly during the Upper Cretaceous. Up to 1200 metres thick chalk series, consisting mainly of coccolith fragments accumulated in the Central Graben and in the southern Viking Graben. Northwards these carbonates grade into an equally thick sequence of marls and calcareous shales. On the highs flanking the central rift system the chalk reaches thicknesses of only 200 to 300 metres. By the close of the Cretaceous most of the rift topography was apparently completely filled in.

During the early Paleocene a last distinct rifting phase, the Laramide Phase, resulted once more in a rapid subsidence of the graben system. This was parallelled by uplifting and erosion of the rift flanks. At the same time regional paleogeographic changes resulted in a return to a clastic depositional regime. Flanking the North Atlantic rift zone the Scottish Highlands and the western Shetland Platform were apparently uplifted. This gave rise to an eastward directed drainage system which deposited, during the Paleocene and Eocene, a thick deltaic and barrier bar sand-sequence on the eastern Shetland Shelf and in the area of the outer Moray Firth basin. Slope failure of this foreset sequence triggered density currents which transported sands into the Central and Viking Graben where they now constitute major hydrocarbon reservoirs.

In the southern North Sea both the Subhercynian and the Laramide phases resulted in an inversion of the Marginal Troughs, as well as of the southern parts of the Central Graben. Similar inversion movements occurred along the Fennoscandian border zone in the Kattegat, in southern Sweden and in Poland (Fig 5e).

Inversion of these basins resulted in uplifting of their axial parts in response to compressive and/or transcurrent movements. These deformations are thought to be caused by compressive stresses which were exerted by the Alpine Orogene on its fragmented foreland in which the above Mesozoic troughs acted like shear pins in an otherwise rigid platform. Once inverted, these basins bacame largely inactive (**P. A. Ziegler** 1975).

From the Eocene till the present day the North Sea area has been subjected to regional subsidence. During this time span a large symmetrical saucer shaped basin developed in which Tertiary series reach a maximum thickness of some 3500 metres in the central North Sea (Fig 5f and 6). The axis of this basin corresponds closely to the axis of the Mesozoic North Sea rift system. Towards the south where the Central Graben loses its identity this basin shallows out. Tertiary sediments consist mainly of clays and minor sands.

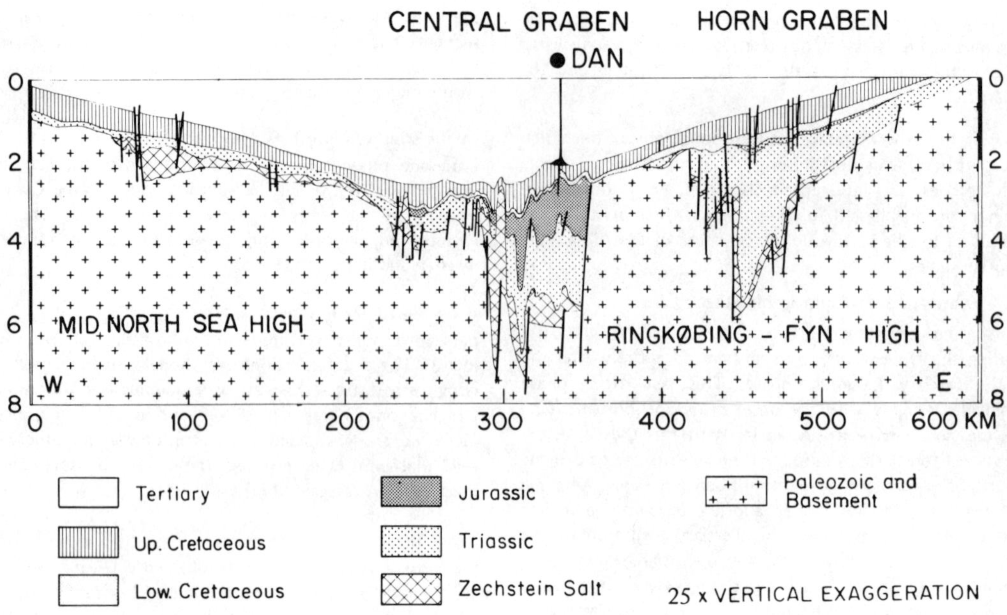

Fig 6 Cross-section Denmark

[*Editor's Note:* Material has been omitted at this point.]

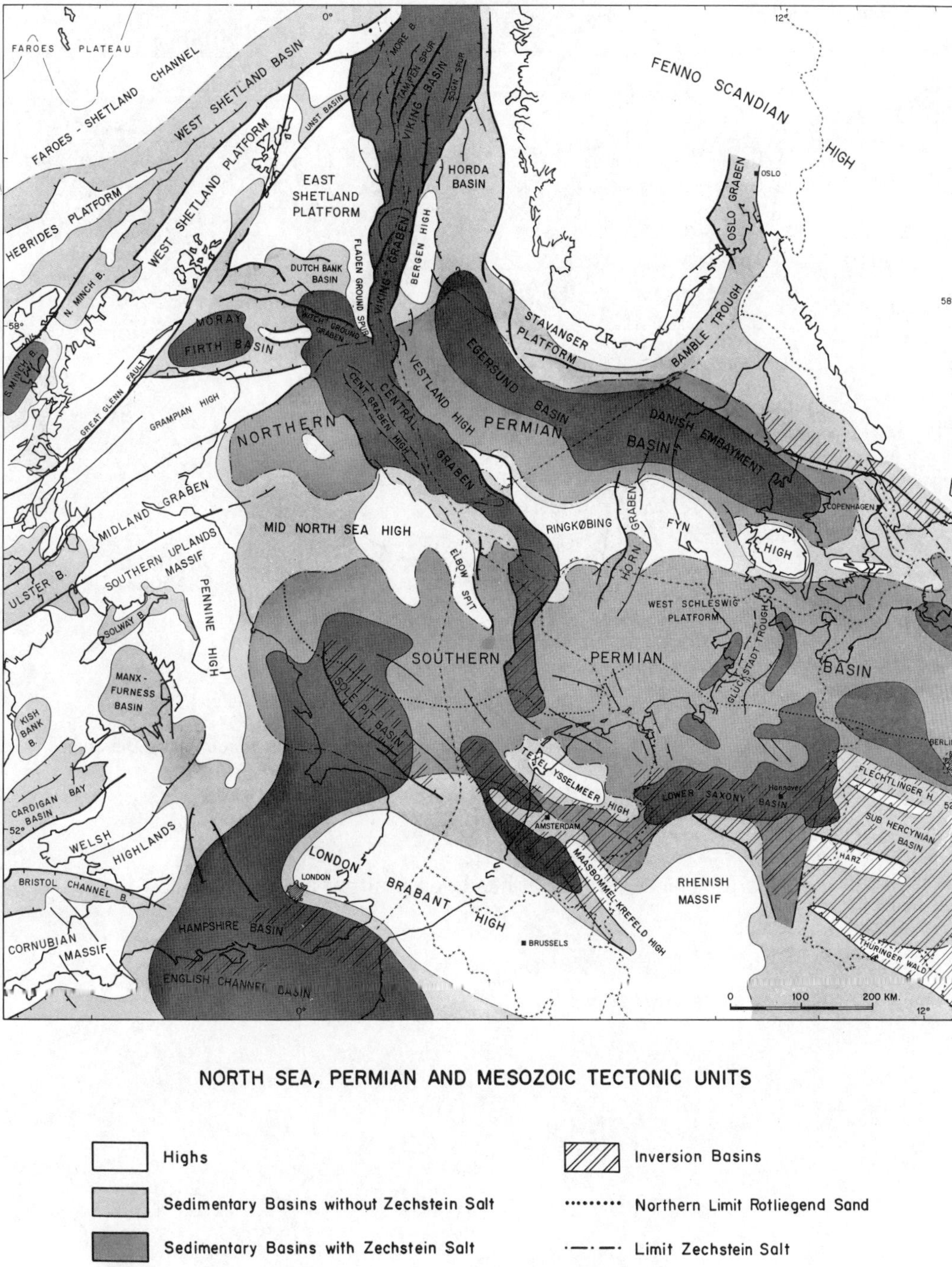

Fig. 8. Mesozoic tectonic units North Sea

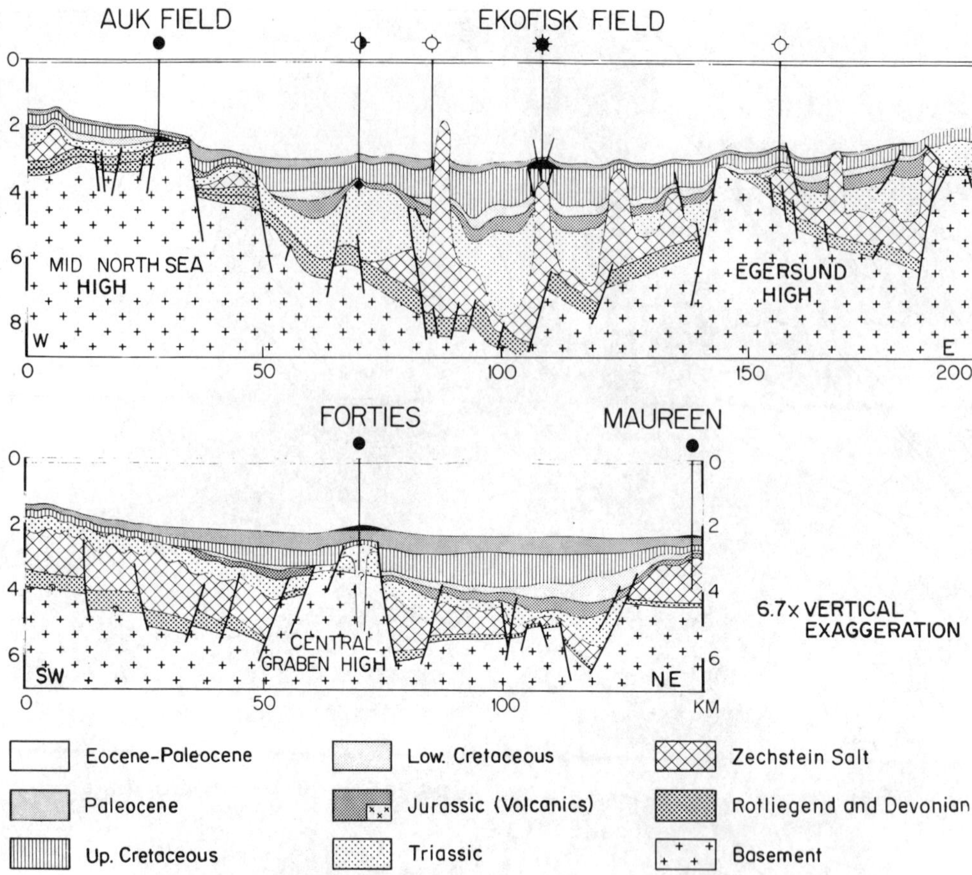

Fig 13 Structural X-sections Ekofisk and Forties

[*Editor's Note:* Material has been omitted at this point.]

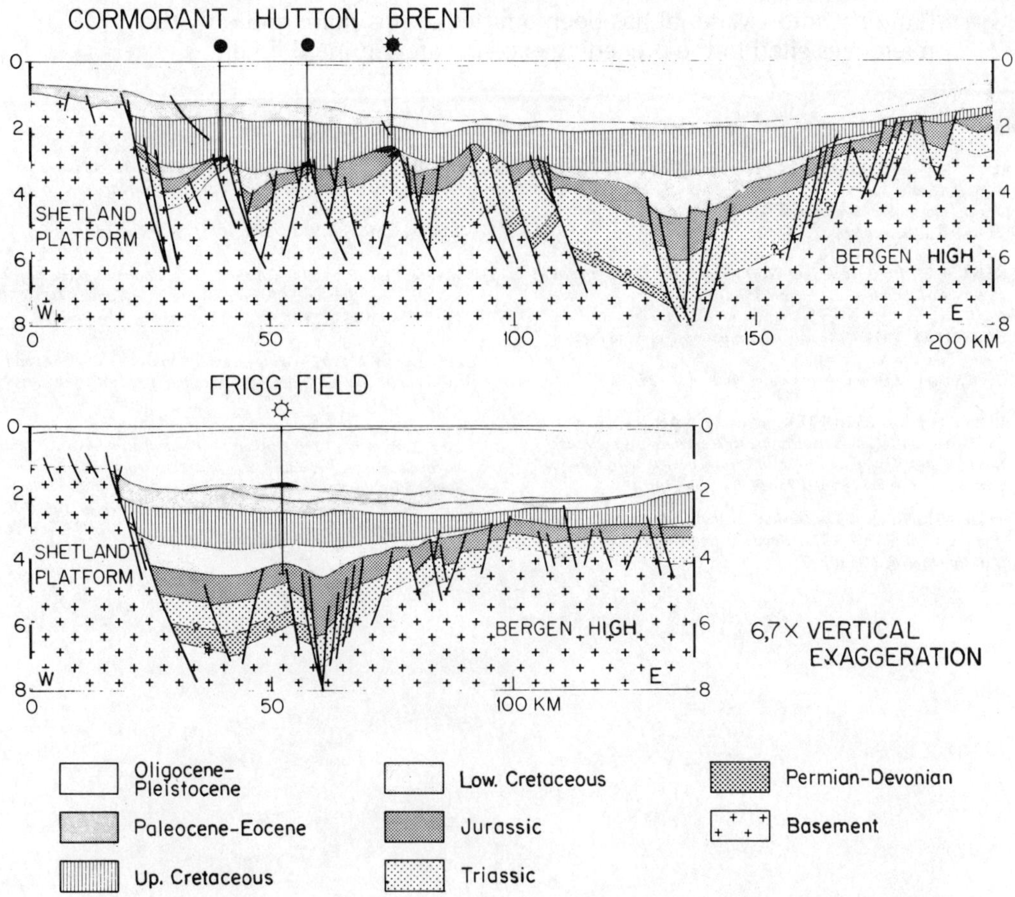

Fig 15 Structural X-section Brent and Frigg

[*Editor's Note:* Material has been omitted at this point.]

[*Editor's Note:* Material has been omitted at this point and only the references cited in the preceding excerpts are reprinted here.]

References

1. AL–KASSIM, F., RØNNEVIK, H.C. and ULLEBERG, K., *1975 Review of the Jurassic offshore Norway. Proceedings Jurassic North Sea Symposium 1975, Stavanger, Norwegian Petroleum Society, JNNSS/3, 1–18*

12. KENT, P.E., *1975 Review of North Sea Basin developments. Jl. geol. Soc. London vol. 131 p 435–468*

13. LARSEN, O., *1971 K/Ar age determination from the Precambrian of Denmark. Geol. Survey of Denmark 2nd series No 97 p 1–34*

14. LUTZ, M., KAASSCHIETER, J.P.H. & VAN WIJKE, D.H. *1975 Geological factors controlling Rotliegendes gas accumulations in the Mid-European Basin. Proceed. 9th World Petro. Congress Tokyo 5/12–16/1975 vol. 3 p 131–140*

21. THOMAS, J.B., *1975 The Geology of the Southern North Sea. Aberdeen Conference 1975, Handout Spearhead Publications Ltd. No OE–75, 216, p 1–7*

22. VOIGHT, E., *1962 Über Randtröge der Schollenränder und ihre Bedeutung im Gebiet der Mitteleuropäischen Senke und angrenzender Gebiete. Deutsch. Geol. Gesell. Zeitschr. vol. 114, 378–418*

23. WALMSLEY, P.J., *1975 Oil and Gas development in the North Sea. Proceed. 9th World Petroleum Congress Tokyo 5/12–15/ 1975, vol 3 p 131–140*

27. ZIEGLER, P.A., *1975 Geologic evolution of the North Sea and its tectonic framework. Am. Ass. Petrol. Geol. Bull 59(4) p 1073–1097*

28. ZIEGLER, W.H., *1975 Outline of the geological history of the North Sea; in: Woodland A.W. (ed): Petroleum and the Continental Shelf of Northwest Europe; the geology and the environment vol. I Geology, Applied Science Publishers p 165–190*

Editor's Comments on Papers 12 and 13

12 KENNEDY
Excerpts from *The Great Glen Fault*

13 FRANCIS
Excerpts from *The Midland Valley as a Rift, Seen in Connection with the Late Palaeozoic European Rift System*

Paper 12 consists of extracts from the classical paper on the Great Glen by W. Q. Kennedy, whose hypothesis of major transcurrent faulting replaced Geikie's idea that it is a normal fault of large displacement with erosion of the zone of crushing. Kennedy's interpretation was based largely on the relative displacement of granite intrusives and other bodies that suggest a strike slip of sinistral sense, and this appears to be confirmed by other evidence. Some other writers have suggested dextral movement (Garson and Plant, 1972). There has been some difference of opinion as to the course the fault takes to the north, whether it passes east or west of the Shetlands. However, Kennedy's paper is included as a classic that was influential in introducing the concept of great horizontal displacements of adjoining continental masses along conservative boundaries.

The Great Glen sinistral strike-slip faulting, with its major downthrow to the southeast and the formation of the graben system of the North Sea, arose from the closing of the Iapetus (the proto-Atlantic Early Paleozoic) ocean with formation of the Iapetus suture in Mid-Paleozoic times (Phillips, Stillman, and Murphy, 1976).

The formation of the Midland valley of Paper 14 (by Professor E. H. Francis of the University of Leeds, U.K.) is closely related to the closure of the Iapetus seaway along a ENE-WSW directed zone that extends across Ireland. The paper continues with an account of the subsequent history of the Graben and the suture. There are great contrasts in petrology and lithology between the Lewisian to the north and the very different (and younger) low-grade, Paleozoic oceanic formations to the south. The question of subduction, whether to north or south or both, is unresolved. The paper goes on to describe the volcanism, the fracturing, and the development of the graben-filling formations. Many age determinations have been made. This rift valley is certainly one that belongs to the penultimate and ultimate stages of the Wilson cycle (Wilson, 1969), as described in the Introduction. The account by Kennedy (1958) describes the geology and the evolutionary history but of course without the benefit of those newer geotectonic concepts.

REFERENCES

Bacon, M., and J. Chesher, 1975, Evidence against post Hercynian transcurrent movement on the Great Glen fault, *Scottish Jour. Geology* **11:**79-82.

Baumann, H., 1981, Regional stress field and rifting in western Europe, in *Mechanisms of Graben Formation,* J. H. Illies, ed., ICG Symposium, Australia, 1979, *Tectonophysics* **73**(1-3) (special issue), pp. 105-111.

Beach, A., 1984, Structural evolution of the Witch Ground Graben, *Geol. Soc. London Jour.* **141:**621-628.

Bluck, B. J., 1980, Evolution of a strike-slip fault-controlled Basin, Upper Old Red sandstone, Scotland, *Spec. Pub. 4, Internat. Assoc. Sedimentologists,* 63-78.

Bodley, M. E., T. Egeberg, and O. Nipen, 1984, Development of rift basins, *Geol. Soc. London Jour.* **141:**639-649.

Cloos, H., 1939, Hebung, Spaltung, Vulcanismus, *Geol. Rundschau.* **30:** 401-527.

Crough, S. T., 1983, Rifts and swells: geophysical constraints on causality, in *Processes of continental rifting,* P. Morgan and B. H. Baker, eds., *Tectonophysics* **94:**23-37.

de Lapparent, A., 1886-1887, Conference sur le sens des movements de l'ecorce terrestre, *Soc. Géol. France Bull.* (ser. 3),**15:**220.

de Lapparent, A., 1898, Soulevements et affaissements, *Rev. Questions Sci.* **14.**

de Martonne, E., 1897, Die Hydrographie des oberen Nilsbeckens, *Zeitschr. Gesellesch. Erdkunde* **32:**315.

Dewey, J. F., 1982, Plate tectonics and the evolution of the British Isles, *Geol. Soc. London Jour.* **139:**371-412.

Elie de Beaumont, 1827, Observations géologiques, *Annales Mines,* ser. 2, pp. 5-82.

Flinn, D., 1961, Continuation of the Great Glen fault beyond the Moray Firth, *Nature* **199:**589-591.

Flinn, D., 1970, The Great Glen fault in the Shetland area, *Nature* **227:**268-269.

Flinn, D., 1975, Evidence for post Hercynian transcurrent movement on the Great Glen fault in the Moray Firth, *Scottish Jour. Geology* **11:**266-267.

Fuchs, K., K. P. Bonjer, and C. Prodehl, The continental rift system of the Rhinegraben—structure, physical properties and dynamical processes, in Mechanisms of Graben Formation, J. H. Illies, ed., *Tectonophysics* **73** (special issue), 79-90.

Garson, M. S., and J. Plant, 1972, Possible dextral movements on the Great Glen and Minch faults in Scotland, *Nature Phys. Sci.* **240:**31-35.

Gibbs, A. D., 1984, Structural evolution of extensional basin margins, *Geol. Soc. London Jour.* **141**:609–620.

Heybrock, P., 1975, On the structure of the Dutch part of the central North Sea graben, in *Petroleum and the Continental Shelf of North-west Europe,* vol. 1, A. W. Woodland, ed., Applied Science, London, pp. 339–351.

Holgate, N., 1969, Paleozoic and Tertiary transcurrent movements on the Great Glen fault, *Jour. Geology* **5**:97–139.

Illies, J. H., 1967*a,* Development and tectonic pattern of the Rhinegraben, in *Rhinegraben Progress Report,* J. P. Rothé and K. Sauer, eds., International Rhinegraben Research Group, Freiburg and Strasbourg pp. 7–9.

Illies, J. H., 1967*b,* An attempt to model the Rhinegraben tectonics, *Rhinegraben Progress Report,* J. P. Rothé and K. Sauer, eds., International Rhinegraben Research Group, Freiburg and Strasbourg, pp. 10–12.

Illies, J. H., 1970, Graben tectonics as related to crust-mantle interaction, in *Graben Problems,* J. H. Illies and St. Mueller, eds., Proceedings of an International Rift Symposium, 1968, Karlsruhe, E. Schweizer bart'sche Verlagsbuchhandlung, Stuttgart, pp. 4–26.

Illies, J. H., 1974, Taphrogenesis, introductory remarks, in *Approaches to Taphrogenesis,* J. H. Illies and K. Fuchs, eds., Proceedings of an International Rift Symposium, Karlsruhe, E. Schweizer bart'sche Verlagsbuchhandlung, Stuttgart, pp. 1–12.

Illies, J. H., 1978, Two stages Rhinegraben rifting, in *Tectonics and Geophysics of Continental Rifts,* I. B. Ramberg and E.-R. Neumann, eds., E. Reidel, Dordrecht, pp. 63–71.

Illies, J. H., and G. Greiner, 1978, Rhinegraben and the Alpine System, *Geol. Soc. America Bull.* **89**:770–782.

Jacob, J., R. D. Russell, and J. T. Wilson, 1974, *Physics and Geology,* 2nd ed., McGraw-Hill, New York, 622p.

Kennedy, W. Q., 1958, The tectonic evolution of the Midland Valley of Scotland, *Geol. Soc. Glasgow Trans.* **23**:106–133.

McConnell, R. B., 1974, Taphrogenic lineaments and plate tectonics, in *Approaches to Taphrogenesis,* J. H. Illies and K. Fuchs, eds., Proceedings of an International Rift Symposium, Karlsruhe, E. Schweizer bart'sche Verlagsbuchhandlung, Stuttgart.

Oftedhal, C., 1978, Main geologic features of the Oslo graben, in *Tectonics and Geophysics of the Continental Rifts,* I. B. Ramberg and E.-R. Neumann, eds., D. Reidel, Dordrecht, pp. 149–166.

Oliver, J., F. Cook, and L. Brown, 1983, COCORP and the continental crust, *Jour. Geophys. Res.* **88**:3329–3347.

Phillips, W. E. A., C. J. Stillman and T. Murphy, 1976, A Caledonian plate tectonic model, *Geol. Soc. London Jour.* **132**:579–609.

Quennell, A. M., ed., 1982, *Rift Valleys: Afro-Arabian,* Benchmark Papers in Geology, vol. 60, Hutchinson Ross, Stroudsburg, Pa. 419 pp.

Russell, M. J., and D. K. Smythe, 1983, Origin of the Oslo Graben in relation to the Hercynian-Alleghenian orogeny and lithosphere rifting in the north Atlantic, *Tectonophysics* **94**:459–472.

St. Mueller, 1970, Geophysical aspects of graben formation in continental rift systems, in *Graben Problems,* J. H. Illies and St. Mueller, eds., Proceedings of an International Rift Symposium, 1968, Karlsruhe, E. Schweizer bart'sche Verlagsbuchhandlung, Stuttgart, pp. 27–37.

St. Mueller, E. Peterschmitt, K. Fuchs, and J. Ansonge, 1969, Crustal structure beneath the Rhinegraben from seismic refraction and reflection measurements, *Tectonophysics* **8:**529-542.

Sittler, C., 1974, Le Fosse Rhenan ou la plaine d'Alsace, in *Géologie de la France,* J. Debelmas, ed., Doin, Paris, pp. 78-104.

Speight, J. M., and J. G. Mitchell, 1979, The Permo-Carboniferous dyke-swarm of northern Argyll and its bearing on dextral displacement of the Great Glen fault, *Geol. Soc. London Jour.* **136:**3-11.

Thorpe, R. S., R. D. Beckinsale, P. J. Patchett, D. A. Piper, G. R. Davies, and J. A. Evans, 1984, Crustal growth and late Precambrian-early Palaeozoic plate tectonic evolution of England and Wales, *Geol. Soc. London Jour.* **141:**521-536.

Whiteman, A., D. Naylor, R. Pegrum, and G. Rees, 1975, North Sea troughs and plate tectonics, *Tectonophysics* **26:**39-54.

Willis, B., 1936, *East African plateaus and Rift Valleys,* Carnegie Inst. of Washington Pub. 470, 358p.

Wilson, J. T., 1969, Static or mobile earth: the current scientific revolution (abstract), *Tectonophysics* **7:**599-601.

Ziegler, P. A., 1975, North Sea Basin history in the tectonic framework of northwest Europe, in *Petroleum and the Continental Shelf of Northwest Europe,* A. W. Woodland, ed., Applied Science, London, pp. 131-149.

Ziegler, P. A., 1978, North Sea rift and basin development, in *Tectonics and Geophysics of Continental Rifts,* I. B. Ramberg and E.-R. Neumann, eds., D. Reidel, Dordrecht, pp. 249-277.

Ziegler, P. A., 1981, Evolution of the sedimentary basins in north-west Europe, in *Petroleum Geology of the Continental Shelf of Northwest Europe,* Institute of Petroleum, London, pp. 3-39.

Ziegler, W. H., 1975, Outline of the geological history of the North Sea, in *Petroleum and the Continental Shelf of Northwest Europe,* A. W. Woodland, ed., Applied Science, London, pp. 165-185.

12

Reprinted from pages 41–43, 48–49, 52–54, 67–68, and 70–71 of *Geol. Soc. London Quart. Jour.* **102:**41–76 (1946)

THE GREAT GLEN FAULT

BY WILLIAM QUARRIER KENNEDY, D.SC. F.G.S.

Read 8 February, 1939

[PLATE III]

CONTENTS

SUMMARY

The powerful dislocation which intersects Scotland along the line of the Great Glen has, in the past, been regarded by most geologists as a normal or dip-slip fault with a predominant vertical downthrow to the south-east. A reconsideration of the entire problem now suggests that this view is no longer tenable and that the dislocation is, in reality, a lateral-slip or wrench fault with a horizontal displacement of approximately 65 miles. Such an interpretation is supported by several independent lines of evidence, as follows :—

(1) The dislocation possesses physical characters unlike those of most normal faults but similar to the great strike-slip shears of the California Coast Range.

(2) It belongs to the same system as the Strathconon, Ericht-Laidon and Loch Tay faults, all of which have proved lateral displacements of up to 5 miles.

(3) It displaces the great belt of regional injection which affects the Moine Schists of the northern and Grampian Highlands, the nature and amount of the displacement being consistent with lateral shift but not with vertical downthrow.

(4) It similarly displaces the metamorphic zones of the Highlands in an equally significant manner.

(5) It truncates the Strontian Granite, the southern portion of which, according to the detailed structural evidence, is missing. The missing portion, moreover, can be identified in the Foyers mass which outcrops on the other side of the fault-line some 65 miles to the north-east and is similarly truncated by the fault. These two major Caledonian intrusions consist of identical rock types and are structurally homologous.

(6) Finally, the occurrence of Lewisian and Torridonian rocks in Islay and Colonsay and the presence of the Moine Thrust-plane in the former island are more readily explained on the assumption of a lateral rather than a vertical displacement along the fault.

Although the dislocation is still active, the available evidence indicates that the main lateral movement was accomplished prior to the deposition of the Upper Carboniferous sediments of Lochaline and subsequent to the intrusion of the Strontian and Foyers (Lower Old Red Sandstone) granites. Middle Old Red Sandstone strata along the Great Glen have, moreover, suffered intense crushing and deformation during the faulting, which must, therefore, be referred partly if not wholly to a post-Middle Old Red Sandstone epoch.

The sinistral nature of the displacement, i.e. towards the south-west on the north-west side of the fracture and towards the north-east on its south-east side, implies the operation of a stress system involving regional compression acting in a general north-and-south direction accompanied by an east-and-west relief of pressure. This is regarded as evidence of the fact that the Hercynian forces, to which the formation of the Great Glen Fault is ascribed, were already in operation during Upper Old Red Sandstone or Lower Carboniferous times.

I. Introduction

The Great Glen Fault, which intersects Scotland from coast to coast along the deep trench-like hollow of Glen More,[1] has long attracted attention by virtue of its length, its singular straightness, and the magnitude of its topographic effect. As early as 1841 Hugh Miller (1841, p. 121) considered that the Great Glen depression must represent " a foot-track, hollowed by the frequent tread of earthquakes, to mark the course in which they journeyed ". It was not until 1861, however, that the existence of a coincident line of disruption was clearly recognized. In that year Murchison and Geikie (1861, p. 207) referred briefly to the Great Glen Fault but, being unable to determine the nature of the displacement, concluded that, " though the fracture is probably more extensive than any other in the country, it has not been attended, so far at least as we have yet been able to ascertain, with any marked upheaval or depression of the rocks on either side." They regarded it as " a fracture without a throw, or at least, with such a throw as not at all to interfere with the regularity and perspicuity of the section ". Four years later, Geikie (1865, p. 177) discussed the origin of the Great Glen in some detail, emphasizing in particular the important part played by erosion in its formation. He correctly interpreted the Loch Linnhe–Loch Lochy–Loch Ness depression as a fault-line feature excavated along the shatter-belt of a powerful dislocation, and pointed out that, along the western shore of the Moray Firth, the fault throws down to the south-east, bringing Jurassic rocks against the Middle Old Red Sandstone. Subsequent work carried out by the Geological Survey during the early part of the present century provided two additional important items of information. In the first place, Horne (Horne and Hinxman 1914, p. 68) showed that south-west of Inverness, where Middle Old Red Sandstone beds on the south-east side of the Great Glen Fault are thrown against schists of the Moine Series to the north-west, the minimum (apparent) vertical displacement must amount to at least 6000 feet. It was also recognized by the Survey geologists (Bailey 1916, p. 215) that in the Loch Linnhe region the schists on the north-west side of the fracture are of notably higher metamorphic grade than those to the south-east, a distinction which does not apply in the districts around the northern end of Loch Ness.

Apart from these observations, no other material evidence has been adduced which might serve to indicate the nature of the fault and the direction and amount of its displacement. Only the extremities of the line have, so far, been mapped in detail by the Geological Survey, and as regards the central sector, from the head of Loch Linnhe to the north-eastern end of Loch Ness, little more is known to-day than is shown on McCulloch's geological map of Scotland published in 1836. Most geologists have, therefore, followed Geikie and Horne in regarding the dislocation as a " normal " or dip-slip fault with a true vertical displacement measurable in thousands of feet. Such an interpretation is, however, inconsistent with certain important features of Highland geology. In particular, it fails to explain the effects of the dislocation on certain

[1] In Gaelic *Gleann Mor na h-Alba*, i.e. the Great Glen of Scotland.

major structures which it intersects. These difficulties have led the writer to reconsider the entire problem and it is now suggested that the dislocation is, in reality, a lateral-slip or wrench fault with a *horizontal* displacement of approximately 65 miles.

The evidence on which these conclusions are based is discussed in detail below. The data have been derived partly from the published and unpublished work of the Geological Survey, and partly represent original observations made by the writer over a period of years. It has been necessary to carry out a rapid survey of the Foyers Granite, for which purpose a grant was received from the Clough Memorial Fund. This is here gratefully acknowledged. Thanks are also due to colleagues in the Scottish branch of the Geological Survey who have assisted materially with information, criticism, and advice during the course of the work, and particularly to Dr. E. M. Anderson, who has discussed at all stages the development of the ideas expressed below. Finally, it may be pointed out that a striking view of the Great Glen from Loch Oich to Loch Ness, taken by Mr. R. M. Adam of the Royal Botanic Gardens, Edinburgh, forms plate xli in Geikie's *Structural and Field Geology*, revised by R. Campbell and R. M. Craig, 1940. The same publication contains a brief reference to the present writer's views regarding the nature of the fault, and a simple figure (fig. 56) based on his original drawings.

[*Editor's Note:* Material has been omitted at this point.]

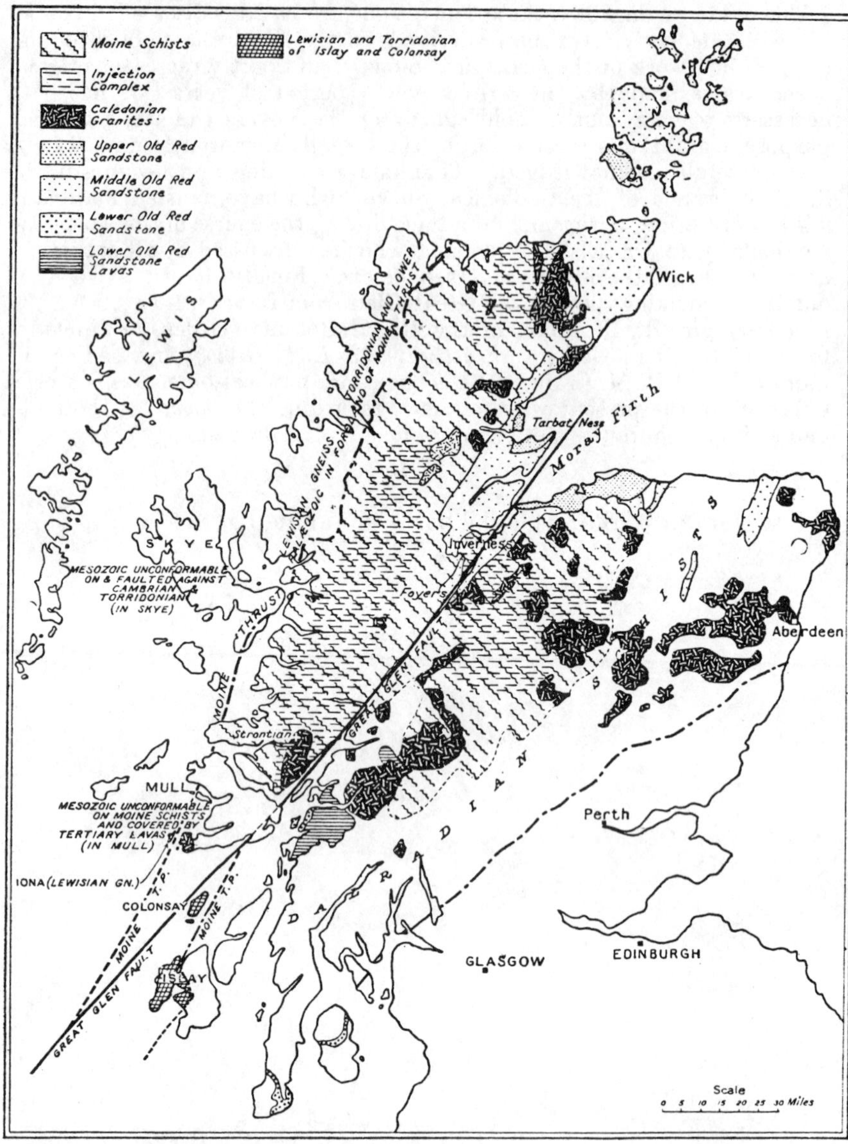

FIG. 2.—Geological map of the Scottish Highlands to show the present position of the Moine injection complexes, the Strontian and Foyers granites, and the Moine Thrust-plane.

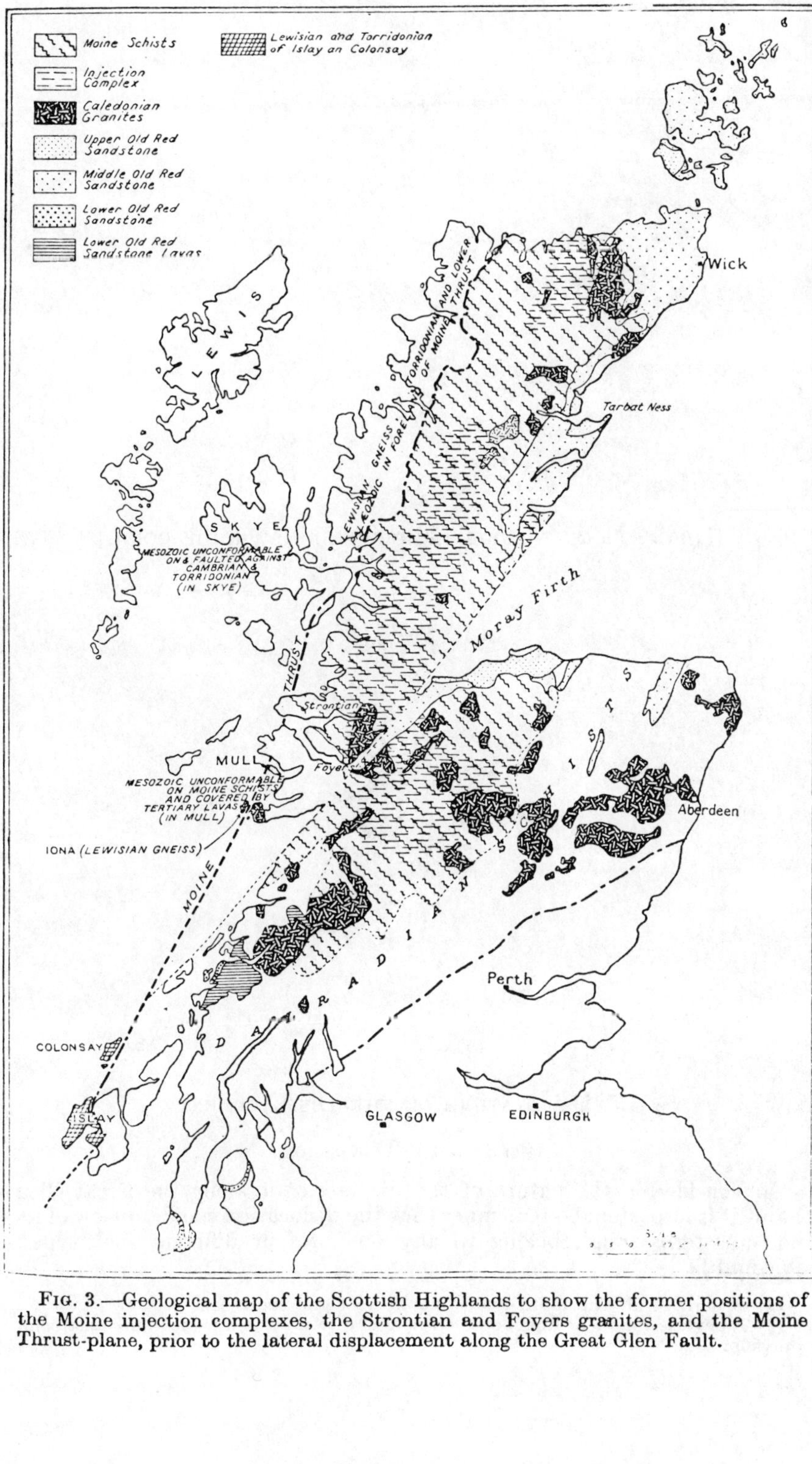

FIG. 3.—Geological map of the Scottish Highlands to show the former positions of the Moine injection complexes, the Strontian and Foyers granites, and the Moine Thrust-plane, prior to the lateral displacement along the Great Glen Fault.

[*Editor's Note:* Material has been omitted at this point.]

IV. The Nature of the Displacement

(a) General Discussion

In considering the nature of the displacement along the Great Glen Fault it is important to remember that the dislocation could, in view of its north-easterly trend, belong to any one of four different fault-types. It could be :—

(1) A normal, dip-slip fault, due to north-west and south-east tension.
(2) A reversed fault or high-angle thrust, due to north-west and south-east compression.

(3) A lateral-slip or wrench fault, involving relative horizontal displacement of the northern fault-block towards the *south-west*.

(4) A lateral-slip or wrench fault, involving relative horizontal displacement of the northern fault-block towards the *north-east*.

Previous writers, with two exceptions, have, in fact, accepted the first of these alternatives, and have regarded the dislocation as a normal fault, with predominant vertical displacement involving downthrow to the south-east. George Barrow (in Tilley 1925, p. 111), however, considered the fracture to be a thrust-plane, whereas Cunningham Craig (in Horne and Hinxman 1914, p. 70) apparently included it in his general statement to the effect that most, if not all, the north-easterly dislocations in the Highlands are characterized by predominant horizontal displacement. The most widely accepted view as to the nature of the dislocation actually involves two distinct conceptions: (a) that the fault is "normal", and (b) that the displacement is vertical. These may be considered briefly, as the one does not necessarily follow from the other.

(*a*) "*Normal character*" *of the fault.*—A normal fault is, by definition, one which hades to the side of the apparent downthrow. Conversely, a reversed fault hades to the upthrow side. In the case of the Great Glen Fault, however, the direction of hade is not known (see p. 45) and it is consequently impossible to tell whether it is normal or reversed. Either is possible so far as direct evidence is concerned.

(*b*) "*Vertical nature*" *of the displacement.*—Previous writers have not claimed specifically that the predominant displacement is vertical, but have merely implied, on the assumption of "normal faulting", that such must be the case. Normal and reversed faults can, however, be produced by horizontal displacement, and the fact that younger sedimentary formations and metamorphic rocks of lower grade on the south-east side of the Great Glen are, at certain localities, thrown against coarsely crystalline schists and gneisses to the north-west, does not, therefore, by itself, constitute proof that the main movement was vertical. The following statement by Ransome (1906, p. 785) is significant in this connexion :—

"... along a plane of dislocation, within which the horizontal component of movement exceeds that corresponding to the dip, apparent normal faults may occur at some points and apparent reverse faults at others. Whether a purely horizontal displacement appears as a normal fault or as a reverse fault at any one place depends merely upon the attitude of the beds (or other structures used as datum planes) relative to the plane of the fault. Furthermore 'normal faults' and 'reverse faults' may both be produced simultaneously by compression and 'normal faulting' is not necessarily a proof of tension.

"In short, the terms normal and reverse as commonly applied to faults have very little real significance. They express merely a local relation of the faulted beds, dikes, or other structures to the plane of the fault. They do not, unless we make an unwarranted assumption, indicate the kind or direction of the movement or whether tension or compression was the dominant stress at the time of dislocation."

Certain features of the Great Glen Fault, moreover, appear to be inconsistent with an assumption of predominantly vertical displacement. The following may be mentioned :—

(1) The dislocation intersects two vertical stocks of granite (the Strontian and Foyers masses) which have been faulted in such a manner that the displaced portions do not appear, on the downthrow side, in the immediate vicinity. In the case of the Strontian Granite at least, it is unlikely that the missing south-easterly portion has been thrown vertically downwards, out of sight, for the downthrow side of the fault still lies within the level of Caledonian granite intrusion, as shown by the presence of many masses of similar age and character in the district.

(2) At the northern end of Loch Ness, where the Middle Old Red Sandstone is

thrown against schists of the Moine Series, the minimum (apparent vertical) displacement is at least 6000 feet (see p. 51). Immediately to the north-east, and to the south-west, the displacement must be much less as, in those areas, Old Red Sandstone strata occur on both sides of the fault (Pl. III). Still farther south, in the Loch Linnhe district, the apparent vertical downthrow cannot be less than 3 to 4 miles (see p. 59) and might be considerably more. Such irregular behaviour is uncommon in dip-slip faults but is characteristic of dislocations in which the principal component of displacement is horizontal.

(3) Within any one area, the deduced direction of throw may vary according to the particular reference datum employed. Thus, at the northern end of Loch Ness, the downthrow is to the south-east when measured with respect to the displaced Middle Old Red Sandstone but to the north-west when measured with respect to the Foyers Granite (Pl. III). In order to explain this relationship by vertical movement it is necessary to postulate either that the Foyers Granite was intruded from above downwards, or that the downthrow has been in different directions at different times. Either case is improbable.

It is clear, therefore, that adequate proof of vertical movement is not available.

On the other hand, positive evidence can be adduced to show that the main movement was horizontal, and amounted to some 65 miles. This is discussed in the following section. It must be emphasized, however, that, although the writer considers the dislocation to be a lateral-slip or wrench fault, this does not exclude the possibility of an important vertical component of displacement. The latter may, in fact, be considerable, but cannot easily be determined in view of the predominant and much greater lateral shift.

(b) Evidence of Lateral Displacement

The conception of a predominant lateral displacement along the Great Glen Fault, amounting to approximately 65 miles, is based on several independent lines of evidence. These belong to different orders of validity and may, for convenience, be referred to the three following categories: (i) indirect evidence, involving certain general comparisons and relationships, (ii) direct evidence, based on a correlation of homologous geological structures intersected and displaced by the fault, and (iii) contributory evidence, mainly of a confirmatory character.

[*Editor's Note:* Material has been omitted at this point.]

(b) Dynamics of Movement

The dynamical conditions involved in faulting of the type under consideration, i.e. in which there has been predominantly horizontal displacement along a vertical or highly inclined fracture, are comparatively simple, and have been discussed very fully by E. M. Anderson (1905, 1942). Such a fracture is essentially a vertical plane of shearing formed by the application of a pressure directed at an angle of less than 45° to the plane of the fault. As Anderson has pointed out (1905, p. 387), the

[1] There is still doubt as to the position of the Pass of Brander rocks in the Carboniferous sequence. They were originally referred to the Calciferous Sandstone Series of the Lower Carboniferous (Kynaston 1908, p. 154) but recent work has shown that some of the evidence on which this conclusion was based is not valid (Pringle and Macgregor 1940, p. 76). The possibility still remains that they may be coeval with the rocks of the Inninmore outlier.

forces acting within a rock-mass can be resolved into three pressures or tensions (or both combined) which act mutually at right angles to one another. There is no tangential stress across the planes normal to these pressures or tensions, but tangential stress is developed to a greater or lesser degree across all other intermediate planes and, for certain cases, attains a maximum. It is these planes of maximum tangential stress which determine the directions of faults. In any given system involving three unequal pressures or tensions there are two such sets of planes across which tangential stress is at a maximum. These planes are mutually perpendicular and bisect the angles between the directions of greatest and least pressure. The actual planes of slipping in rocks do not, however, coincide with the planes of maximum tangential stress but, owing to the effects of frictional resistance, deviate from these positions so as to form smaller angles (less than 45°) with the direction of greatest pressure. The following conditions, therefore, hold with regard to faulting in general :—

(1) Reversed faults and thrust-planes originate when the greatest pressure is horizontal, and the least pressure vertical. They strike in a direction perpendicular to that of the greatest pressure, and dip in either direction at angles of less than 45°.

(2) Normal (dip-slip) faults originate when the greatest pressure is vertical, and the least pressure is in some horizontal direction. They strike in a direction perpendicular to that of the least pressure, and dip in either direction at angles of more than 45°.

(3) Lateral-slip or wrench faults originate when the greatest pressure is in one horizontal direction, and the least pressure in another horizontal direction, necessarily at right angles to the first. They strike in two possible directions, forming acute angles which are bisected by the direction of greatest pressure. Their hade is theoretically vertical. The displacement along one set of shear-planes is sinistral, and along the other set is dextral.

The Great Glen Fault belongs to class (3) above, and has a sinistral displacement in which the relative movement is towards the south-west on the north-west side of the fracture, and towards the north-east on the south-east side. It follows, therefore, that the dislocation must have originated through the operation of a stress system involving regional compression acting in a general north-and-south direction accompanied by an east-and-west relief of pressure (Fig. 8a).

[*Editor's Note:* Material has been omitted at this point.]

(c) Tectonic Significance

The evidence adduced in the preceding sections leads to certain interesting conclusions regarding the tectonic significance of the faulting movements. Thus, it has been established that :—

(1) The main lateral displacement along the line of the Great Glen was almost certainly accomplished during Upper Old Red Sandstone or Lower Carboniferous times.

(2) The faulting represented the response of the more or less rigid, crystalline, Highland mass to the application of a stress system involving regional compression in a north-and-south direction accompanied by an east-and-west relief of pressure.

The main phase of movement is, therefore, seen to have occurred during the interval between the Caledonian and Hercynian orogenies : and so far as *time relations* are concerned the stresses might equally well have been genetically related to either of these major crustal revolutions. That the faulting is not a residual or posthumous effect of the Caledonian movements, *but represents an early manifestation of the forces which culminated later in the main Hercynian folding,* is clearly indicated, however, by the orientation of the principal pressure relative to the direction of the Hercynian front in Britain (Fig. 8b).

This conclusion is in accordance with observations made by Evans, who has pointed out (1926, p. lxxviii) that "the first effective action of such compressive forces usually antedates by a considerable period the epoch of maximum disturbance", and, furthermore, that "the earliest influence of the Hercynian movement seems to have affected South Wales and the south-western peninsula of England long before the close of the Devonian Period". It also raises issues which have an important bearing on certain fundamental problems of orogenic evolution. Discussion of these must, however, be deferred until some future occasion. In the meantime, attention is directed to the following considerations :—

(1) That, during the development of the Hercynian geosyncline, the foreland area, of which the Highlands formed part, was in a state of *compression.*

(2) That the Caledonian folding apparently did not serve to relieve the crustal stresses for any prolonged period but was followed almost immediately by the onset of the Hercynian pressure.

It must be emphasized that the formation of the Great Glen fracture system represents a major diastrophic event comparable in magnitude and significance to a phase of orogenic movement. Lateral displacement along the faults has, in fact, resulted in a meridional shortening of the crust in Scotland by some 50 miles (compare Figs. 2 and 3), a result which, in the case of less rigid rocks, would presumably have been accomplished by a considerable degree of folding and thrusting. There can be little doubt, therefore, that the relief of stress effected by the faulting within the crystalline foreland must have influenced, to a greater or lesser extent, the structural and stratigraphical development of the regions to the south.

[*Editor's Note:* Only the references cited in the preceding excerpts are reprinted here.]

VI. List of Works to which Reference is Made

Anderson, E. M. 1905. The dynamics of faulting. *Trans. Edin. Geol. Soc.* viii, p. 387.

——. 1942. The dynamics of faulting. Edinburgh.

Bailey, E. B. 1916. *In* The geology of Ben Nevis and Glen Coe. *Mem. Geol. Surv.*

Evans, J. W. 1926. Regions of compression. *Q.J.G.S.* lxxxii, p. lx.

Geikie, A. 1865. The scenery of Scotland. 1st edition. London.

Geikie, J. 1912. Structural and field geology. 3rd edition. Edinburgh.

Horne, J., & L. W. Hinxman. 1914. The Geology of the Country round Beauly and Inverness. *Mem. Geol. Surv.*

Kynaston, H. 1908. *In* The geology of the country near Oban and Dalmally. *Mem. Geol. Surv.*

Miller, H. 1841. The Old Red Sandstone. 1st edition. Edinburgh.

Murchison, R. I., & A. Geikie. 1861. On the altered rocks of the Western Islands of Scotland, and the North-Western and Central Highlands. *Q.J.G.S.* xvii, p. 171.

Perry, E. L. 1935. Flaws and tear faults. *Am. Journ. Sci.* ser. 5, xxix, p. 112.

Pringle, J., & M. Macgregor. 1940. The outlier of Carboniferous rocks at Bridge of Awe, Argyllshire. *Trans. Geol. Soc. Glasgow*, xx, p. 73.

Ransome, F. L. 1906. The directions of movement and the nomenclature of faults. *Econ. Geol.* i, p. 777.

Richey, J. E. 1939. The dykes of Scotland. *Trans. Edin. Geol. Soc.* xiii, p. 393.

Tilley, C. E. 1925. A preliminary survey of metamorphic zones in the Southern Highlands of Scotland. *Q.J.G.S.* lxxxi, p. 100.

13

Reprinted from pages 133-135 of *Tectonics and Geophysics of Continental Rifts,* I. B. Ramberg and E.-R. Neumann, eds., D. Reidel, Dordrecht, 1978, 444p.

THE MIDLAND VALLEY AS A RIFT, SEEN IN CONNECTION WITH THE LATE PALAEOZOIC EUROPEAN RIFT SYSTEM

E.H. Francis

Institute of Geological Sciences, Leeds, England*

ABSTRACT. The Midland Valley is first recognisable as a graben after the closure of Iapetus. NE-SW structural lineaments parallel to the closure then controlled deposition of early Devonian intermontane sequences and the siting of some of the calc-alkaline volcanism within the rift. The phase, which included similar volcanism and related plutonism beyond the rift, was ended by a mid-Devonian orogeny. Carboniferous volcanism was alkaline and mid-continental; its location, like the distribution of sediments, indicates continuing NE-SW basement controls. These probably owed less to subduction, hot-spots or plate collision than to adjustment-fracturing and underlying partial melting in an inhomogeneous continental plate drifting northwards.

Late in the Carboniferous new stresses led to emplacement of E-W tholeiite dykes, along extensional fractures transgressing all the earlier structures. They may continue eastwards beneath the North Sea, reappearing along ESE to SE trends in southern Sweden and are possibly comagmatic with the earliest lava flows in the Oslo Graben. This orogenic phase may have been initiated by rotation of a South Europe microplate during the closure of a mid-European ocean. After this brief interruption, continental alkali-basaltic magmatism became re-established and continued into early Permian times. It was more silica-undersaturated than before - a change which may represent a new stress regime, corresponding with NW or NNW basement controls recognised in the western part of the Midland Valley and correlatable with the N-S rifts of the North Sea and the Oslo Graben.

* Dept. of Earth Sciences, University of Leeds, from Oct. 1977.

1. INTRODUCTION

In Scotland, the Midland Valley Graben is about 80 km wide and 200 km long extending from the firths of Tay and Forth south-westward to the Firth of Clyde. The downfaulted rocks within the structure are mainly Devonian and Carboniferous in age. They are separated from folded Ordovician and Silurian rocks to the south by the Southern Uplands Fault and from even older metamorphosed Dalradian rocks to the north by the Highland Boundary Fault (Figs. 1,2). The graben can be traced for a further 300 km or so across Ireland, where it is narrower (c. 50 km) and less clearly defined, partly because the boundary faults are obscured by younger rocks.

Previous syntheses [1-3] have shown that the Midland Valley was not recognisable as a rift until the end of the Caledonian

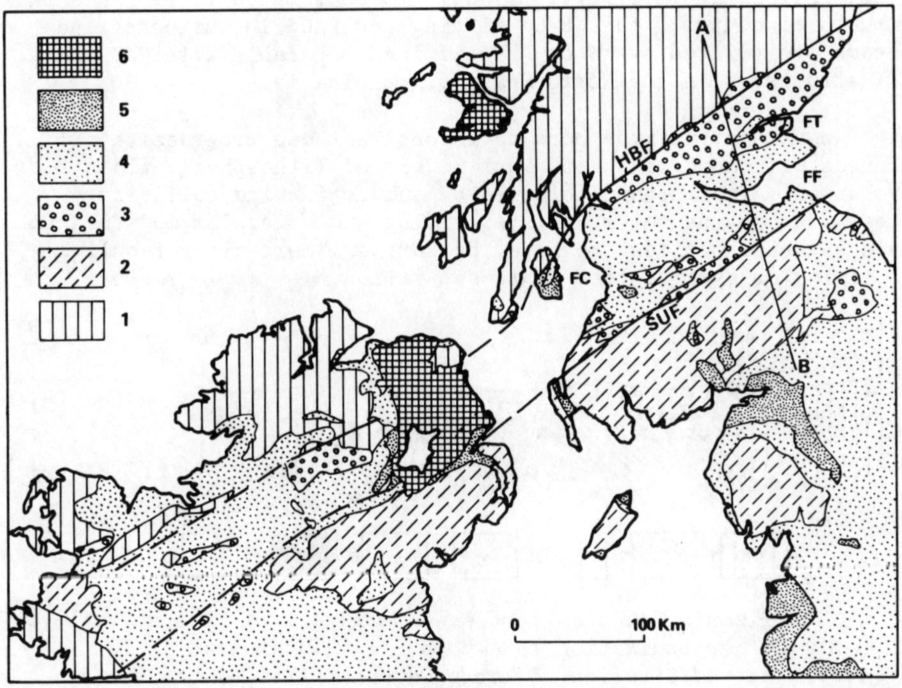

Fig. 1. Geological sketch map of the Midland Valley Graben. 1 Dalradian and older; 2 Ordovician and Silurian; 3 Early Devonian (L.ORS); 4 Late Devonian (U.ORS) and Carboniferous; 5 Mainly Permian; 6 Mesozoic and Tertiary. HBF Highland Boundary Fault; SUF Southern Uplands Fault; FC Firth of Clyde; FF Firth of Forth; FT Firth of Tay. A-B see Figs. 2 and 3.

orogeny and that it was most active as a structure during the Devonian and Carboniferous. However, because it is now believed that the Caledonian orogeny was caused by the closure of a Lower Palaeozoic proto-Atlantic Ocean (Iapetus) some reappraisal of both the timing and foundations of the graben is desirable. Moreover, although the mechanisms governing the main structure and the parallel to sub-parallel internal rifts appear to have operated continuously throughout Upper Palaeozoic times there were profound changes with time in the form taken by penecontemporaneous volcanism. Thus during the early Devonian, the products of activity were extrusive and plutonic calc-alkaline rocks located both within the rift and well to the north and south of it. In early to mid-Carboniferous times, by contrast, volcanism was essentially alkali-basaltic and extrusive and was virtually confined to the rift or along hinge lines of newly formed graben farther south. Finally, the late Carboniferous to early Permian activity, though continuing to be alkali-basaltic except for a brief tholeiitic interruption, seems to be related to a change in structural control: the NE-SW rifting declined in influence and became overprinted by NW or NNW rifting comparable with N-S and NNE structures in the North Sea and Scandinavia.

An attempt is made here to reconcile these progressive changes with some of the current models of lithospheric plate evolution. After oceanic closure by subduction and collision there was continental mid-plate rifting and volcanism modified in late Palaeozoic times, perhaps, by further, more distant oblique collision or by incipient plate separation.

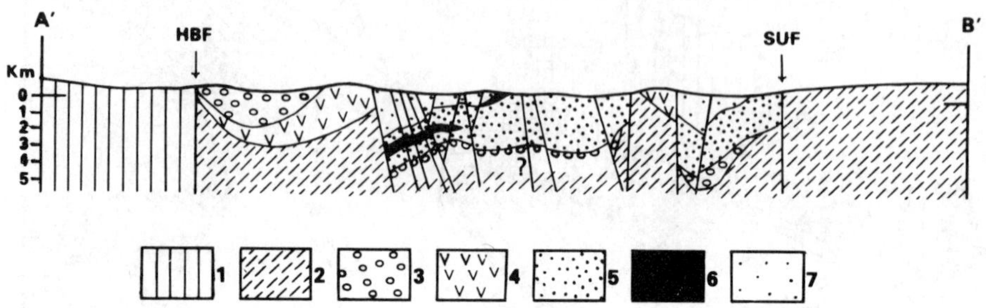

Fig. 2. Horizontal section across eastern part of Midland Valley. Line A′ - B′ approximating to A-B on Fig. 1. 1 Dalradian; 2 Ordovician and Silurian; 3 Early Devonian (L.ORS) sediments; 4 Early Devonian lava and tuffs; 5 Late Devonian (U.ORS) and Lower Carboniferous sediments; 6 Carboniferous basalt lavas; 7 Upper Carboniferous sediments.

[*Editor's Note:* Material has been omitted at this point and only the references cited in the preceding excerpt are reprinted here.]

REFERENCES

1. M. Macgregor and A.G. MacGregor, The Midland Valley of Scotland (2nd Edit.), British Regional Geology, 1948.
2. W. Q. Kennedy, Trans. geol. Soc. Glasg., 23, 106, 1958.
3. T. N. George, Trans. geol. Soc. Galsg., 24, 32, 1960.

Part IV

ASIA

Editor's Comments on Papers 14 Through 17

14 MOLNAR, TAPPONNIER, and CHEN
Extensional Tectonics in Central and Eastern Asia: A Brief Summary

15 FLORENSOV
Excerpts from *Rifts of the Baikal Mountain Region*

16 KING
The Baikal Rift

17 KING
Ode to the Baikal Rift

Paper 14, by Molnar of M.I.T., Tapponnier of Université Pierre et Marie Curie, and Chen of M.I.T., is included because it considers the extensional phenomena accompanying intercontinental collision and contains a useful map of central Asia, which can serve as a key to the relevant localities—Tibet; the Baikal rift system; the Shansi graben field; and Yunnan. The authors link the normal and graben faulting with the uplift of Tibet and the probable latitudinal spreading. Transmitted via the northeasterly Inner Asiatic belt of "rejuvenated mountains" (Zonenshain and Sarostin, 1981), the compressional effects of the collision had an influence on the tectonics of the Baikal region and the Stanovoi fault zone beyond. The possibility of "hiving off" of southwest Asia in the manner postulated for East Africa, can be envisaged.

Papers 15 and 16 are almost wholly devoted to the geology of the Baikal rift system. Paper 15 by Professor Florensov of Irvrutsk, contains the most appropriate map of Lake Baikal and its environs. The accompanying text contains an account of the geology and the geomorphology, but the geophysical data are omitted because more recent treatments are contained in the listed references. The tectonics do not include reference to the role played in the Baikal rifting by the collision of the Indian plate with the Eurasian one, which had not been analyzed when the paper was written. Papers 16 and 17 are both by Professor B. C. King of University of London, who made personal observations of the geology of the Baikal rift. The first of the two papers is an easily digested summary of the extensive Russian work, which is largely inaccessible; the second is his "Ode to the Baikal Rift."

As a background to the tectonics as understood at the time these accounts were written, some early references are included that are still quite relevant today, notably Salop and Scheinman (1969) and Salop (1967). Florensov (1966) is the first (in English) of a series on the rift zone itself. Artemjev and Artyushkov (1971) question the dominance of the wedge-subsidence theories of Cloos, Vening-Meinesz and others; this paper is one of the early ones to appreciate the active part played by the asthenosphere (see also Artyushkov, 1983). From about 1973 onward there is a realization of the importance of geophysics, especially seismic and thermal studies. This was no doubt partly due to the Upper Mantle Project, which was then succeeded by the International Geodynamic Project.

REFERENCES

Artemjev, M. E., and E. V. Artyushkov, 1971, Structure and isostasy of the Baikal Rift and the mechanism of rifting, *Jour. Geophys. Research* **76:**1197-1211.

Artyushkov, E. V., 1983, *Developments in Geotectonics,* Geodynamics, Vol. 18, Elsevier, Amsterdam, 1983.

Florensov, N. A., 1965, The Baikal rift zone, in *The World Rift System,* Report of the Upper Mantle Committee Symposium on the World Rift System, T. N. Irvine, ed., Geol. Surv. Canada Paper 66-14, pp. 307-329.

Florensov, N. A., ed., 1968, *The Baikal Rift Zone,* Nauka, Moscow, 181p. [in Russian].

Huang, T. K., 1980, An outline of the tectonic characteristics of China, in *Continental Tectonics,* Studies in Geophysics, B. C. Burchfiel, panel chairman, National Academy of Sciences, Washington, pp. 184-197.

Logatchev, N. A., and N. A. Florensov, 1978, The Baikal system of rift valleys, *Tectonophysics* **45:**1-13.

Logatchev, N. A., V. A. Rogozhina, V. P. Solenenko, and Y. A. Zorin, 1978, Deep structure and evolution of the Baikal Rift zone, in *Tectonics and Geophysics of Continental Rifts,* I. B. Ramberg and E.-R. Neumann, eds., D. Reidel, Dordrecht, pp. 49-61.

Logatchev, N. A., Y. A. Zorin, and V. A. Rogozhina, 1983, Baikal Rift: active or passive—comparison of the Baikal and Kenya rift zones, *Tectonophysics* **94:**223-240.

Molnar, P., and P. Tapponnier, 1975, Cenozoic tectonics of Asia: effect of a continental collision, *Science* **189:**419-426.

Muratov, M. V., 1977, Principal tectonic features of the USSR, *Internat. Geology Rev.* **21:**39-50.

Puzyrev, N. N., M. M. Mandelbaum, S. V. Krylor, B. P. Mishenkin, G. V. Krupkaya, and G. V. Petrick, 1973, Deep seismic investigation in the Baikal rift zone, *Tectonophysics* **20:**85-95.

Ruzhich, V. V., S. I. Sherman and S. I. Tarasevich, 1972, New data on thrusts in the southwestern part of the Baikal rift zone, *Doklady Akad Nauk. SSSR* **205:**68-71.

Salop, L. I., 1967, *Geology of the Baikal Mountain Region,* Nyedra, Moscow, vol. **1,** 515p.; vol. 2, 699p. [in Russian].

Sherman, S. I., 1978, Faults of the Baikal rift zone, *Tectonophysics* **45:**31-39.

Tapponier, P., and P. Molnar, 1977, Active faulting and tectonics in China, *Jour. Geophys. Research* **82:**2095-3005.

Zonenshain, L. P., and L. A. Savostin, 1981, Geodynamics of the Baikal rift zone and plate tectonics of Asia, *Tectonophysics* **76:**1-45.

Zorin, Y. A., 1981, The Baikal rift: an example of the intrusion of asthenospheric material into the lithosphere as the cause of disruption of lithospheric plates, *Tectonophysics* **73:**91-104.

Zorin,Y. A., and V. A. Rogozhina, 1978, Mechanism of rifting and some features of the deepseated structure of the Baikal rift zone, *Tectonophysics* **45:**23-30.

14

Reprinted from *Royal Soc. London Philos. Trans.* **A300:**403-406 (1981)

Extensional tectonics in central and eastern Asia: a brief summary

By P. Molnar,† P. Tapponnier‡ and W.-P. Chen†

† *Department of Earth and Planetary Sciences, Massachusetts Institute of Technology, Cambridge, Massachusetts* 02139, *U.S.A.*

‡ *Institut de Physique du Globe, Université Pierre et Marie Curie, 4 Place Jussieu*, 75230 *Paris cedex* 05, *France*

Although most of the neotectonics of eastern and central Asia seems to be due to the collision and continued convergence between India and Eurasia, extensional tectonics prevails in four portions of this region: Tibet, the Baikal rift system, the Shansi graben and Yunnan. We think that the convergence between India and Eurasia is responsible for crustal extension in all four regions, but the physical mechanisms responsible for the extension are not the same in all of these cases.

Fault plane solutions of earthquakes, and clearly defined escarpments and adjacent basins seen on the Landsat imagery, indicate widespread normal faulting in Tibet (Molnar & Tapponnier 1975, 1978; Ni & York 1978). The direction of extension appears to be east–west, but the poorly determined fault plane solutions and the lack of field studies allow some variation in the direction of extension. The normal faulting seems to be confined to the areas of highest elevation and some of the normal faults cross the Indus–Zangbu suture into the high Himalaya.

Most of the earthquakes seem to have occurred at shallow depths, but at least one event, that of 14 September 1976, occurred at 90 ± 10 km depth (Chen *et al.* 1980). The fault plane solution for it shows normal faulting with an east–west T axis. That this event occurred in the mantle beneath the thick Tibetan crust corroborates the inference that the mantle immediately beneath Tibet is not unusually hot (Chen & Molnar 1980). If the orientation of the T axis for this event is representative of the direction of the least compressive stress in the uppermost mantle beneath all or most of Tibet, then it implies that the entire outer 100 km of the Earth beneath Tibet is being stretched in an east–west direction.

We infer that the normal faulting results from the gravitational body forces acting on the high altitudes and the crustal root of Tibet (Molnar & Tapponnier 1975, 1978). For equilibrium, this force must be balanced by a horizontal compressive stress in the surrounding regions or horizontal extensional stress within Tibet. Tibet's elevation is then analogous to a pressure gauge (Molnar & Tapponnier 1978) and it measures the average pressure that India applies to the rest of Asia (Tapponnier & Molnar 1976). Presumably India applies sufficient pressure to prevent north–south extension within Tibet, but if eastern China is free to move eastwards relative to both India and Eurasia, extension can occur in an esat–west direction.

This current style of tectonics is probably quite young and presumably it did not occur until the high elevations and thick crust had formed. One mechanism for creating the thick Tibetan crust is very large-scale crustal shortening (Dewey & Burke 1973). Reconstructions of the relative positions of India and Eurasia since the late Cretaceous require 2000–3000 km of convergence since the collision 40–50 Ma ago (Molnar & Tapponnier 1975). Palaeomagnetic studies in southern Tibet also indicate this amount of convergence between southern Tibet and

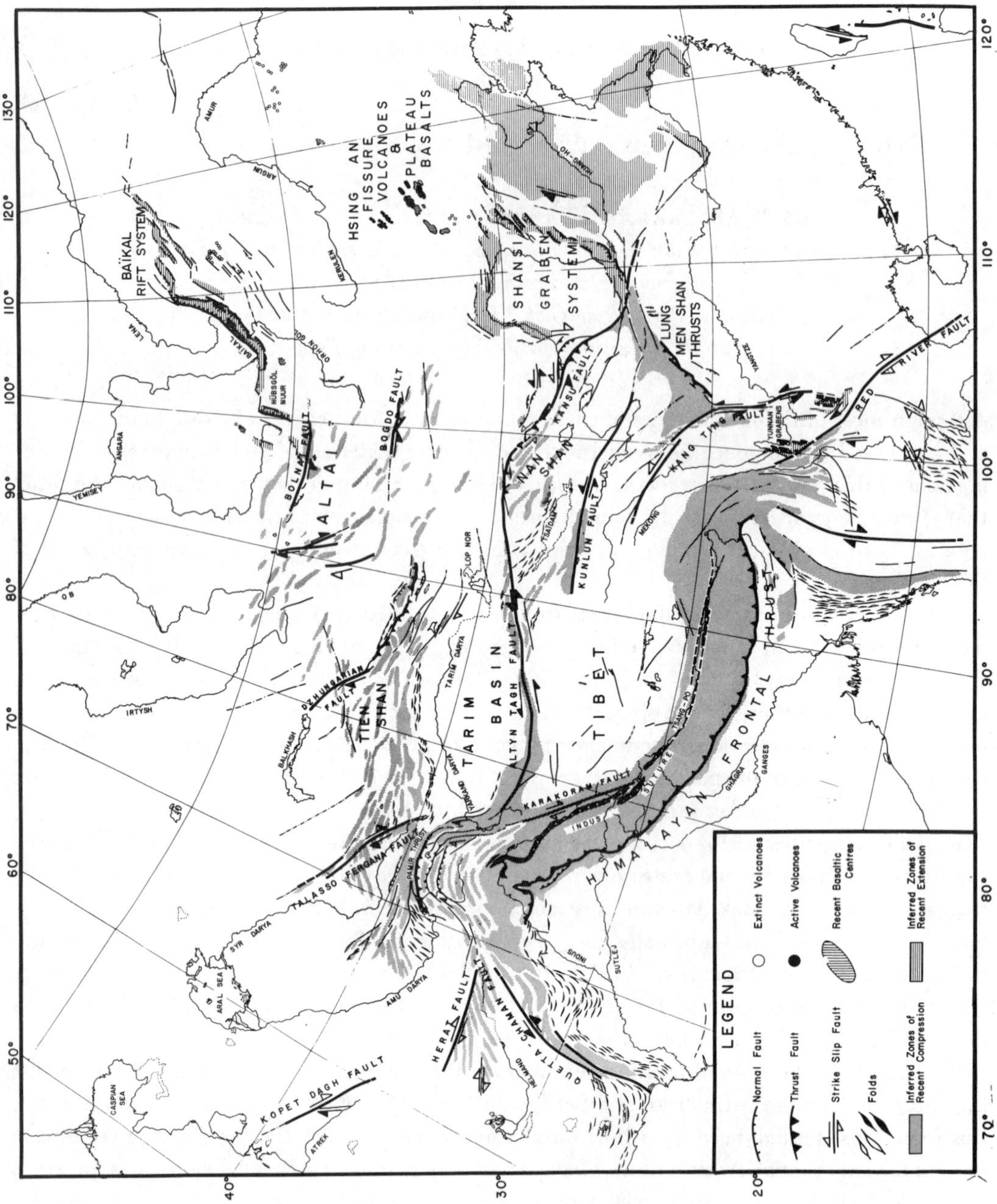
LEGEND
Normal Fault
Thrust Fault
Strike Slip Fault
Folds
Inferred Zones of Recent Compression
Extinct Volcanoes
Active Volcanoes
Recent Basaltic Centres
Inferred Zones of Recent Extension
BAÏKAL RIFT SYSTEM
HSING AN FISSURE VOLCANOES & PLATEAU BASALTS
SHANSI GRABEN SYSTEM
LUNG MEN SHAN THRUSTS
KANSU FAULT
NAN SHAN
KUNLUN FAULT
KANG TING FAULT
RED RIVER FAULT
YUNNAN GRABENS
BOLNAI FAULT
BOGDO FAULT
ALTAI
DZHUNGARIAN FAULT
TIEN SHAN
TARIM BASIN
ALTYN TAGH FAULT
TIBET
KARAKORAM FAULT
HIMALAYAN FRONTAL THRUST
TALASSO FERGANA FAULT
PAMIR THRUST
HERAT FAULT
CHAMAN FAULT
QUETTA
KOPET DAGH FAULT
CASPIAN SEA
ARAL SEA
BALKHASH
LOP NOR
TSAIDAM
INDUS
GANGES
YANGTZE
MEKONG
IRTYSH
OB
YENISEY
ANGARA
LENA
AMUR
ARGUN
KERULEN
HUANG HO
SYR DARYA
AMU DARYA
TARIM DARYA
HELMAND
ATREK
SUTLEJ

Siberia (Molnar & Chen 1978). These results militate against large-scale underthrusting of India beneath Tibet, but are consistent with large-scale shortening within Tibet. The existence of normal faulting within Tibet at present implies that shortening there has stopped, although the continuing convergence of India and Eurasia does seem to cause thrust faulting and crustal shortening on the margins of Tibet (Molnar *et al.* 1973, 1977), so that as these regions are elevated, the plateau seems to grow in area but not in altitude (Molnar & Tapponnier 1978). The convergence also seems to cause strike-slip faulting, primarily north and east of Tibet, by means of which parts of Asia are displaced eastwards with respect to the rest of Asia.

The normal faulting in the Baikal rift system, in the Shansi graben and in Yunnan, which are well studied by Soviet and Chinese geologists, is clear on the Landsat imagery and, except for the Shansi graben, is confirmed by fault plane solutions (Tapponnier & Molnar 1977, 1979). We think that there is an intimate relation between the normal faulting and some of the major strike slip faults: the Khangai fault in Mongolia (Tapponnier & Molnar 1979), the Kansu fault in eastern China (Molnar & Tapponnier 1977; Tapponnier & Molnar 1977), and (less clearly) the Red River fault in southwest China (Tapponnier & Molnar 1977), respectively. The relation of the normal and strike-slip faults is crudely analogous to that of tension cracks near the ends of shear zones. Moreover, the stress field associated with these extensional zones is approximately that which one would expect if Asia behaved like a plastic solid indented by a rigid indenter (Tapponnier & Molnar 1976).

Thus we associate the extensional tectonics in all of these areas with the convergence of India and Eurasia, but the mechanisms by which the convergence causes normal faulting seem to be different. The buoyancy forces that apparently drive the extension in Tibet are probably small in Baikal and Shansi. These grabens seem instead to open in response to a regional stress field that tends to pull the crust apart, but that nevertheless results from the convergence between India and the rest of Eurasia. The grabens in Yunnan probably result from a combination of these two causes.

This summary is not intended to be all-encompassing, but rather to give the reader an annotated bibliography of our work. The papers cited give more complete reference to the data upon which it is based and to the work of others.

The work summarized here has been made possible by continued support from the National Science Foundation. Molnar also acknowledges support from the Simon Guggenheim Foundation during the preparation of the text.

FIGURE 1. Preliminary map of recent tectonics in Asia. Bold lines represent faults of major importance, usually seismic and with very sharp morphology. Bold arrows indicate sense of motion, corroborated by fault plane solutions or surface faulting of earthquakes. Open arrows indicate sense inferred from analysis of photographs. For Tertiary folding, bold symbols indicate more prominent, more recent folds. The dotted areas indicate region of inferred recent vertical motion associated with thrust faulting and compressional tectonics. Areas shaded by dashed lines are covered by thick recent alluvial deposits and are dominated by horizontal extension and subsidence. The names of faults are not official names but purely for reference in this article.

References (Molnar *et al.*)

Chen, W. P. & Molnar, P. 1980 *J. geophys. Res.* (Submitted.)
Chen, W. P., Nabelek, J. L., Fitch, T. J. & Molnar, P. 1980 *J. geophys. Res.* (In the press.)
Dewey, J. F. & Burke, K. C. A. 1973 *J. Geol.* **81**, 683–692.
Molnar, P. & Chen, W. P. 1978 *Nature, Lond.* **273**, 218–220.
Molnar, P. & Tapponnier, P. 1975 *Science, N.Y.* **189**, 419–426.
Molnar, P. & Tapponnier, P. 1977 *Geology* **5**, 212–216.
Molnar, P. & Tapponnier, P. 1978 *J. geophys. Res.* **83**, 5361–5375.
Molnar, P., Chen, W. P., Fitch, T. J., Tapponnier, P., Warsi, W. E. K. & Wu, F. T. 1977 In *Himalaya: sciences de la terre*, pp. 269–294. Paris: Centre National de la Recherche Scientifique.
Molnar, P., Fitch, T. J. & Wu, F. T. 1973 *Earth planet. Sci. Lett.* **19**, 101–112.
Ni, J. & York, J. 1978 *J. geophys. Res.* **83**, 5377–5387.
Tapponnier, P. & Molnar, P. 1976 *Nature, Lond.* **264**, 319–324.
Tapponnier, P. & Molnar, P. 1977 *J. geophys. Res.* **82**, 2905–2930.
Tapponnier, P. & Molnar, P. 1979 *J. geophys. Res.* **84**, 3425–3459.

15

Reprinted from pages 443-452 of *Tectonophysics* **8**:443-456 (1969)

RIFTS OF THE BAIKAL MOUNTAIN REGION

N. A. Florensov

SUMMARY

Large and small graben comprising the Baikal rift zone are connected with an extensive Pliocene–Pleistocene uplift with amplitude 3,000–4,000 m above sea level. These grabens inherit the strike of the Early Neogene subsidences and they cannot be considered as the result of fracturing or sinking of the crust in the arch of the uplift. The long and discontinued basalt volcanism, which preceded and accompanied the rift formation, as well as active mountain formation and the character of fracture tectonics point to the leading role of stretching of the Earth's crust during the formation of this zone; the zone has no evident connection with the world rift system; moreover, it has a specific strain field in which, according to seismological data, near-horizontal stretching forces are dominant. They are directed across the major tectonic and orographic lines.

A reverse strain pattern in earthquakes foci in the western region of the zone (near-horizontal compression), in the presence of the outer structure-morphological monotony of the latter, can testify to the growing influence on this region of conditions which are inherent in the neighbouring region of the central-Asiatic depressions (Mongolia) where earthquake foci have similar strains and seismic activity is much higher.

Numerous signs of increase of rift depressions in width and depth, the presence of embryonal but active rift structures within the Stanovoi Mountains indicate the continuation of the development of the zone in the modern epoch. Its rift structures are only partially connected with the system of the boundary seam of the Siberian Platform. They also utilize the inner ancient faults in the Baikalides and Early Caledonides belt. Contrary to the opinions of some authors, the horizontal displacement component in the structure of the Baikal Rift has a clearly subordinate role. Among the existing interpretations of the gravimetrical data related to the Baikal zone the hypothesis suggested by Borisov (1965) should be particularly noted. The hypothesis proposes "a crust–mantle mixture" comprising the root of the general structure of the Baikal Rift, while the cause of considerable mobility of the crust in this zone can be found in highly heated upper layers of the upper mantle. This is supported by recent researches in Pribaikalie carried out by the magneto-telluric sounding.

INTRODUCTION

Large and small graben form the 2,500 km long mountainous Baikal rift zone (Fig.1). Situated in the inner part of Asia this isolated zone has no evident connection with the remaining parts of the World Rift System. At the same time geological age, structural and morphological features and some geophysical characteristics of the Baikal zone do not allow us to consider it a feature, genetically and historically alien to the world rift system. On the contrary, all these features make us consider it one of the elements of this worldwide system. Its continental isolation from this point of view demands particular attention.

The grabens of the Baikal rift zone, as well as the grabens of the other rift zones of the globe, are connected with a vast but distinctly linear uplift with an amplitude to 3,000–4,000 m above sea level. This uplift obtained essential geomorphological expression in Pliocene and Pleistocene. At the same time, as is already well known in the middle, deepest part of the Baikal zone (the Baikal, Tunka, Bargusin depressions), the rift structures inherit the position and extension of earlier Oligocene and Early Neogene basins, which were much flatter and probably more extensive than the later rift depressions. This fact excludes the possibility of the development of the Baikal zone by arching and rift depressions, because in the places of maximum depth of their bottoms sediment accumulation went on more or less uninterruptedly, having begun before the vast warping and consequent mountain building. It would not be correct to date the beginning of rifting as Oligocene or Miocene because the structures and deposits of this age are spread over the vast territory of east Siberia and not only within the boundaries of the Baikal rift zone. However, the biggest thickness of Oligocene–Miocene deposits is in the grabens of the Baikal zone. Since Pliocene the composition of beds of the later Cenozoic Period sharply changes everywhere. Thus it is possible to distinguish two stages in the newest geological history of the Baikal zone: (1) pre-rift stage, i.e., Pliocene and Pleistocene; and (2) rift stage, i.e., Pliocene and Pleistocene.

Some authors believe that the age of the layers which are found in the lower part of the sedimental section of rift structures determines well enough the age of rifting itself. The author cannot agree with this statement. He believes that it is necessary to qualify as rift originated deposits only those, which accumulated in connection with the existing marginal faults, which are responsible for the origin of typical rifts. If not, any basins of sedimentation may be attributed to rift depressions.

On the other hand, geological data do not allow us to represent the Baikal Pliocene and Pleistocene uplift as an indivisible simple flat-topped arch. On the contrary, these data prove that the Sayan–Baikalian arched construction consists of separate arched uplifts, which are large recent ranges. Its structural surface, if we do not take into consideration its denudative transformation, strongly resembles a relatively flat anticlinorium, but not a simple anticlinal arch. The depressions in the body of the complicated Baikalian uplift are older than the uplift itself. This causes difficulties in constructing a general mechanical model of the Baikal rift zone.

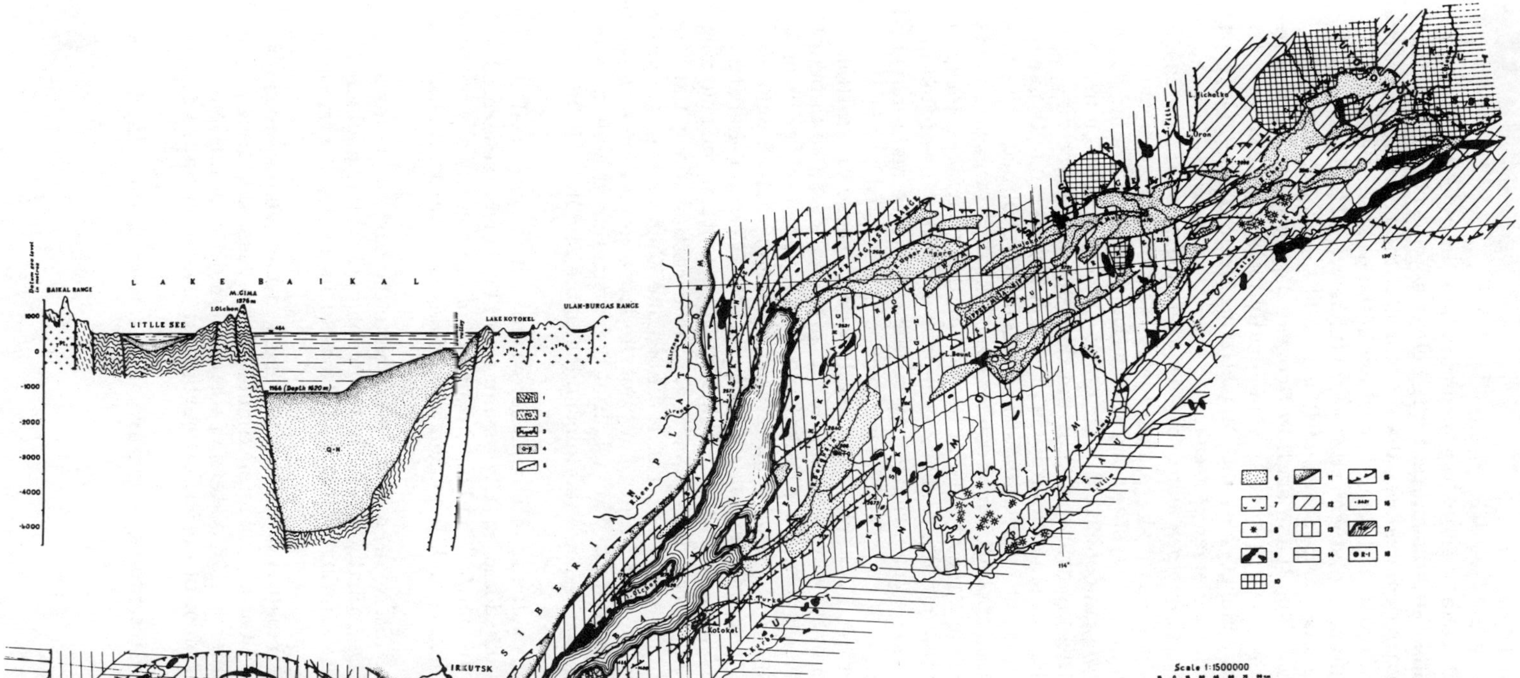

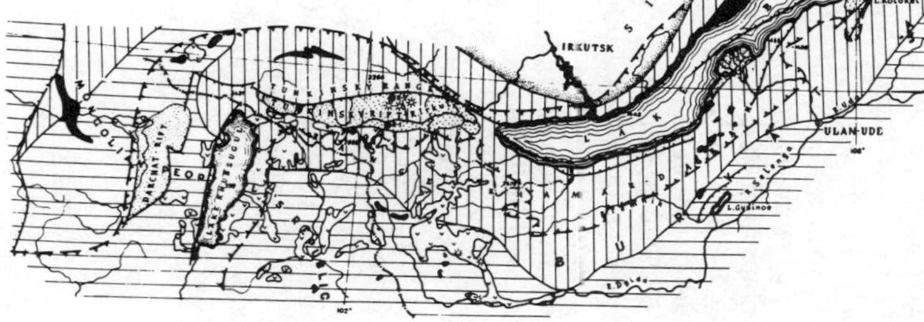

Fig.1. The tectonic scheme of the Baikal rift zone. *1* = Marble, gneiss, hornblende schist; *2* = volcanic sedimentary metamorphic formation; *3* = granitoids; *4* = Neogene–Quaternary formations; *5* = faults; *6* = rift-valleys; *7* = Neogene–Quaternary plateau-basalts; *8* = extinct volcanoes, Pleistocene–Holocene; *9* = ultramafic and basic rocks (Precambrian); *10* = Aldan Shield and Archean tectonic blocks; *11* = edge of the Siberian Platform; *12* = region of the Preriphean folding; *13* = region of the Baikalian folding; *14* = region of the Caledonian folding; *15* = seismic belt, intensity up to M = 6.5; *16* = sea level datum in metres; *17* = depth of the lake in metres; *18* = some deep bore holes. (After V.P. Solonenko and N.A. Florensov.)

pp. 445–450

GEOLOGICAL CHARACTERISTICS

The Baikal rift zone overlies the mountain belt limiting the Siberian Platform on the south. On the southeast this belt is adjoined by the region of Jurassic–Cretaceous depressions, the structure of the region being very much like that of the Basin and Range province in the western United States. In the Miocene and Pleistocene this region underwent but a very slight rejuvenation. On the side of the Siberian Platform at the foot of the mountain belt there are only small and shallow Neogene depressions. This is indicative of the fact that the rift zone is geologically independent of the tectonic development undergone by the neighbouring territories - the trans Baikalian area and mid-Siberian Plateau.

The Baikal rift valleys present two-sided and single-sided grabens. All the big grabens possess a sharp asymmetry - their northern and southwestern slopes are more steep and linear than the southern and southeastern ones. Here we can observe different degrees of expression in the relief of faults limiting the grabens. Only the Holocene faults are directly expressed but they have a small amplitude. The more ancient faults are to a varying extent blotted out by erosion and accumulation, but are, nonetheless, well detected in mapping. The steepness and linearity of escarpments are characteristics indicative of normal faults with their surfaces abruptly inclined (60°–70°) down into the grabens. The larger grabens are found in the central part of the rift zone where they closely adjoin the main marginal fault of the Siberian Platform. The biggest graben is occupied by Lake Baikal whose bottom lies 1,200 m. below sea level and whose crystalline foundation, as shown by geophysical data lies approximately 5,000 m below sea level. The deepest part of Lake Baikal and of the graben lying closer to the southern boundary of the Siberian Platform is surrounded by the relatively low mountains of the Primorsky Ridge and the Kharmar–Daban. The height of the graben floors and the height of the mountains surrounding them, increases from Lake Baikal in the same direction along the rift zone to the west and northeast. The level of Lake Khubsugul in the graben of the same name on the western extremity of the rift zone is almost 1,200 m higher than that of Lake Baikal. The height of the mountains at both ends of the rift zone exceeds the height of the mountains in the middle of it by at least 1,000 m.

The Baikal Depression is of complex structure. It is composed of two large closely adjoined and equally asymmetrical grabens, separated by a narrow discontinuous uplift. The latter consists of the sharply asymmetric tectonic block of Olkhon Island, the underwater Podvodny Academic Sill and the Ushkan Islands.

The Svjatoy Nos (St. Nose) massif, situated to the east, occupies an isolated position in the Baikal graben. In its morphological and geological features this massif is similar to Ruvensori massif in the East-African rift zone.

The present distribution of maximum depths in Lake Baikal does not coincide with the distribution of the maximum thicknesses of sediments which are associated with the shallow waters opposite the mouths of the main rivers flowing into the lake. Most of the Baikal earthquake epicenters are associated with them.

The sediments filling the grabens are unconsolidated, their thickness in big grabens amounts to 2,500–3,000 m, and in the Baikal graben probably to 5,000–6,000 m. They are subdivided into two series - that of the pre-rift stage and that of the rift stage. Both series are represented by the deposits of ancient rivers, lakes and swamps, the upper series showing deposits of glacial origin. The composition and structure of the upper series reflect a fast uplift and intensive erosion of the adjacent mountains. Simultaneously the basalt floors that are very common at the western flank of the rift zone are distributed all over the section of the sedimentary series. The commencement of rifting (Mid-Pliocene) was accompanied by strong volcanic explosions and massive accumulation of basalt tuffs.

[*Editor's Note:* Material has been omitted at this point and only the references cited in this excerpt have been reprinted here.]

REFERENCES

Borisov, A.A., 1965. Deep Structure of the Territory of the U.S.S R., according to Geophysical Data. Nauka, Moscow, 301 pp.

16

Reprinted from *Geol. Soc. London Jour.* **132:**348–349 (1976)

THE BAIKAL RIFT

B. C. King

The Baikal system of rift valleys shows striking analogies with that of East Africa, not least in comparability of scale. It consists of numerous fault-bounded troughs of varying size which are partly infilled with sediments. The Baikal system of rift faulting is essentially Cainozoic in age and is closely associated with zones of contemporary uplift or arching, but its pattern has been largely determined by much earlier grains and trends. Partly for this reason its extent is a matter of somewhat arbitrary definition. The system is generally described as extending for a distance of about 2500 km from north-western Mongolia through the mountains of south-eastern Siberia into Yakutia and the the neighbourhood of River Olekma (*c.* 97°E 50°N to 120°E 57°N), but not dissimilar structures extend westwards towards Tuva and Kazakhstan (Ubsa Nor) and it is increasingly recognized that the system continues eastwards to the Stanovoy range and possibly, indeed, to the sea of Okhotsk (a total distance of approaching 3500 km).

In general terms the rift system follows the complex orogenic belt that flanks the southern margin of the Siberian platform. The sharply convex southern extension of the platform known as the Irkutsk amphitheatre is defined by major dislocations which date back to late Precambrian and are referred to as the 'marginal suture' of the Siberian platform.

The latest formations to enter into the foundations of the Baikal rift system are Jurassic to Cretaceous in age. They are typical continental deposits, sandstones and clays, and are coal-bearing.

In the Baikal region numerous but sporadic areas of Mesozoics occur; they have been disconnected partly by Cainozoic movement, whereby they lie at widely different elevations, but mostly by erosion so they are found as remnants on uplifted belts. Evidence for reconstructing Mesozoic palaeogeography is therefore fragmentary.

The rift structures consist of linear systems of intermontane depressions, essentially graben, along the crestal belts of arched uplifts. The maximum altitudes of mountain ranges are thus very close to the margins of depressions. The depressions are mostly dry, but in some places are occupied by lakes, of which Lake Baikal (670 km long) is by far the largest; they either form direct extensions, one to another, or relay each other 'en echelon'. The rift system comprises some twelve large depressions ranging from about 100 to 700 km in length and up to 15 to 18 km in width, together with very many smaller depressions. All are characteristically floored by Cainozoic sediments and some have intercalated volcanics.

The rift depressions mostly show rather similar morphologies; they are generally asymmetric in cross section with one margin steep and determined by a major dislocation, the other, more gently sloping, with minor faults or merely downwarped; they are thus one-sided graben. Usually the steep margin is to the north or west of the trough; Lake Baikal is a typical example.

The nature and distribution of the Cainozoic formations of the Baikal rift zone are the result of its development and consequently provide evidence for its course of evolution. They occupy separate basins, which are thus genetically distinct, and consist of continental deposits of Oligocene, Miocene, Pliocene and Pleistocene ages. The mamimum thicknesses in the largest basins are from 1500 to 2500 m, as judged from geophysical evidence and deep borings. The thickness of sediments in the Baikal depression appears to be exceptional; several bore-holes in the Selenga delta penetrated Oligocene at around 3000 m, but geophysical data suggest that the maximum thicknesses are 5000 to 5500 m.

In the immediate *pre-rift history of the Baikal region* the lack or paucity of sediments of later Cretaceous and early Cainozoic age reflects conditions of tectonic stability with sub-aerial weathering and erosion.

The *Cainozoic formations of the Baikal rift zone* consist of continental sediments and volcanics. While both are broadly developed on the scale of the Baikal arch as a whole, they show only limited association in detail and largely occur in separate areas. Indeed, the maximum volumes of lavas were erupted at the ends of the rift system and over the basement outside the grabens.

Although the sediments accumulated independently in separate troughs, they show great uniformity of types and sequences, with virtually no local variations. In vertical succession the sediments are divisible into two major units corresponding to two different tectonic environments.

The *lower group* is mainly of shallow lake, swamp and river deposits of Oligocene, Miocene and possibly lower Pliocene age and has thicknesses of from 1000 to 3000 m.

The *upper group* consists of extremely diverse assemblages of sediments of the later Pliocene, Pleistocene and Holocene, with thicknesses of the order of 500 to 1200 m which accumulated under cold climatic conditions. They were evidently produced from mountain relief, the height and contrast of which increased in the course of time, and were strictly confined to the developing intermontane troughs or depressions of the present day.

Vulcanism in the Baikal region is largely independent of the sedimentary troughs and mostly occurs on the uplifted blocks and arches.

In all the volcanic areas lavas greatly predominate; intrusives mainly consist of dykes within the troughs or on uplifts, although sheets and laccoliths occur in the thick sediment–lava sequence in the Tunkinsky depression. Compositions of the volcanics show a limited range and are mostly rather uniform. Olivine basalt and andesite-basalts occur at intervals in the sequences. Minor occurrences of andesite and trachyte are recorded among late effusions (e.g. Udokan). The suites are generally only slightly alkaline, intermediate between tholeiites and alkali olivine basalts.

The Baikal rift system is narrowest and deepest where it swings around Irkutsk amphitheatre and widens both to the east and the west. The height of the arched uplift is also at its minimum on the margins of the south Baikal trough and rises eastwards and westwards; the difference in altitude is 1500 to 2000 m. There is thus a relationship between the dimensions of the rift depressions and those of the arched uplift; the vertical displacements have the same general order of magnitude throughout the rift zone despite the fact that in the central sector the prevailing sense of movement must have been subsidence of the troughs, while on the flanks it was vertical uplift, with the floors of the graben lagging behind.

The continuing development of the Baikal rift system is attested by its high level of seismic activity at the present day. Epicentres of earthquakes correlate closely with the position of known faults.

The gravity field of the rift Ione is sufficiently well known in the region of Lake Baikal itself. Two intense gravity minima are closely localised within the limits of the Baikal trough and are evidently conditioned primarily by the great thicknesses of light sediments deposited within the south Baikal and north Baikal depressions. Quantitative interpretation indicates a thickness of as much as 6000 m in the former and 4500 m in the latter.

17

Reprinted with permission of the author

ODE TO THE BAIKAL RIFT

B. C. King

O Baikalian Rift in the heart of Siberia
you're in no way inferior
To the rifts that developed elsewhere;
You're so far from the oceans,
You show no plate motions
To give you false notions
And yet you're undoubtedly *there*.

You show crustal convulsions, but no
triple junctions,
And your course follows archaic trends;
There's been no occasion
Of mantle invasion
Though, by seismic persuasion,
You yet could make future amends.

O Baikalian Rift, with continent drift,
You could widen with greatest of ease;
A gulf of salt water
Along the right quarter
Is something that ought to
Please Russians as well as Chinese.

October, 1975

Part V

NORTH AMERICA

Editor's Comments on Papers 18 Through 21

For a general picture of rifting in North America, works by King (1969, 1976) can be consulted. For general and regional information about North and South America, as well as Australia and Antarctica, see the *Encyclopedia of World Regional Geology, Part I* (Fairbridge, 1975). Papers 18 through 21 cover only rifts and rift systems lying within the western North American region (Fig. 1-2, in Paper 18).

East of the Rocky Mountains, however, there are also many paleorifts and aulacogens of Proterozoic to Cenozoic age (Burke, 1980), set generally in an intracontinental environment. In particular, there was the Triassic/Early Jurassic rifting associated with the separation of Eurasia and North America, with the formation of the Atlantic type continental margin. This extensional zone has been traced to the northwest as far as the McKenzie delta (Easton, 1983). It included two principal zones of faulting of mid-Mesozoic age. The St. Lawrence rift system has at least four almost identical graben in its western part (which may reflect earlier aulacogens), the eastern part being submerged by the St. Lawrence delta (Kumarapeli, 1978). Most important of all is the belt paralleling the Atlantic coast from Nova Scotia southwestward to the Carolinas and subsurface to the Gulf of Mexico (Philpotts, 1979), which are the type examples of the "taphrogeosynclines," a term used by Kay (1945, 1951). Many are in the form of "half-grabens."

Hoffman, Dewey, and Burke (1974) described Precambrian

aulacogens from Canada and gave criteria for their general recognition. In mid-continent is the Keweenawan rift (Halls, 1978; Green, 1983), and south of this the Oklahoma aulacogens (Gilbert, 1983, who also describes their history). The central North American rift system, which is chiefly a geophysically determined feature, is described by Ocolu and Meyer (1973). A useful account of aulacogens and paleo-rifting of all ages east of the Rocky Mountains is given by Keller et al. (1983).

For the western North American continent, good accounts are to be found in Gilluly (1963) and in Smith and Eaton (1978).

Before turning to the subjects of the reprinted papers—Basin Range province, Rio Grande rift zone, and the San Andreas rift—other structures within the western cordillera may be noted. The most striking is the Rocky Mountain trench (or trenches), a chain of erosional and faulted valleys extending parallel to the Canadian cordillera in British Columbia to Montana for 2500 km, and initiated in the early Tertiary (Leech, 1965; Crickmay, 1964). Northwestward it appears to be continuous with the Tintina valley. Early accounts and opinions are provided by Daly (1912) and Evans (1932). The Republican graben that lies in northeastern Washington may also be part of this system.

The classical account of the history of research on the Basin and Range province is by Nolan (1943), but theories of its origin date from the 1870s. Paper 18, by John Harris Stewart of the U.S. Geological Survey, and Stewart (1971) are modern reviews. An earlier and more detailed account of the geology is provided by Cook (1969). Detailed descriptions and application of dynamics are by Thompson (1972), Sheridan (1978), Eaton (1980), and Effimoff and Pinezich (1981). Many other papers have considered the Great Basin's plate tectonic implications.

The San Andreas fault system has received almost as much attention from American earth scientists as has the Basin and Range province, especially since the 1906 earthquake. Hill (1981) records the "history of concepts" and furnishes a good bibliography. Classical accounts are by Noble (1926), Hill and Diblee (1953), and Crowell (1962, as well as Paper 19). John C. Crowell is with the University of California, Santa Barbara. Regarding Southern California, a historically interesting paper is by Davis (1927).

The southern extension of the San Andreas fault reaches the Gulf of California (Crowell, 1981). Paper 20, (by W. A. Elders at the University of California, Riverside, and colleagues), in excerpts, gives an account of the Salton Sea and the merging of the fault zone with the Gulf of California spreading zone (Larson et al., 1968).

Paper 21, by Charles E. Chapin (of the New Mexico Bureau of

Mines in Socorro) is a summary of the evolution of the Rio Grande rift during late Paleogene times. It lies within the Rockies, east of the Basin and Range province and Colorado Plateau, and it is separated in time and space from the regions discussed in the other papers. Publication 16 in the Bibliography is the symposium of papers of IUCG 1978 conference. Cordell (1978) reviews the geophysics. Golombek, McGill, and Brown (1983) give an up-to-date account of the tectonic evolution and nature of the Rio Grande rift.

REFERENCES

Bridwell, R. J., 1978, The Rio Grande rift and a diapiric mechanism for continental rifting, in *Tectonics and Geophysics of Continental Rifts,* I. B. Ramberg and E.-R. Neumann, eds., D. Reidel, Dordrecht, pp. 73–80.

Burke, K., 1980, Intracontinental rifts and aulacogens, in *Continental Tectonics,* National Research Council, National Academy of Sciences, Washington, D.C., pp. 42–49.

Cook, K. L., 1965, Rift system in the Basin and Range province, in *the World Rift System,* Report of the Upper Mantle Committee Symposium on the World Rift System, T. N. Irvine, ed., Geol. Surv. Canada Paper 66-14, pp. 246–279.

Cook, K. L., 1969, Active rift system in the Basin and Range province, *Tectonophysics* **8:**469–511.

Cordell, L., 1978, Regional geophysical setting of the Rio Grande rift, *Geol.* Soc. America Bull. **89:**1073–1090.

Crickmay, C. H., 1964, The Rocky Mountain Trench: a problem, *Canadian Jour. Earth Sci.* **1:**184–205.

Crowell, J. C., 1962, *Displacement along the San Andreas Fault, California,* Geol. Soc. America Spec. Paper 71, 61p.

Crowell, J. C., 1974, Origin of late Cenozoic basins in southern California, *Soc. Econ. Paleontologists and Mineralogists Spec. Pub.* 22, Tulsa, pp. 190-204.

Crowell, J. C., ed., 1975, *San Andreas Fault in Southern California,* Calif. Div. Mines and Geology Spec. Rept. 118.

Crowell, J. C., ed., 1981, Juncture of San Andreas transform system and Gulf of California rift, *Oceanologica Acta 1981,* Actes 26th Congrès Int. de Géologie, Coll. Géologie des Marges Continentales, Paris, 1980, pp. 137–141.

Daly, R. A., 1912, *Geology of the North American Cordillera at the 49th Parallel,* Canada Geol. Survey Mem. 38.

Davis, W. M., 1925, The Basin and Range problem, *Natl. Acad. Sci. Proc.* **11:**387-392.

Davis, W. M., 1927, The rifts of southern California, *Am. Jour. Sci.,* 5th ser., **13:**55–72.

Decker, E. R., and S. B. Smithson, 1975, Heatflow and gravity interpretation across the Rio Grande Rift in Southern New Mexico and West Texas, *Jour. Geoph. Research.* **80:**2542–2552.

Eardley, A. J., 1951, *Structural Geology of North America,* Harper, New York, 624p.

Easton, R. M., 1983, Crustal structure of rifted continental margins: geological constraints from the Proterozoic ricks of the Canadian shield, *Tectonophysics* **94:**371-390.

Eaton, G. P., 1980, Geophysical and geological characteristics of the crust of the Basin and Range province, in *Continental Tectonics,* National Research Council, National Academy of Sciences, Washington, pp. 96-113.

Effimoff, I., and A. R. Pinezich, 1981, Tertiary structural development of selected valleys based on seismic data: Basin and Range province, northeast Nevada, *Royal Soc. London Philos. Trans.* **A300:**435-442.

Evans, C. S., 1932, Bisco-Dogtooth map area, B. C., in *Canada Geol. Survey Summary Rept., Part AII.,* pp. 106-176.

Fairbridge, R. W., ed., 1975, *The Encyclopedia of World Regional Geology, Part I,* Dowden, Hutchinson, and Ross, Stroudsburg, Pa., 704p.

Gilbert, G. K., 1928, *Studies of Basin and Range Structure,* U.S. Geol. Survey Prof. Paper 153.

Gilbert, M. C. 1983, Timing and chemistry of igneous events associated with the Southern Oklahoma aulacogens, *Tectonophysics* **94:**439-455.

Gilluly, J., 1963, The tectonic evolution of the western United States, *Geol. Soc. London Quart. Jour.* **119:**133-174.

Golombek, M. P., G. E. McGill, and L. Brown, 1983, Tectonic and geologic evolution of the Espanola Basin, Rio Grande rift: Structure and rate of extension, and relation to the state of stress in the western United States, *Tectonophysics* **94:**483-507.

Green, J. C., 1983, Geologic and geochemical evidence for the nature and development of the Middle Proterozoic (Keweenawan) mid-continent rift of North America, *Tectonophysics* **94:**1-10.

Halls, H. C., 1978, The late Precambrian Central North American rift system, in *Tectonics and Geophysics of Continental Rifts,* I. B. Ramberg and E.-R. Neumann, eds., D. Reidel, Dordrecht, pp. 111-123.

Hill, M. L., 1981, San Andreas fault: history of concepts, *Geol. Soc. America Bull.,* Part 1, **92:**112-131.

Hill, M. L., and T. W. Dibblee, 1953, San Andreas, Garlock and Big Pine Faults, California, *Geol. Soc. America Bull.* **64:**443-458.

Hoffman, P., J. F. Dewey, and K. Burke, 1974, Aulacogens and their genetic relation to geosynclines with a Proterozoic example from Great Slave Lake, Canada, in *Soc. Econ. Paleontologists and Mineralogists Spec. Pub.* No. 19, pp. 38-55.

Joestling, H. R., J. E. Case, and L. E. Cordell, 1961, The Rio Grande trough near Albuquerque, New Mexico, *U.S. Geol. Survey Prof. Paper* 424D, pp. 282-286.

Kay, M., 1945, North American geosynclines—their classification (abstract), *Geol. Soc. America Bull.* **56:**1172.

Kay, M., 1951, North American geosynclines, *Geol. Soc. America Mem.* **48:**60, 86-87, 107.

Keller, G. R., E. G. Likiak, W. J. Hinze, and L. W. Braile, 1983, The role of rifting in the tectonic development of the Midcontinent, U.S.A., *Tectonophysics* **94:**391-412.

King, P. B., 1969, Tectonic map of North America, scale 1:5m, 2 sheets, *U.S. Geol. Surv. Prof. Paper* 628, pp. 1-94.

King, P. B., 1976, *Geological Evolution of North America,* Princeton University Press, Princeton, 190p.

Kovack, R. L., ed., 1973, *Tectonic Problems of San Andreas Fault System,* Stanford Univ. Pubs. Geol. Sci. 13.

Kumarapeli, P. S., 1978, The St. Lawrence Paleo-Rift System—a comparative study, in *Tectonics and Geophysics of Continental Rifts,* I. B. Ramberg and E.-R. Neumann, eds., D. Reidel, Dordrecht, pp. 367-386.

Larson, R. L., H. W. Menhard, and S. M. Smith, 1968, Gulf of California: a result of ocean floor spreading and transform faulting, *Science* **161:**781-784.

Leech, G. B., 1965, Rocky Mountain Trench, in *The World Rift System,* Report of the Upper Mantle Committee Symposium on the World Rift System, T. N. Irvine, ed., Geol. Surv. Canada Paper 66-14, pp. 307-329.

Noble, L. F., 1926, The San Andreas rift and some other active faults in the desert region of south-eastern California, *Carnegie Inst. Washington Year Book* 25, pp. 415-428.

Nolan, T. B., 1943, The Basin and Range province in Utah, Nevada and California, *U.S. Geol. Survey Prof. Paper* 197D, pp. 141-196.

Ocolu, L. C., and R. P. Meyer, 1973, The Central North American rift system, *Tectonophysics* **20:**215.

Philpotts, A. R., 1978, Rift-associated igneous activity in eastern north America, in *Petrology and Geochemistry of Continental Rifts,* E.-R. Neumann and I. B. Ramberg, eds., D. Reidel, Dordrecht, pp. 153-154.

Scholz, C. H., 1977, Transform fault systems of California and New Zealand: similarities in their tectonic and seismic styles, *Geol. Soc. London Jour.* **133:**215-228.

Sheridan, M. F., 1978, Owens valley—a major rift between the Sierra Nevada batholith and Basin and Range province, U.S.A., in *Tectonics and Geophysics of Continental Rifts,* I. B. Ramberg and E.-R. Neumann, eds., Reidel, Dordrecht, pp. 81-88.

Smith, R. B., and G. R. Eaton, eds., 1978, Cenozoic tectonics and regional geophysics of the western Cordillera, *Geol. Soc. America Mem.* 152, 370p.

Stewart, J. H., 1971, Basin and range structure. A system of horsts and grabens produced by deep-seated extension, *Geol. Soc. America Bull.* **82:**1019-1044.

Stewart, J. H., 1978, Rift systems in the Western U.S., in *Tectonics and Geophysics of Continental Rifts,* I. B. Ramberg and E.-R. Neumann, eds., D. Reidel, Dordrecht, pp. 89-110.

Thompson, G. A., 1972, Cenozoic Basin Range Tectonism, in *Rept. 24 Int. Geol. Cong. 1972 Sec. 3,* pp. 84-90.

Zoback, M. L., R. E. Anderson, and G. A. Thompson, 1981, Cainozoic evolution of the state of stress and style of tectonism of the Basin and Range province of the Western United States, *Royal Soc. London Philos. Trans.* **A300:**407-434.

18

Reprinted from pages 1-2 and 13-30 of *Geol. Soc. America Mem.* **152:**1-30 (1978)

Basin-range structure in western North America: A review

JOHN H. STEWART
U.S. Geological Survey
345 Middlefield Road
Menlo Park, California 94025

ABSTRACT

For more than 1,500 km along the western Cordillera of North America, late Cenozoic extensional faulting has produced block-faulted basin-range structure characterized by alternating elongate mountain ranges and alluviated basins. The faulting follows older geologic patterns, particularly those of Mesozoic and early Tertiary deformation and of early and middle Tertiary igneous activity. Basin-range structure is commonly inferred to represent either (1) blocks tilted along downward-flattening (listric) faults in which the upslope part of an individual rotated block forms a mountain and the downslope part a valley or (2) alternating downdropped blocks (grabens) that form valleys and relatively upthrown blocks (horsts) that form mountains. Such structure has been produced by extension estimated to be from 10% to 35% of the original width of the province and as much as 100% in specific areas. The province is characterized by anomalous upper mantle, thin crust, high heat flow, and regional uplift.

Current theories on the origin of basin-range structure can be grouped loosely into four main categories. In the first, the structure is presumed to be related to oblique tensional fragmentation within a broad belt of right-lateral movement and distributed extension along the west side of the North American lithospheric plate. This motion was initiated by the collision of the East Pacific Rise with the North American plate, which brought together the North American and Pacific plates to form the right-lateral San Andreas transform fault system. The second theory relates extension to spreading caused by upwelling from the mantle behind an active subduction zone (back-arc spreading). The third theory relates the basin-range structure to spreading that resulted from presumed subduction of the East Pacific Rise beneath part of North America. The fourth theory relates the basin-range structure to plate motion caused by deep-mantle convection in the form of narrow mantle plumes. The combination of anomalous upper mantle, thin crust, high heat flow, regional uplift, and extension with a previous history of high heat generation can best be related to back-arc spreading. The spreading may have been accelerated by slackening of confining pressure after the destruction of the subduction system along western North America and may have been accompanied by right-lateral shear because of the development of the transform western margin of North America.

INTRODUCTION

Basin-range structure consists of block-faulted mountain ranges and intervening alluviated valleys. It is one of the most distinctive geologic features of western North America and is more extensively developed there than in any other part of the world. Knowledge of its origin is critical in understanding the structural development of western North America, as well as of other regions where similar structures occur.

In this report, I will outline the geologic setting and characteristics of basin-range structure and current theories of origin. The subject involves many facets of geology that I can only summarize here. A review of the historical development of ideas concerning basin-range structure has been given by Nolan (1943) and Roberts (1968).

DISTRIBUTION OF BASIN-RANGE STRUCTURE

High-angle extension faulting extends throughout much of the western Cordillera of North America from Canada to northern Mexico (Fig. 1-1). Basin-range structure consisting of alternating mountains and valleys is best developed in the Basin and Range physiographic province (Fig. 1-2) that extends from southern Oregon and Idaho through most of Nevada and parts of California, Utah, Arizona, and New Mexico to northern Mexico—a total distance of more than 1,500 km. In the United States, the province is about 500 to 800 km across in the Great Basin region of Nevada and western Utah. The elevations of the valleys in the Great Basin are generally 1,300 to 1,600 m; mountain crests are commonly 2,000 to 3,000 m and locally about 3,600 m. Elevations of valleys and mountains are commonly 500 to 1,000 m lower in most of southeastern California, southern Arizona, and southwestern New Mexico. In the United States, basin-range structure extends around much of the Colorado Plateau province, a tectonically stable region underlain by relatively undeformed Paleozoic, Mesozoic, and Cenozoic sedimentary rocks, with a general surface elevation of 1,200 to 1,800 m. Extensional faulting along the east margin of the Colorado Plateau and southward to near the international boundary forms the Rio Grande rift valley (Chapin and Seager, 1975). In Mexico, basin-range structure extends southward along either side of the Sierra Madre Occidental, a high plateau (1,800 to 3,000 m) of relatively undeformed Cenozoic volcanic rocks. Elevations west of the Sierra Madre Occidental range from sea level along the coast to 1,000 to 2,000 m in some of the higher mountains inland. East of the Sierra Madre Occidental, elevations are generally higher; they range from about 1,000 to 1,500 m in valleys to 2,000 to 3,000 m in mountain ranges. Basin-range structure also occurs in Mexico along the east side of Baja California, north of lat 28°N. Some extensional faulting and possible basin-range structure also occur in the southern part of the Sierra Madre Occidental and in the so-called Central Mesa (really a physiographic basin) to the west.

[*Editor's Note:* Material has been omitted at this point.]

CHARACTER OF BASIN-RANGE STRUCTURE

Basin-range structure consists of a highly complex system of normal faults along which movement has resulted in the relative uplift of linear segments of the crust to form mountains and the relative sinking of adjacent segments to form valleys (Fig. 1-8). The mountains are usually about 15 to 20 km across and are separated by alluviated valleys of comparable width. In detail, the patterns of mountains and valleys are highly complex (Pl. 1-1, in pocket) and the ranges, although they are generally elongate, are locally equidimensional or elliptical in plan view.

Basin-range structure has generally been related to block faulting, whereby ranges are formed by vertical movements along profound faults on one or both sides of the mountain block. This theory accounts for the gross pattern of mountains and valleys, and for the uniform internal structure of some of the mountain blocks. The concept of block faulting, on the other hand, can imply that the structure is simple, whereas in reality, faulting is not merely confined to the sides of mountains but is distributed throughout the mountain areas and presumably in the suballuvial rocks of the valleys as well. Major structural blocks, therefore, should not be viewed as rigid coherent masses, but rather as aggregate structural units that generally move in a more or less uniform manner relative to adjacent structural units.

Basin-range structure is typically developed in the Great Basin region of Nevada, western Utah, and parts of Oregon, Idaho, California, and Arizona. Here the general distribution of major basin-range faults, the tilting of fault blocks, and the general attitudes of Tertiary rocks are well known (Pl. 1-1). A small-scale map, however, cannot adequately portray the complexities of the structure. Numerous known extension faults have been omitted because they are too small to show. Also omitted are faults that are inferred to bound major blocks but are covered by alluvium and poorly located.

In the Great Basin, basin-range structure is clearly the expression of a system of major blocks, many with significant tilts. In places, the mountains (horsts) and valleys (grabens) are bounded on both sides by normal faults. Tilts of the mountain ranges are generally from a few degrees to about 30°, although higher tilts are known and locally Tertiary units are overturned (Anderson, 1971; Proffett, 1972). Within the Great Basin as a whole, about two-thirds of the tilts are to the east and one-third to the west (Fig. 1-9). Patterns of east- or west-directed tilts are complex, but near the eastern margin of the province, most tilts are to the east; near the western margin, most are to the west (Proffett, 1972; Ekren and others, 1974), although less consistently so. The pattern of tilts near the margins of the Great Basin extends into adjacent provinces; to the east, adjacent parts of the Rocky Mountains and Colorado Plateau are tilted east, and to the west the Sierra Nevada is tilted west; thus, the Great Basin is a central, highly extended part of a giant uplift (Le Conte, 1889). The westward tilts in the western part of the province and eastward tilts in the eastern part contribute to an overall bilateral symmetry of the Great Basin that is also expressed in regional topography and Bouguer gravity, thickness of lithosphere, and pattern of inception of fundamentally basaltic volcanism (G. P. Eaton, 1976, written commun.; Best and Hamblin, this volume).

The amount of valley fill in major valleys in the Great Basin is from a few hundred metres to more than 3,000 m. The structural relief between the lowest bedrock areas under valleys

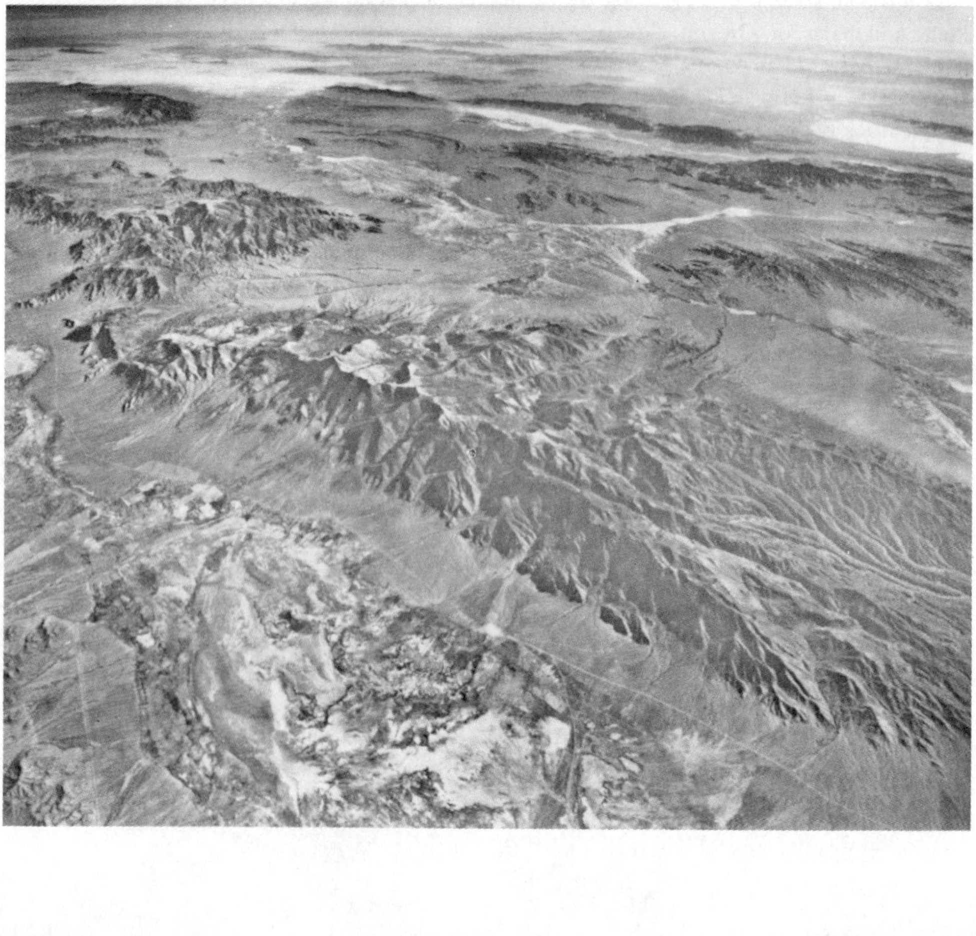

Figure 1-8 (facing page). Aerial photographs showing basin-range structure in central Great Basin. Upper photograph, low-altitude oblique photograph in central Nevada looking northwest across the Toquima Range at the Toiyabe Range in the background. Fault-bounded eastern margin of Toquima Range is well defined. Photograph by the author. Lower photograph, high-altitude oblique aerial photograph looking northeast at Snake Range in easternmost Nevada. Western Utah in background. Block-like character of range and fault-bounded western margin are clearly visible. U.S. Geological Survey-U.S. Air Force photograph.

to the highest adjacent mountains is generally from 2,000 to 5,000 m. Structural relief along the eastern escarpment of the Sierra Nevada in California and the western escarpment of the Wasatch Mountains in Utah is locally about 6,000 m (Christiansen, 1966).

Strike-slip faults are locally an important part of the tectonic framework of the Great Basin. The most conspicuous group of these faults occurs in a northwest-trending belt (the Walker lane) of right-lateral displacement and disrupted structure in the western part of the Great Basin (Albers, 1967). Total right-lateral displacement along this belt is probably 130 to 190 km and occurs in part as fault slip and in part as a more pervasive large-scale drag (Stewart

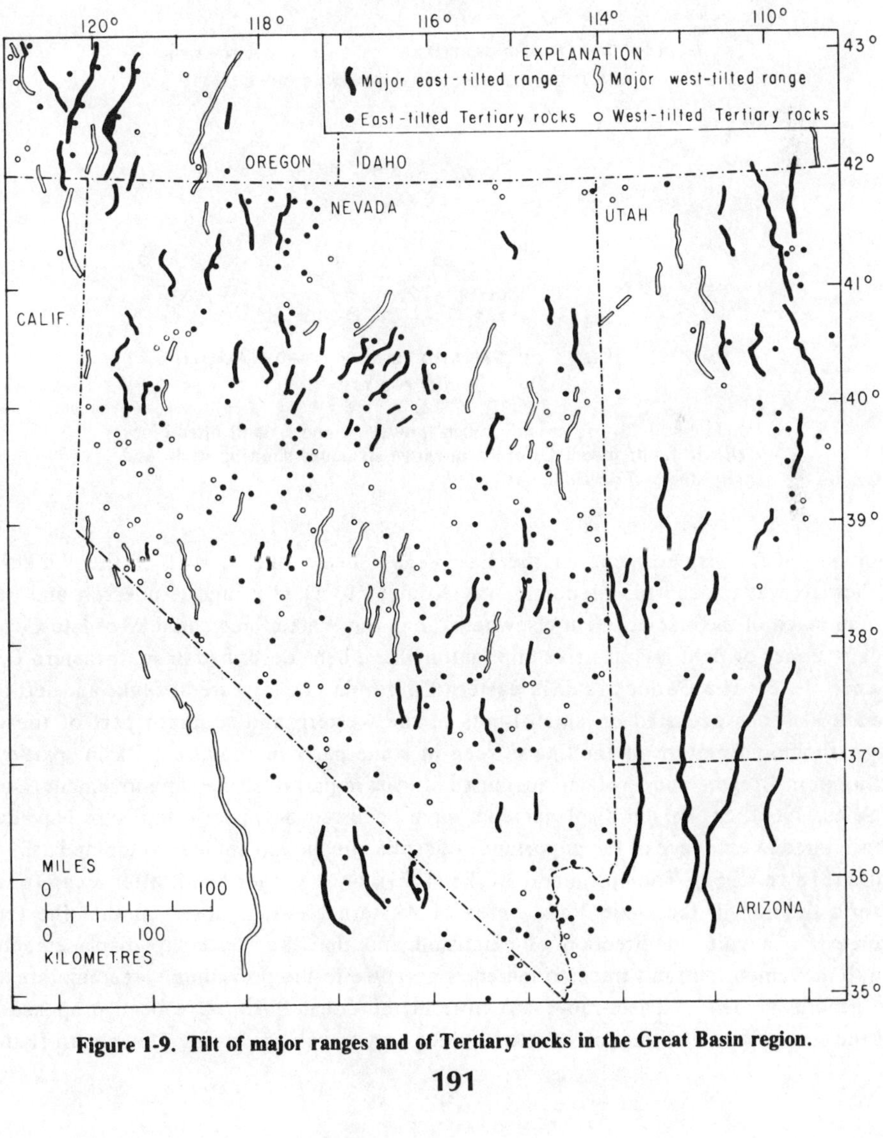

Figure 1-9. Tilt of major ranges and of Tertiary rocks in the Great Basin region.

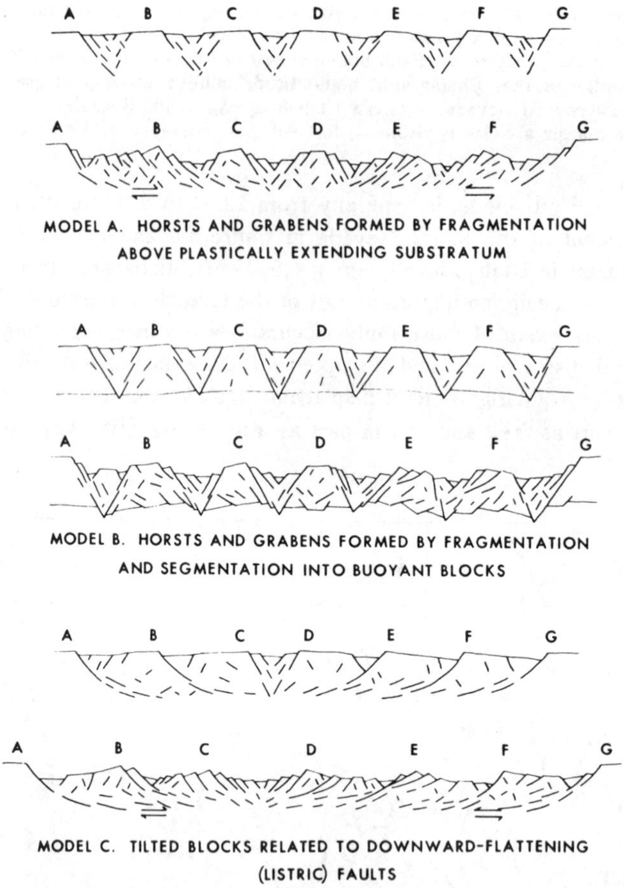

Figure 1-10. Horst-and-graben (models A and B) and tilted-block (listric fault; model C) of basin-range structure showing early and late stages of development.

and others, 1968). Displacement on the Las Vegas shear zone, a part of the Walker lane in southern Nevada, occurred predominantly from 15 to 11 m.y. ago (Anderson and others, 1972), and much of the displacement elsewhere along the Walker lane could be of late Cenozoic age. Major zones of right-lateral strike-slip faulting have been described in southeastern Oregon (Lawrence, 1976). The Garlock fault in eastern California is a left-lateral fault, and left-lateral movement also has occurred on small faults in the western and southern part of the Great Basin. A rhomboid pattern of faulting is seen in some parts of the Great Basin, particularly in southeastern Oregon, and has been attributed at least in part to strike-slip movement (Donath, 1962; Sales, 1966). Strike-slip displacements have occurred on historic faults in Nevada and have been cited as evidence of the importance of strike-slip movement in developing basin-range structure (Shawe, 1965). Thompson and Burke (1973), on the other hand, after a careful study of historic faulting in the Dixie Valley area of western Nevada, observed that the faulting was related to a uniform direction of extension and that the strike-slip displacements are related to movement on fault traces oriented at an angle to the prevailing spreading direction.

Two general models of basin-range structure in the Great Basin have been proposed. One relates the structure to a system of structural blocks rotated along curving, downward-flattening normal faults, sometimes called listric faults (for example, Lowell and others, 1975). The uptilted part of an individual block forms a mountain, and the downtilted part a valley (Fig. 1-10, model C). The second model, the horst-and-graben model, relates the structure to a system of relatively downthrown blocks (grabens) that form valleys, and relatively upthrown blocks (horsts or tilted horsts) that form mountains (Fig. 1-10, models A and B). Evaluating the relative merits of these two models is not easy, and structure in the Basin and Range province probably involves elements of both models.

In a previous article (Stewart, 1971), I discussed the two models of basin-range structure and indicated a preference for the horst-and-graben model. In this model, basin-range structure is produced by the fragmentation of an upper crustal slab over a plastically extending substratum. Extension of the substratum causes the basal part of the slab to be pulled apart along narrow, systematically spaced zones; this, in turn, causes the downdropping of complex horizontal prisms (grabens) of the brittle upper crust. Tilting is produced either by (1) the development of asymmetric grabens in which the mountain block on one side of the graben slumps downward, (2) the rotation of large blocks along downward-flattening faults after the initial formation of the grabens, or (3) rotation of large buoyant blocks due to mass differences. In support of the horst-and-graben model, I outlined geologic and geophysical evidence indicating that many of the valleys in the Great Basin are underlain by grabens and that the alluvial fill in valleys is in many cases much too thick to be accounted for in a simple tilted-block (listric fault) model. In southeastern Oregon, the horst-and-graben structure is particularly well developed, and tilting relatively minor. Structure in this area may represent an initial stage in the development of basin-range structure, before widespread tilting occurs. Studies of first motions of earthquakes and of the distribution of microearthquakes generally indicate high-angle faults (Sbar and others, 1972; Smith and Sbar, 1974) compatible with the horst-and-graben model but not the tilted-block model, in which low-angle faults are predicted at depth.

The horst-and-graben model can be related to the fragmentation of an upper crustal slab above a plastically extending substratum (model A, Fig. 1-10) or to the fragmentation and segmentation of the upper crustal slab into buoyant blocks that float on the substratum (model B, Fig. 1-10; Fig. 1-11). The buoyant block model (Taber, 1927; de Sitter, 1959; Sales, 1976)

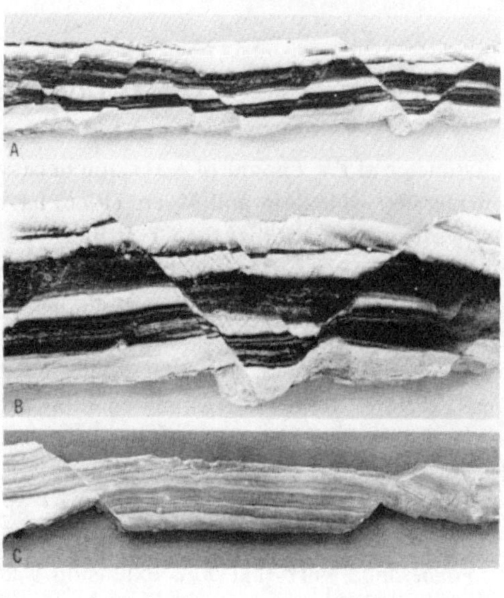

Figure 1-11. Plaster-barite mud models of basin-range structure made by John K. Sales (see Sales, 1976). The photographs were supplied by Sales and are reproduced here with his permission. The models were produced by pouring a layer of molding plaster over a layer of soupy barite drilling mud and then, after the plaster reached the proper consistency, allowing the plaster to extend by backing off a movable piston at one end of the box that holds the model. The plaster is less dense than the barite and floats on it. The photograph shows the plaster after it has hardened. The plaster slab is about 10 cm thick. The now-disrupted markings (representing layering) were originally horizontal.

Photograph A shows widespread high-angle normal faulting. Photograph B is a detail of A showing a keel caused by the sinking of the graben prism. Photograph C shows a graben with elevated marginal blocks that tilt outward. Such geometry is caused by the greater buoyancy of the marginal blocks adjacent to the graben (Taber, 1927; de Sitter, 1959; Sales, 1976). Several grabens in Oregon show this general structure (see Pl. 1-1).

is particularly attractive because it allows for the rotation of large blocks on a scale that could cause entire ranges to tilt. Because of the shape of the large blocks, excess buoyancy and deficient mass is created in blocks adjacent to grabens, thus accounting for elevated margins and the tilt of mountain blocks in opposite directions away from a central graben (Fig. 1-11). Such oppositely directed tilts away from a central graben are observed at several places in southeastern Oregon (Pl. 1-1).

Advocates of the tilted-block (listric fault) model have noted that (1) major mountain blocks in the Great Basin are commonly tilted by as much as 10° to 30° or more and (2) the simplest way to explain this tilting is by rotation of large structural blocks along downward-flattening faults. This concept is supported by surface observations that some faults do decrease in dip downward (Longwell, 1945; Hamblin, 1965; Proffett, 1971, 1977) and that low-angle faults are numerous locally (Anderson, 1971; Wright and Troxel, 1973). In some areas, however, these low-angle faults may have been originally steep, and their present low dip is due to rotation of major structural blocks (Hunt and Mabey, 1966). Additional evidence for the rotation of large structural blocks above downward-flattening faults has been suggested by Moore (1960), who noted that the plan view of many fault traces are crescent-shaped in the Great Basin, indicative of spoon-shaped faults in the subsurface. Crescent-shaped traces are evident only locally on Plate 1-1, however.

The two models of basin-range structure are similar in some respects. Both probably involve the fragmentation of an upper-crustal slab over a plastically extending substratum. The thickness of this slab may be about 12 to 14 km, judged from the greatest depth of most earthquakes in the Great Basin. Both models require substantial internal readjustments along many small faults. These faults may tend to have mostly high angles; new faults may develop as older ones are rotated into low-angle attitudes (Proffett, 1972; Chamberlain, 1976). In both models, tilting of blocks has been related to (1) gravity sliding off regional highs (J. G. Moore, in Wallace, 1964), (2) lateral flow in the crust below the fragmenting slab (Proffett, 1972), or (3) mass differences that cause the rotation of individual blocks.

With present data, the character of basin-range structure at depth cannot be firmly established. A better understanding will come from integrated studies of surface geology, deep seismic profiles, first-motion studies of earthquakes, and the spatial distribution of individual earthquake swarms and aftershocks.

AMOUNT OF EXTENSION

Estimates of the amount of extension necessary to produce basin-range structure have varied considerably. Hamilton and Myers (1966) have suggested between 50 and 100 km of generally east-west extension across the Great Basin, an 8% to 18% increase in the width of the province, in late Cenozoic time on the basis of an average estimate for horizontal extension on each of 25 major range-bounding faults along lat 40°N. They and Elston (1976, 1977) have suggested that extension may have occurred in the early and middle Cenozoic Era as well, a concept not generally accepted, and that total Cenozoic extension in this case could have been as great as 100%. In 1971, I estimated total extension across the Great Basin of about 72 km (13%) in late Cenozoic time on the basis of the "graben rule" of Hansen (1965), in which the cross-sectional area of a graben can be related to the amount of lateral motion (Stewart, 1971). Also in 1971, Proffett, on the basis of detailed geologic work in the Yerington mining district in western Nevada, indicated local extension of more than 100%. He suggested from reconnaissance work that 50% extension was likely in the western and eastern parts of the

Great Basin and that less occurred in the central part. Wright and Troxel (1973) and Wright (1976) indicated 30% to 50% extension in part of the Death Valley area of eastern California, on the basis of detailed mapping. In 1974, Thompson and Burke estimated the total amount of extension across Dixie Valley to be 5 km on the basis of geophysical exploration of the valley and, using this as a typical amount of extension for major grabens, suggested about 100 km of total extension (18% increase in width) across the Great Basin. An extension of 30% to 50% (Thompson, 1972) is required if the thinness of the crust in the Basin and Range province is due entirely to lateral spreading, but as Thompson (1972) noted, this estimate may not be meaningful because of the possibility of phase conversions between crust and mantle.

Variations in estimates of extension are the result, in part, of uncertainties in interpreting basin-range structure at depth. If the structure is related to a system of downward-flattening faults, then the total extension necessary to produce the structure is probably much larger than if it is related to generally steep faults. Even extension related to steep faults, however, may lead to a large amount of spreading. This is illustrated by the Turnagain Heights translatory landslide produced by the 1964 Alaskan earthquake. During this slide, total extension was about 100%, and displacement apparently took place on generally steep faults (Hansen, 1965). The complex surface patterns of elongate ridges and valleys produced during the landslide are similar, except in scale, to those of the Basin and Range province.

The amount of extension can also be studied in relation to the thinning of an upper crustal slab and related regional uplift. These relations (shown in Figs. 1-12 and 1-13, Table 1-1) apply best to models A and C in Figure 1-10 where basin-range structure occurs above a specific level in the crust.

In Figure 1-12, the lower diagram represents in cross section a slab of the upper crust with dimensions *a* and *b*. In the upper diagram, the slab is shown with new dimensions *c* and *d* after extension and the development of basin-range structure. Some erosion of material from the uplifted blocks has taken place, and the material has been deposited in the basins. In the model, all of the material eroded is considered to be trapped in the basins, an assumption approximately true in the Great Basin where drainage is predominantly internal. Thus the total eroded area of the ranges is approximately equal in cross section to the total depositional area of the basins, not counting the relatively small increase in volume as a result of disaggregation

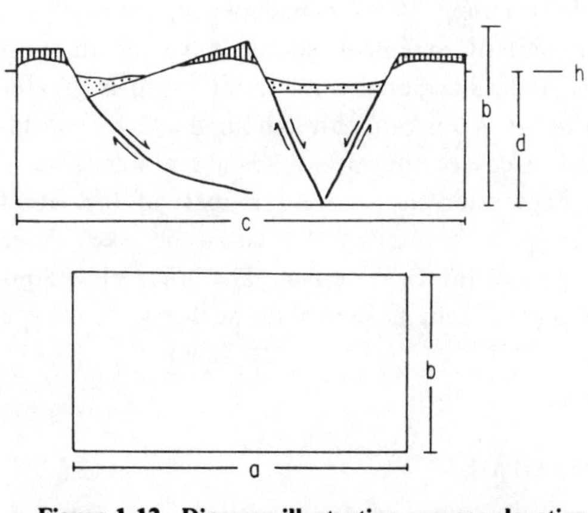

Figure 1-12. Diagram illustrating average elevation (*h*) in relation to basin-range structure. Vertically lined pattern, eroded areas; dotted pattern, depositional areas. Further explanation in text.

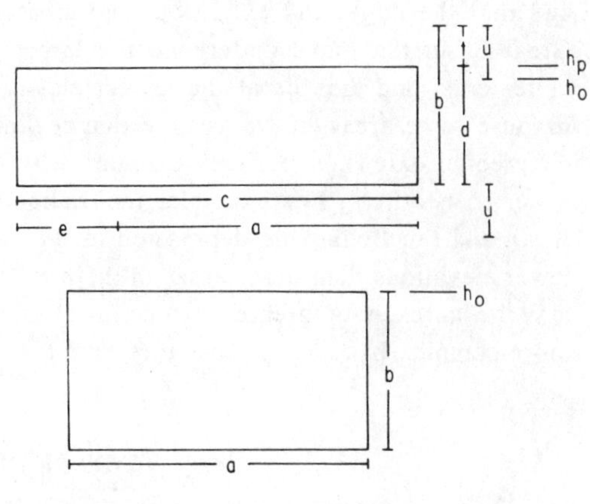

Figure 1-13. Diagram illustrating extension and uplift during development of basin-range structure. Further explanation in text.

TABLE 1-1. RELATION OF EXTENSION TO UPLIFT

Percent crustal extension $(e/a) \times 100\%$ (%)	Amount of crustal extension $e = c - a$ (km)	Thinning of slab $t = b - d$ (m)	Total uplift $u = h_p - h_o + t$ (m)		
			$h_o = 0$ m	$h_o = 700$ m	$h_o = 1{,}200$ m
5	30	639	2,539	1,839	1,339
10	58	1,296	3,196	2,496	1,996
15	83	1,937	3,837	3,137	2,637
20	107	2,610	4,510	3,810	3,310
25	128	3,250	5,150	4,450	3,950
30	148	3,910	5,810	5,110	4,610
35	166	4,553	6,453	5,753	5,253
40	183	5,206	7,106	6,406	5,906

Note: Geometric relations between variables are shown in Figures 1-12 and 1-13; description is in text.

and deposition. If these relations are approximately correct, then the average elevation h in the upper diagram (Fig. 1-12) can be used to calculate the total cross-sectional area of the upper diagram (cd). As no material has been gained or lost from the system, the cross-sectional area of the lower diagram (ab) is equal to that of the upper diagram (cd).

In Figure 1-13, the relations derived from Figure 1-12 are applied to the Great Basin along lat 40°N. Here, the average present elevation h_p is approximately 1,900 m, the width c of the Great Basin is 640 km, and the thickness d of the slab is approximately 13 km (based on average maximum focal depth of earthquakes). The average original elevation h_o of the Great Basin at lat 40°N before development of basin-range structure is poorly known, but probably was between sea level and about 1,200 m from the types of early Tertiary flora in California and Nevada (Axelrod, 1966, Fig. 4). Axelrod (1957) estimated that the average elevation of western Nevada was about 600 to 750 m in Miocene-Pliocene time, also on the basis of fossil flora. These estimates of elevations are uncertain, but nonetheless appear to indicate the extremes that should be considered. The relations between the amount of extension e, the thinning t of the slab, and the amount of uplift u, (see Fig. 1-13) are listed in Table 1-1.

Table 1-1 indicates that extension is accompanied by thinning of the upper crustal slab and that the larger the extension, the greater the thinning. As a consequence, areas of the Great Basin that have undergone the largest amount of extension should have the thinnest upper crust and thus be at the lowest elevations, unless regional uplift of the slab is greater in these same areas. If we assume that regional uplift is uniform throughout the Great Basin, differential extension perhaps explains why some areas of the Great Basin are significantly lower than others. For example, the Lahontan depression in the western part of the Great Basin and the Bonneville depression in the eastern part, both sites of Pleistocene lakes, have lower elevations than other areas in the northern part of the Great Basin. The lower elevations may be a result of greater extension at these places, which seems to be borne out by a more complex basin-range structure here (Stewart, 1971).

TIME OF DEVELOPMENT

Basin-range structure is clearly related predominantly to late Cenozoic tectonic events. In many areas of western North America, extensional faults cut widespread tuffs of early and middle Cenozoic age that formed as highly mobile ash flows, which tend to fill troughs much like water. The occurrence of individual ash-flow units at many different topographic and structural levels can only be explained by faulting.

Two types of evidence have been used to date precisely the development of basin-range structure. The first is indirect and is based on the assumption that the transition from calc-alkalic volcanic rocks in early and middle Cenozoic time to fundamentally basaltic volcanic rocks in late Cenozoic time marks the change from predominantly compressional tectonics (related to a subduction zone) to extensional tectonics (related to wrench faulting, back-arc spreading, or some other factor). The second type of evidence is direct and based on the first appearance of fault-controlled sedimentary basins and topographic forms approximately resembling those seen today.

The volcanic transition in most areas of the Great Basin was about 17 m.y. ago (McKee and others, 1970; McKee, 1971). It was much later than this, however, along the Cascade belt of calc-alkalic volcanism parallel to the western margin of North America in the westernmost Great Basin and the Sierra Nevada of California. Snyder and others (1976) have carefully documented that calc-alkalic volcanism along this belt ended perhaps as much as 20 m.y. ago at lat 35°N and progressively more recently northward to near lat 40°N, where it is still active. In southern Arizona and New Mexico, the transition may have been somewhat older than in the Great Basin. Christiansen and Lipman (1972) indicated that the transition date there was between 23 and 37 m.y. ago; Armstrong and Higgins (1973) suggested a general range of 25 to 30 m.y.; McKee and Noble (1974) suggested about 21 m.y. Several authors (Christiansen and Lipman, 1972; Proffett, 1972; Best and Hamblin, this volume) have suggested a general northward migration of the time of inception of basin-range faulting in the Western United States. Data are not complete enough to generalize about the time of transition from calc-alkalic to basaltic rocks in Mexico, although basaltic rocks in Baja California (Gastil and others, 1975) have the same spread in ages as those in most areas of the Western United States.

Direct evidence of the time of development of basin-range structure is sparse and more difficult to evaluate. In the Great Basin, large sedimentary basins, presumably formed by extensional faulting, were well defined about 11 to 13 m.y. ago (late Barstovian to Clarendonian) (Axelrod, 1957; Robinson and others, 1968; Gilbert and Reynolds, 1973). In a study in western Nevada, Gilbert and Reynolds (1973) described one such basin that was in existence from about 12.5 to 8 m.y. ago, but noted that it was much more extensive than present basins in the area. Sedimentary basins that have a distribution similar to those of today did not develop in that area until approximately 7.5 m.y. ago. In the Nevada Test Site area of southern Nevada, evidence of the late development of basin-range structure has also been described (Ekren and others, 1968). In this area, two systems of faulting have been recognized. The earlier system consists of two sets of faults—one striking northeast and the other northwest—and is developed only in rocks older than 17 m.y. The younger system, a single set, strikes north and cuts 14-m.y.-old tuff that is not cut by the older system of faults. Thus, the north-trending system started to develop sometime between 17 and 14 m.y. ago, but high relief related to this faulting did not develop rapidly. That the area had low relief about 11 m.y. ago is indicated by a widespread tuff of that age that does not vary significantly in thickness between present-day valleys and mountains, a situation that would be impossible if present-day topographic relief had developed by then. By the time another tuff had erupted about 7 m.y. ago, however, the topographic grain was much as it is today. This younger tuff lapped up against some of the ranges and in places flowed into valleys that are the sites of present-day streams.

In summary, the present basins and ranges clearly are a late Cenozoic structure. Extensional

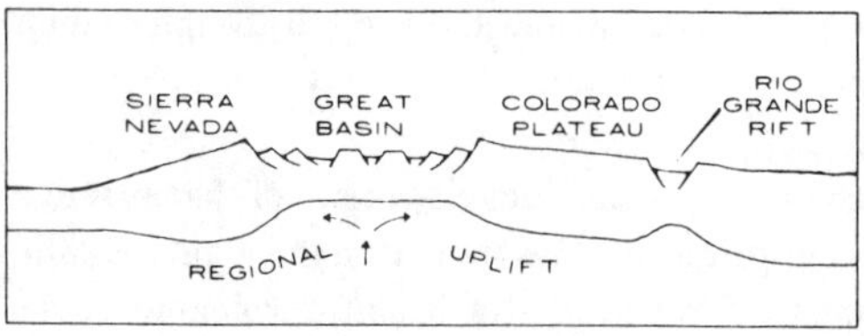

Figure 1-14. Schematic diagram illustrating regional uplift in western North America.

faulting probably started about 17 m.y. ago in much of the Great Basin, but may have started slightly earlier in southern Arizona and New Mexico. The grain of present-day topography, however, probably did not develop before 10 m.y. ago, at the earliest.

REGIONAL UPLIFT

The Basin and Range province lies within a broad region of late Cenozoic uplift (Fig. 1-14). This uplift is documented by mid-Cenozoic floras in the Cordilleran region that indicate elevations at least 1 km below those at present (see discussion and references in Proffett, 1972; Suppe and others, 1975). More precise data on uplift are available west of the Great Basin, where regional uplift of more than 1,800 m in the past 9 m.y. has been described by Christiansen (1966). The Colorado Plateau east of the Great Basin also was uplifted in late Cenozoic time, primarily between 5 and 10 m.y. ago, according to McKee and McKee (1972). Chapin and Seager (1975), however, indicated that the earliest uplift may be as old as 24 m.y. Within the Great Basin, an uplift of 2,000 to 3,000 m in late Cenozoic time seems likely (Table 1-1).

THEORIES OF ORIGIN OF BASIN-RANGE STRUCTURE

Current theories of the origin of basin-range structure can be loosely grouped into four main categories: wrench faulting, back-arc spreading, subduction of the East Pacific Rise, and mantle plumes.

Wrench Faulting

The wrench-faulting concept relates the development of basin-range structure to oblique tensional fragmentation within a broad belt of right-lateral movement along the western side of the North American lithospheric plate. This theory is based on concepts developed by Carey (1958), Wise (1963), Shawe (1965), Hamilton and Myers (1966), Sales (1966), and Slemmons (1967) and has been put in terms of plate-tectonics theory by Atwater (1970) and Christiansen and McKee (this volume). According to this view, western North America is within a broad belt of right-lateral movement related to differential motion between the North American and Pacific plates. Some of the right-lateral movement is taken up on the San Andreas fault and related zones of right-lateral shear, such as the Walker lane in the western Great Basin. The movement is also thought to produce distributed extension and tensional crustal fragmentation (including basin-range structure) along trends oriented obliquely to the trend of the San Andreas fault. This concept is illustrated in Figure 1-15, a map drawn on a Mercator projection about the pole of relative rotation between the Pacific and North American plates (Atwater, 1970). On the map, pure strike-slip motion occurs along horizontal lines, pure tension on vertical lines, and oblique tensional fragmentation on lines having other orientations. Many basin-range faults, as shown on the map, have an oblique orientation consistent with the concept of wrench faulting.

The concept of a broad zone of right-lateral displacement in the Western United States also is consistent with the concept that the broad dextral-curving pattern of Paleozoic eugeosynclinal rocks and of Mesozoic plutonic rocks in the Western United States is due to large-scale oroclinal folding (Mendocino and Idaho oroclines). The oroclinal folding from this point of view is part of the same tectonic pattern that developed basin-range structure (Hamilton and Myers, 1966).

The wrench-faulting concept is particularly attractive when accounting for major zones of strike-slip faulting in the Basin and Range province and for strike-slip first motions of earthquakes in some regions of western North America. The concept is also supported by the contemporaneous development of a transform boundary and of basin-range structure, and the similarity in latitudinal extent (Fig. 1-1) of both structures. The concept does not easily account for regional uplift that has affected the western Cordillera, nor does it entirely explain why shear stress was transmitted over such a broad zone of western North America rather than being confined to the edge of the plate, or why the Basin and Range province should extend (increase in area) instead of simply shearing without an increase in area. These and other factors of plate interactions possibly related to the development of basin-range structure are discussed by Christiansen and McKee (this volume).

Back-Arc Spreading

The back-arc–spreading theory relates basin-range structure to mantle upwelling and associated spreading above a subduction zone (Fig. 1-16). The concept was originally developed to explain the origin of island arcs in the western Pacific that, according to the theory, drifted out from the Asian continent by spreading (back-arc spreading) on the continent side of the arc at the same time that the arc was being underthrust by the Pacific plate (Karig, 1971, 1974; Matsuda and Uyeda, 1971). The theory relates spreading to high heat generated along a Benioff zone by friction and the upwelling of mantle material that causes spreading in the crust and, in the case of the western Pacific, leads to the development of oceanic crust in back-arc

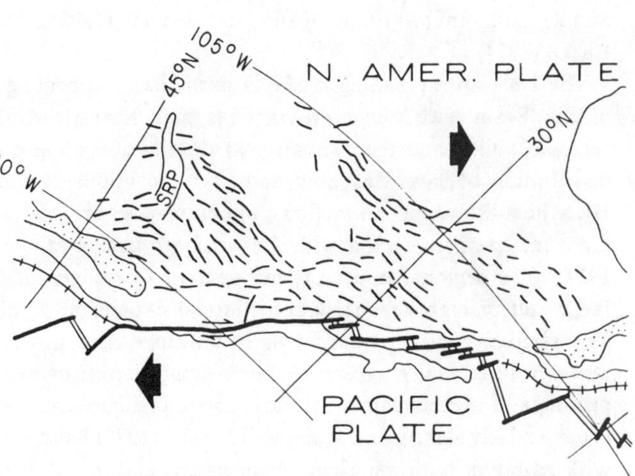

Figure 1-15. Map projected about the pole of relative rotation of Pacific and North American plates, showing relation of extensional faulting to shear in western North America. After Atwater (1970). Double lines, spreading ridges; lines with cross bars, trenches; broad lines, transform faults; narrow, short lines, extensional faults and a few strike-slip faults; dotted areas, late Cenozoic calc-alkalic volcanic rocks; SRP, Snake River Plain.

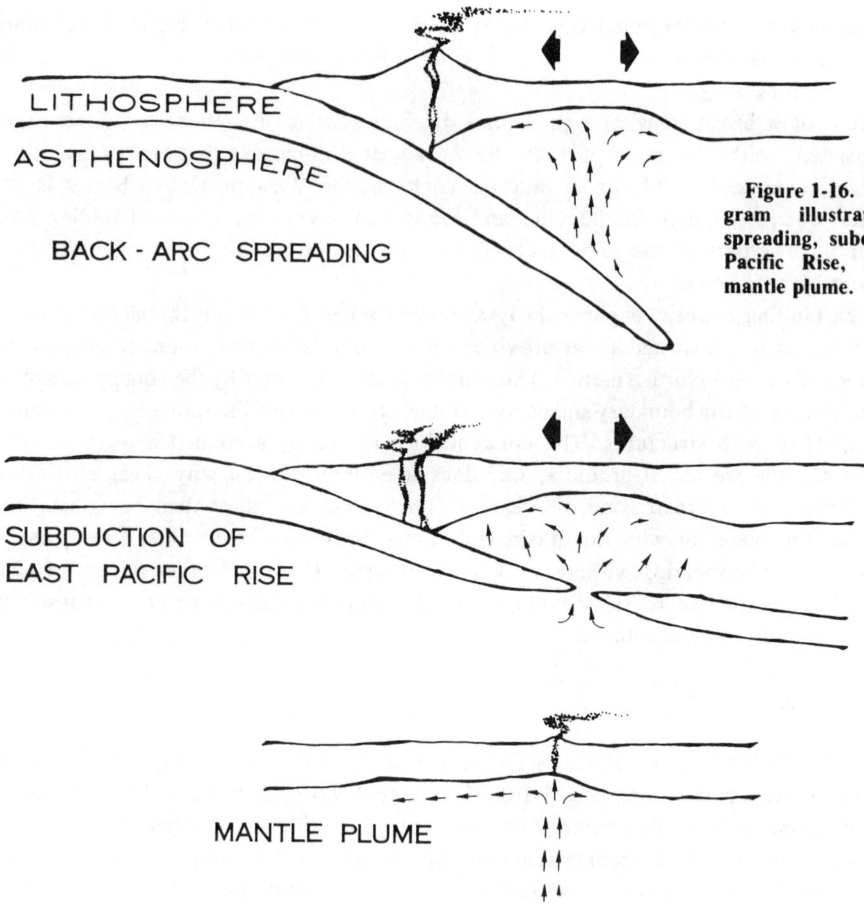

Figure 1-16. Schematic diagram illustrating back-arc spreading, subduction of East Pacific Rise, and action of mantle plume.

basins. The process is slow; calculation by Hasebe and others (1970) suggests that a subduction system would have to be active for 100 m.y. to reach the heat flow observed in the Japan arc system. The back-arc–spreading concept has been applied to western North America by Scholz and others (1971), Karig and Jensky (1972), Thompson (1972), and Thompson and Burke (1974).

The back-arc–spreading model is particularly appealing when explaining such characteristics of the Basin and Range province as high heat flow, thin crust, anomalous upper mantle, regional uplift, and the similarity in distribution of late Cenozoic extensional faulting to the distribution of late Mesozoic and early to middle Cenozoic igneous and tectonic features. High heat flow and anomalous upper mantle are characteristics that the Basin and Range province has in common with the back-arc basins of the western Pacific (Scholz and others, 1971). The regional uplift of the western Cordillera of North America can be explained as the result of high heat flow and thermal expansion or phase changes in the mantle or crust. The similarity of distribution of basin-range structure and Mesozoic and Cenozoic igneous and deformational events indicates a long history of heat generation in the Basin and Range province. The abundance of mid-Cenozoic silicic volcanism in the Basin and Range province is particularly significant. Karig and Jensky (1972) have noted the association of silicic volcanism with rifting in both marginal ocean basins and in volcano-tectonic rift zones. The broad zone of Mesozoic and Cenozoic igneous activity in the Western United States and northern Mexico is anomalous in comparison with other parts of western North and South America; this suggests that somehow the anomalously broad zone and spreading are related. Uyeda and Miyashiro (1974) also noted an association of spreading in the western Pacific with a broad zone of igneous activity.

Why did back-arc spreading, if this concept is correct, start in late Cenozoic time? Perhaps the generation of heat along a subduction system had reached some critical point where mantle upwelling and crustal spreading could start. The spreading could have been triggered by the factors, as yet unknown, that caused an episode of intense volcanic activity in much of the circum-Pacific area during middle Miocene time (McBirney and others, 1974). The voluminous basalts of the Columbia Plateau, themselves indicative of back-arc spreading, were erupted at virtually the same time as this mid-Miocene episode. Scholz and others (1971), on the other hand, have suggested that spreading was accelerated by the cessation of subduction and development of a transform boundary along the western margin of North America. They reasoned that subduction created a confining pressure on western North America and that once this pressure was relieved, the spreading rate would increase. The fact that extention is greatest in regions of North America inland from the transform boundary (Fig. 1-1) supports this concept.

Bott (1973) has outlined how oceanward spreading might occur along a stable Atlantic-type continental margin, and a similar process might take place after termination of compression. Terman (1975) has also suggested that extension is related to lack of confining pressure, although in a somewhat different context than here.

East Pacific Rise

Basin-range structure has been related to convection currents and lateral spreading on the flanks of the East Pacific Rise, which according to this theory, extends under western North America (Menard, 1960; Cook, 1969; McKee, 1971; Gough, 1974). In support of this view, Menard (1960) noted that the East Pacific Rise, if projected northward, would extend under the Great Basin area and that both the Great Basin and the East Pacific Rise are characterized by high heat flow. In addition, Menard pointed out that the ridges and troughs analogous to the basins and ranges also occur on the ocean bottoms on the flanks of the East Pacific Rise.

Present-day plate-tectonics theories discount the possibility that the East Pacific Rise extends under western North America. In Atwater's (1970) model, the upwelling of mantle material along the East Pacific Rise is considered to be a passive response to the separation of plates—not the driving mechanism of plate separation. Upwelling should cease, according to this view, once plate separation ceased because of the development of a transform boundary (Fig. 1-7). Therefore, upwelling related to the East Pacific Rise should not occur beneath western North America. An additional argument against any relation between the Basin and Range province and the East Pacific Rise is the widespread presence of basin-range structures in Mexico, in areas far east of where the ridge system still exists at the mouth of the Gulf of California and locally as spreading centers within the gulf.

A somewhat different concept can be considered if basin-range structure is not related directly to upwelling along the East Pacific Rise but to the presence of hot oceanic lithosphere, a remnant of the Farallon plate, beneath the Great Basin. This lithosphere would be relatively young and therefore hot, because the youngest parts would have been created from upwelling magma immediately prior to the collision of the ridge and trench. Its heat could help drive the spreading in the Great Basin. This concept apparently requires that part of the Farallon plate ceased its descent under North America and remained under the Great Basin. If, on the other hand, the plate did continue to descend, a hole would be left in the plate where no new lithosphere was being created inland of the transform system (Fig. 1-7). If this were true, material from the asthenosphere would presumably ascend into the hole as the plate descended, creating a convection system much like that along a ridge system. Both the "hot plate" and "hole filling" concepts are difficult to apply to Mexico, where basin-range structure lies east of areas where the ridge system still exists.

Mantle Plumes

Deep-mantle convection in the form of narrow rising plumes has been thought to play a major role in the late Cenozoic history of the Western United States (Matthews and Anderson, 1973; Wilson, 1973; Smith and Sbar, 1974; Suppe and others, 1975). According to Morgan (1971, 1972a, 1972b), such mantle plumes provide the driving mechanism for continental drift and cause plates to rift and to be driven apart from one another. The Snake River Plain–Yellowstone area in southern Idaho and northwestern Montana has been suggested (Suppe and others, 1975) as the trace of such a mantle plume. According to this theory, the plume lay under the western part of the Snake River Plain about 15 m.y. ago and has traced a path relatively eastward as the North American plate moved over it. Its present position is under the Yellowstone region of northwest Wyoming. Matthews and Anderson (1973), Smith and Sbar (1974), and Suppe and others (1975) have suggested that this plume may have caused the breakup of a large segment of the Western United States. Others (Eaton and others, 1975) related the volcanism in the Snake River Plain to a major crustal fracture that propagated northeastward, guided by structures in Precambrian rocks.

As of now, mantle plumes similar to those proposed for oceanic areas have not been proved to underlie western North America, nor is the effect of such plumes, if they exist, certain. Seemingly, several widely scattered plumes would be necessary to produce the extensional faulting observed in such a large segment of the Cordillera of western North America. As yet only one area (Yellowstone) can credibly be considered to be underlain by a plume, and even that is a matter of controversy.

CONCLUSIONS

Of the many characteristics of the Basin and Range province that should be considered in evaluating the origin of basin-range structure, the most important appear to be (1) low seismic velocity of the upper mantle, indicative of partial melting; (2) thin crust; (3) high heat flow; (4) regional uplift and extension; (5) previous history of deformation in Mesozoic and earliest Cenozoic time, and of widespread siliceous volcanism in middle Cenozoic time; and (6) position inland from a transform plate boundary.

Considered together, these characteristics generally fit well with the concept of back-arc spreading. In this system, heat generated by friction along a descending slab provides the energy for upwelling of mantle material and the resultant thinning of the crust, regional uplift, and near-surface spreading. High heat generation can be assumed because of long history of subduction along the western margin of North America that led to the widespread emplacement of plutonic rocks in the Mesozoic Era and to the eruption of voluminous volcanic rocks, particularly silicic types, in early and middle Cenozoic time. The back-arc setting is further indicated by the distribution of volcanic rocks in late Cenozoic time (Snyder and others, 1976; Stewart and Carlson, this volume) that clearly shows that an active andesitic magmatic arc existed along the western margin of the Great Basin and adjacent parts of the Sierra Nevada in California at the same time that extensional faulting was occurring farther east.

As proposed by Scholz and others (1971), back-arc spreading may have been accelerated by the slackening of confining pressure owing to the destruction of the subduction system along western North America and the development of a transform boundary. Such a concept explains why basin-range structure is mostly confined to areas inland from the transform boundary of the western margin of North America. In addition, a major component of right-lateral shear related to distributed stress from the transform system can be incorporated in the back-arc model to account for right-lateral motion along the Walker lane and other fault zones in the Western United States.

ACKNOWLEDGMENTS

I am indebted to many geologists for help during the preparation of this report. I am particularly indebted to John K. Sales, who kindly supplied photographs of his plaster-barite models of basin-range structure. My knowledge increased greatly from discussions at the Penrose Conference on Regional Geophysics and Tectonics of the Intermountain West convened by G. P. Eaton, D. R. Mabey, and R. B. Smith. The maps were improved considerably from help by M. G. Best, P. M. Blacet, M. D. Crittenden, Jr., W. A. Duffield, R. K. Hose, F. J. Kleinhampl, and W. J. Moore. The manuscript was improved by thoughtful reviews by D. R. Shawe and G. A. Thompson.

[*Editor's Note:* Only the references cited in the preceding excerpts are reprinted here. The following reference is cited in the text but not included in the list: Stewart, J. H., and Carlson, J. E., 1978, Generalized maps showing distribution, lithology, and age of Cenozoic igneous rocks in the Western United States, *in* Smith, R. B., and Eaton, G. P., eds., Cenozoic tectonics and regional geophysics of the western Cordillera: Geol. Soc. America 152, scale 1:5,000,000 (this volume).]

REFERENCES CITED

Albers, J. P., 1967, Belt of sigmoidal bending and right-lateral faulting in the western Great Basin: Geol. Soc. America Bull., v. 78, p. 143-156.

Anderson, R. E., 1971, Thin skin distension in Tertiary rocks of southeastern Nevada: Geol. Soc. America Bull., v. 82, p. 43-58.

Anderson, R. E., Longwell, C. R., Armstrong, R. L., and Marvin, R. F., 1972, Significance of K-Ar ages of Tertiary rocks from the Lake Mead region, Nevada-Arizona: Geol. Soc. America Bull., v. 83, p. 273-288.

Armstrong, R. L., and Higgins, R. E., 1973, K-Ar dating of the beginning of Tertiary volcanism in the Mojave Desert, California: Geol. Soc. America Bull., v. 84, p. 1095-1100.

Atwater, Tanya, 1970, Implications of plate tectonics for the Cenozoic tectonic evolution of western North America: Geol. Soc. America Bull., v. 81, p. 3513-3535.

Axelrod, D. I., 1957, Late Tertiary floras and the Sierra Nevadan uplift [California-Nevada]: Geol. Soc. America Bull., v. 68, p. 19-45.

——1966, The Eocene Copper Basin flora of northeastern Nevada: California Univ. Pubs. Geol. Sci., v. 59, 125 p.

Best, M. G., and Hamblin, W. K., 1978, Origin of the northern Basin and Range province: Implications from the geology of its eastern boundary, *in* Smith, R. B., and Eaton, G. P., eds., Cenozoic tectonics and regional geophysics of the western Cordillera: Geol. Soc. America Mem. 152 (this volume).

Bott, M.H.P., 1973, Shelf subsidence in relation to the evolution of young continental margins, *in* Tarling, D. H., and Runcorn, S. K., eds., Implications of continental drift to the earth sciences, Vol. 2: New York, Academic Press, p. 675–683.

Carey, S. W., 1958, The tectonic approach to continental drift, *in* Carey, S. W., ed., Continental drift, a symposium: Hobart, Tasmania, Tasmania Univ. Geology Dept., p. 177–355.

Chamberlain, R. M., 1976, Rotated early-rift faults and fault blocks, Lemitar Mountains, Socorro County, New Mexico: Geol. Soc. America Abs. with Programs, v. 8, p. 807.

Chapin, C. E., and Seager, W. R., 1975, Evolution of the Rio Grande rift in the Socorro and Las Cruces areas, *in* Seager, W. R., Clemons, R. E., and Callender, J. F., eds., Guidebook of the Las Cruces country, 1975: New Mexico Geol. Soc. Guidebook 26, p. 297–321.

Christiansen, M. N., 1966, Late Cenozoic crustal movements in the Sierra Nevada of California: Geol. Soc. America Bull., v. 77, p. 163–181.

Christiansen, R. L., and Lipman, P. W., 1972, Cenozoic volcanism and plate-tectonic evolution of the Western United Sates. II, Late Cenozoic: Royal Soc. London Philos. Trans., v. 271, p. 249–284.

Christiansen, R. L., and McKee, E. H., 1978, Late Cenozoic volcanic and tectonic evolution of the Great Basin and Columbia intermontane region, *in* Smith, R. B., and Eaton, G. P., eds., Cenozoic tectonics and regional geophysics of the western Cordillera: Geol. Soc. America Mem. 152 (this volume).

Cook, K. L., 1969, Active rift system in the Basin and Range province, *in* The world rift system—Internat. Upper Mantle Comm., Upper Mantle Project sci. rept. 19: Tectonophysics, v. 8, p. 469–511.

Donath, F. A., 1962, Analysis of basin-range structure, south-central Oregon: Geol. Soc. America Bull., v. 73, p. 1–15.

Eaton, G. P., Christiansen, R. L., Iyer, H. M., Pitt, A. M., Mabey, D. R., Blank, H. R., Jr., Zietz, Isidore, and Gettings, M. E., 1975, Magma beneath Yellowstone National Park: Science, v. 188, p. 787–796.

Ekren, E. B., Rogers, C. L., Anderson, R. E., and Orkild, P. P., 1968, Age of basin and range normal faults in Nevada Test Site and Nellis Air Force Range, Nevada, *in* Nevada Test Site: Geol. Soc. America Mem. 110, p. 247–250.

Ekren, E. B., Bath, G. D., Dixon, G. L., Healey, D. L., and Quinlivan, W. D., 1974, Tertiary history of Little Fish Lake Valley, Nye County, Nevada, and implications as to the origin of the Great Basin: U.S. Geol. Survey Jour. Research, v. 2, p. 105–118.

Elston, W. E., 1976, Tectonic significance of mid-Tertiary volcanism in the Basin and Range province: A critical review with special reference to New Mexico, *in* Elston, W. E., and Northrop, S. A., eds., Cenozoic volcanism in southwestern New Mexico: New Mexico Geol. Soc. Spec. Pub. 5, p. 93–151.

——1977, Rifting and volcanism in the New Mexico segment of the Basin and Range province, southwestern USA, *in* Larson, B. T., and Neumann, E. R., eds., NATO Advanced Study Institute on Paleorift Systems, Oslo, Norway, Proc.: Dordrecht, Holland, D. Reidel (in press).

Gastil, R. G., Phillips, R. P., and Allison, E. C., 1975, Reconnaissance geology of the state of Baja California: Geol. Soc. America Mem. 140, 170 p.

Gilbert, C. M., and Reynolds, M. W., 1973, Character and chronology of basin development, western margin of the Basin and Range province: Geol. Soc. America Bull., v. 84, p. 2489–2509.

Gough, D. I., 1974, Electrical conductivity under western North America in relation to heat flow, seismology, and structure: Jour. Geomagnetism and Geoelectricity, v. 26, p. 105–123.

Hamblin, W. K., 1965, Origin of "reverse drag" on the downthrown side of normal faults: Geol. Soc. America Bull., v. 76, p. 1145–1164.

Hamilton, Warren, and Myers, W. B., 1966, Cenozoic tectonics of the Western United States: Rev. Geophysics, v. 5, p. 509–549.

Hansen, W. R., 1965, Effects of the earthquake of March 27, 1964, at Anchorage, Alaska: U.S. Geol. Survey Prof. Paper 542-A, p.A1–A68.

Hasebe, K., Fujii, N., and Uyeda, S., 1970, Thermal process under island arcs: Tectonophysics, v. 10, p. 335–355.

Hunt, C. B., and Mabey, D. R., 1966, General geology of Death Valley, California—Stratigraphy and structure: U.S. Geol. Survey Prof. Paper 494-A, p. A1–A165.

Karig, D. E., 1971, Origin and development of marginal basins in the western Pacific: Jour. Geophys. Research, v. 76, p. 2542–2561.

——1974, Evolution of arc systems in the western

Pacific: Earth and Planetary Sci. Ann. Rev., v. 2, p. 51–75.

Karig, D. E., and Jensky, Wallace, 1972, The proto-Gulf of California: Earth and Planetary Sci. Letters, v. 17, p. 169–174.

Lawrence, R. D., 1976, Strike slip faulting terminates the Basin and Range province in Oregon: Geol. Soc. America Bull., v. 87, p. 846–850.

Le Conte, Joseph, 1889, On the origin of normal faults and of the structure of the Basin region: Am. Jour. Sci., 3rd ser., v. 38, p. 257–263.

Longwell, C. R., 1945, Low-angle normal faults in the Basin-and-Range province: Am. Geophys. Union Trans., v. 26, pt. 1, p. 107–118.

Lowell, J. D., Genik, G. J., Nelson, T. H., and Tucker, P. M., 1975, Petroleum and plate tectonics of the southern Red Sea, *in* Fischer, A. G., and Judson, Sheldon, eds., Petroleum and global tectonics: Princeton, N.J., Princeton Univ. Press, p. 129–153.

Matthews, Vincent, III, and Anderson, C. E., 1973, Yellowstone convection plume and break-up of the Western United States: Nature, v. 243, p. 158–159.

Matsuda, T., and Uyeda, S., 1971, On the Pacific-type orogeny and its model-extension of the paired belts concept and possible origin of marginal seas: Tectonophysics, v. 11, p. 5–27.

McBirney, A. R., Sutter, J. F., Naslund, H. R., Sutton, K. G., and White, C. M., 1974, Episodic volcanism in the central Oregon Cascade Range: Geology, v. 2, p. 585–589.

McKee, E. D., and McKee, E. H., 1972, Pliocene uplift of the Grand Canyon region; Time of drainage adjustment: Geol. Soc. America Bull., v. 83, p. 1923–1931.

McKee, E. H., 1971, Tertiary igneous chronology of the Great Basin of Western United States—Implications for tectonic models: Geol. Soc. America Bull., v. 82, p. 3497–3502.

McKee, E. H., and Noble, D. C., 1974, Timing of late Cenozoic crustal extension in the Western United States: Geol. Soc. America Abs. with Programs, v. 6, p. 218.

McKee, E. H., Noble, D. C., and Silberman, M. L., 1970, Middle Miocene hiatus in volcanic activity in the Great Basin area of the Western United States: Earth and Planetary Sci. Letters, v. 8, p. 93–96.

Menard, H. W., Jr., 1960, The East Pacific Rise: Science, v. 132, p. 1737–1746.

Moore, J. G., 1960, Curvature of normal faults in the Basin and Range province of the Western United States, *in* Geological Survey research 1960: U.S. Geol. Survey Prof. Paper 400-B, p. B409–B411.

Morgan, W. J., 1971, Convective plumes in the lower mantle: Nature, v. 230, p. 42–43.

——1972a, Deep mantle convection plumes and plate motions: Am. Assoc. Petroleum Geologists Bull., v. 56, p. 203–213.

——1972b, Plate motion and deep mantle convection, *in* Shagam, R., and others, eds., Studies in Earth and space science: Geol. Soc. America Mem. 132, p. 7–22.

Nolan, T. B., 1943, The Basin and Range province in Utah, Nevada, and California: U.S. Geol. Survey Prof. Paper 197-D, p. 141–196.

Proffett, J. M., Jr., 1971, Late Cenozoic structure in the Yerington district, Nevada, and the origin of the Great Basin: Geol. Soc. America Abs. with Programs, v. 3, p. 181.

——1972, Nature, age, and origin of Cenozoic faulting and volcanism in the Basin and Range province (with special reference to the Yerington district, Nevada [Ph.D. thesis]: Berkeley, Univ. California, 77 p.

——1977, Cenozoic geology of the Yerington district, Nevada, and implications for the nature and origin of basin and range faulting: Geol. Soc. America Bull., v. 88, p. 247–266.

Roberts, R. J., 1968, Tectonic framework of the Great Basin, *in* A coast to coast tectonic study of the United States: UMR Jour., no. 1 (V. H. McNutt–Geology Dept. Colloquim Ser. 1, Univ. Missouri, Rolla) p. 101–119.

Robinson, P. T., McKee, E. H., and Moiola, R. J., 1968, Cenozoic volcanism and sedimentation, Silver Peak region, western Nevada and adjacent California, *in* Coats, R. R., Hay, R. L., and Anderson, C. A., eds., Studies in volcanology—A memoir in honor of Howel Williams: Geol. Soc. America Mem. 116, p. 577–611.

Sales, J. K., 1966, Structural analysis of the Basin and Range province in terms of wrench faulting [Ph.D. thesis]: Reno, Nevada Univ., 178 p.

——1976, Model studies of continental rifting: Geol. Soc. America Abs. with Programs, v. 8, p. 1083.

Sbar, M. L., Barazangi, M., Dorman, J., Scholz, C. H., and Smith, R. B., 1972, Tectonics of the Intermountain seismic belt, Western United States: Microearthquake seismicity and composite fault plane solutions: Geol. Soc. America Bull., v. 83, p. 13–20.

Scholz, C. H., Barazangi, M., and Sbar, M. L., 1971, Late Cenozoic evolution of the Great

Basin, Western United States, as an ensialic interarc basin: Geol. Soc. America Bull., v. 82, p. 2979–2990.

Shawe, D. R., 1965, Strike-slip control of basin-range structure indicated by historical faults in western Nevada: Geol. Soc. America Bull., v. 76, p. 1361–1378.

Sitter, L. U. de, 1959, Structural geology: New York, McGraw-Hill Book Co., 552 p.

Slemmons, D. B., 1967, Pliocene and Quaternary crustal movements of the Basin and Range province, USA, *in* Sea level changes and crustal movements of the Pacific—Pacific Sci. Cong., 11th, Tokyo, 1966, Symposium 19: Osaka City Univ. Jour. Geosciences, v. 10, p. 91–103.

Smith, R. B., and Sbar, M. L., 1974, Contemporary tectonics and seismicity of the Western United States with emphasis on the Intermountain seismic belt: Geol. Soc. America Bull., v. 85, p. 1205–1218.

Snyder, W. S., Dickinson, W. R., and Silberman, M. L., 1976, Tectonic implications of space-time patterns of Cenozoic magmatism in the Western United States: Earth and Planetary Sci. Letters, v. 32, p. 91–106.

Stewart, J. H., 1971, Basin and Range structure: A system of horsts and grabens produced by deep-seated extension: Geol. Soc. America Bull., v. 82, p. 1019–1044.

Stewart, J. H., Albers, J. P., and Poole, F. G., 1968, Summary of regional evidence for right-lateral displacement in the western Great Basin: Geol. Soc. America Bull., v. 79, p. 1407–1413.

Suppe, John, Powell, Christine, and Berry, Robert, 1975, Regional topography, seismicity, Quaternary volcanism, and the present-day tectonics of the western United States: Am. Jour. Sci., v. 275A, p. 397–436.

Taber, Stephen, 1927, Fault troughs: Jour. Geology, v. 35, p. 577–606.

Terman, M. J., 1975, Plate-tectonic implications of crustal extension models in east Asia and North America: Geol. Soc. America Abs. with Programs, v. 7, p. 1295.

Thompson, G. A., 1972, Cenozoic basin range tectonism in relation to deep structure: Internat. Geol. Cong., 24th, Montreal 1972, Proc., p. 84–90.

Thompson, G. A., and Burke, D. B., 1973, Rate and direction of spreading in Dixie Valley, Basin and Range province, Nevada: Geol. Soc. America Bull., v. 84, p. 627–632.

——1974, Regional geophysics of the Basin and Range province: Earth and Planetary Sci. Ann. Rev., v. 2, p. 213–238.

Uyeda, Seiya, and Miyashiro, Akiho, 1974, Plate tectonics and the Japanese Islands: A synthesis: Geol. Soc. America Bull., v. 85, p. 1159–1170.

Wallace, R. E., 1964, Structural evolution, *in* Mineral and water resources of Nevada: U.S. 88th Cong., 2nd sess., Senate Doc. 87, p. 32–39; Nevada Bur. Mines Bull. 65, 314 p.

Wilson, J. T., 1973, Mantle plumes and plate motions: Tectonophysics, v. 19, p. 149–164.

Wise, D. U., 1963, An outrageous hypothesis for the tectonic pattern of the North American Cordillera: Geol. Soc. America Bull., v. 74, p. 357–362.

Wright, L. A., 1976, Late Cenozoic fault patterns and stress fields in the Great Basin and western displacement of the Sierra Nevada block: Geology, v. 4, p. 489–494.

Wright, L. A., and Troxel, B. W., 1973, Shallow-fault interpretation of Basin and Range structure, southwestern Great Basin, *in* de Jong, K. A., and Scholten, Robert, eds., Gravity and tectonics: New York, John Wiley and Sons, Inc., p. 397–407.

Reprinted from *Geol. Soc. London Jour.* **136:**293–302 (1979)

The San Andreas fault system through time

J. C. Crowell

SUMMARY: The active San Andreas fault system today lies at the splintered boundary between the Pacific and North American lithospheric plates, a tectonic arrangement that originated in California in the Oligocene. By late Miocene time sedimentary breccias derived from San Andreas scarps were being shifted laterally from their source areas. During the Pliocene the system extended from the Mendocino triple-junction to the head of the Gulf of California. The Big Bend in the San Andreas developed and the Transverse Ranges rose, in association with the opening of the Gulf of California.

The boundary between the Pacific and North American plates is wide and fragmented and during the late Cenozoic, the crust has yielded laterally by simple shear and deformed through pull-apart stretching and fault-block squeezing and uplift and rotation. Splay faults, such as the San Jacinto, Calaveras, and Hayward, branch from the main fault near constraining bends. The strewn out arrangement of displaced slices may account for displacements of 600 km on the northern San Andreas and only 300 km on the southern across a wide mobile boundary zone that has a total displacement exceeding 1000 km. Strike slip may also have occurred along faults parallel to the system in Cretaceous–Palaeocene times when the Kula Plate was moving laterally northwestward with respect to the North American Plate.

The western margin of North America has been the site of interactions between lithospheric plates as far back into geological time as we are able to decipher the record. These interactions have included convergent, divergent, and transform plate movements so that older terrains have been subjected to repeated strong deformation of markedly differing styles. The purpose of this paper is to describe briefly the complexity in structural style and geological history in terrains along the plate boundaries. One tectonic style after another has been overprinted upon already deformed rocks, across a broad boundary zone extending well into each of the impinging plates. Another purpose is to review the history of the San Andreas fault system. Largely by examining successively older rocks, and especially sedimentary facies, events as far back as 165 Ma for which a fairly complete record is preserved, are being disentangled and a tectonic history is being elucidated by many investigators.

Plate tectonic review

Interpretations of plate interactions across the California margin have been made by Atwater (1970) and Atwater & Molnar (1973). These studies, simplified here (Fig. 1), have been based mainly on the interpretation of ocean-floor magnetic anomalies, tied together on a global scale without direct correlation of the North American Plate with the succession of plates on the W. According to these studies and inferences from geological relations, in the late Cretaceous and early Tertiary (Fig. 1A, B, for 80 and 60 Ma) a spreading ridge between the Kula and Farallon plates probably impinged against the margin of the North American Plate. The inferred direction of relative motion between the combined two western plates and the North American Plate was largely transform with major strike-slip faults in the plate boundary zone. When the Kula Plate and its juncture with the Farallon Plate passed northwestward out of the scene, a long interval of oblique convergence ensued. This interval for California lasted from *c.* 55 Ma (the beginning of the Eocene) until *c.* 38 Ma (early Oligocene), and was followed by the impingement of the Pacific Plate against the North American Plate. Two triple junctions then migrated along the margin of the North American Plate: a trench-transform-transform junction on the N and a transform-trench-ridge junction on the S (Dickinson *et al.* 1972). The northern triple junction moved slowly, and probably irregularly, northwestward relative to the North American Plate, and the southern one southeastward, perhaps more rapidly. Complexities at these triple junctions are considerable. Here, near the spreading ridge the plates are thin and plate margins are irregular. The margins do not form simple arcs with respect to plate poles of rotation so that segments are easily sliced off. Moreover, the crust at the edge of the continent is heterogeneous mechanically and flawed with many pre-existing faults. For example, the geology at the present-day triple junction near Cap Mendocino (Fig. 2, CM), a trench-transform-transform junction, is especially complex and has not yet been geologically mapped satisfactorily. In this region the terrain is severely sheared: late Mesozoic strata that were at first pervasively deformed during subduction predominate, and have since been overprinted by strong transform tectonics.

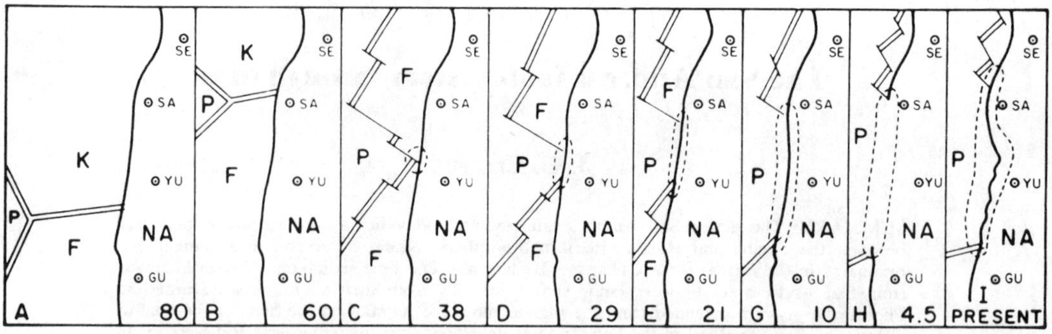

FIG. 1. Schematic diagrams of lithospheric plate arrangements along coast of western North America. Age in Ma given at lower right of each sketch. P = Pacific Plate, K = Kula Plate, F = Farallon Plate, NA = North American Plate, SE = Seattle, SA = Sacramento, YU = Yuma, GU = Guadalajara. Dashed line encloses region of plate boundary tectonic complexity. Adapted from Atwater (1970), Blake *et al.* (1978), and Nilsen (1977*a*).

Fig. 2. (*facing*)

FIG 2. Simplified fault map of coastal California and northwesternmost Mexico. CM = Cape Mendocino, SC = Shelter Cove, PA = Point Arena, GV = Great Valley, BA = San Francisco Bay area, SN = Sierra Nevada, SJ = San Juan Bautista Bend, SB = Salinian Block, SM = Santa Maria Basin, BB = Big Bend of San Andreas Fault, SB = Ventura Basin, MB = Mojave Block, CI = Channel Islands, TR = Transverse Ranges, LA = Los Angeles Basin, GB = San Gorgonio Bend, CO = Southern California offshore borderland, ET = Eastern Transverse Ranges, PR = Peninsular Ranges, ST = Salton Trough, BP = Baja California Peninsula, GC = Gulf of California, BO = Baja California offshore borderland, SO = Sonora. *Faults*: 1 = San Andreas, 2 = Mendocino fracture zone, 3 = Oregon subduction zone, 4 = Hayward, 5 = Calaveras, 6 = San Gregorio, 7 = Hosgri, 8 = Rinconada, 9 = Red Hills–San Juan–Chimeneas, 10 = Nacimiento, 11 = Big Pine, 12 = White Wolf–Kern, 13 = Garlock, 14 = Santa Ynez, 15 = San Gabriel, 16 = Santa Monica–Malibu Coastal, 17 = East Santa Cruz Basin, 18 = Newport–Inglewood, 19 = Elsinore, 20 = Pinto, 21 = San Jacinto, 22 = Salton Creek, 23 = Agua Blanca, 24 = Sonoran faults, 25 = A Gulf of California transform fault. Base from King (1969).

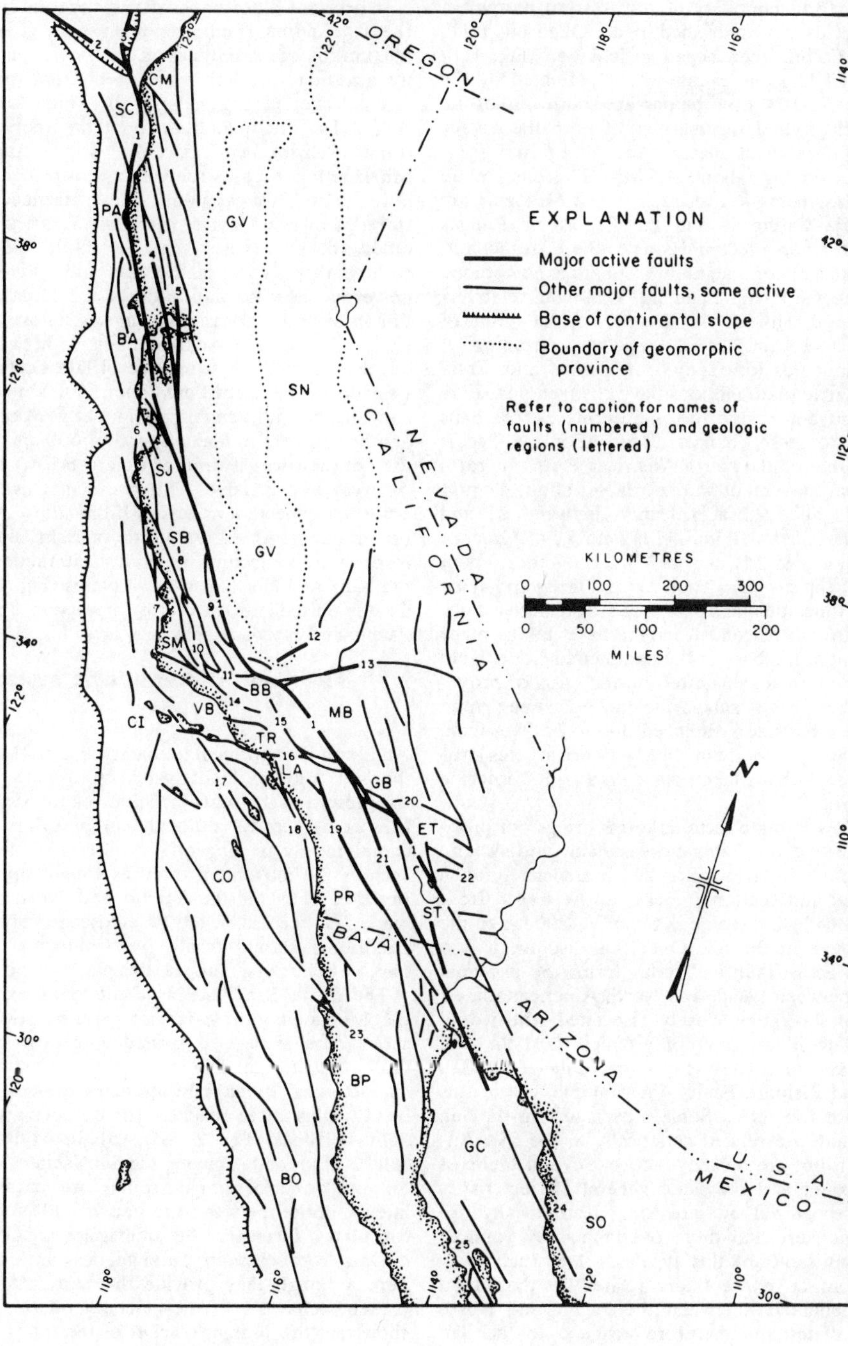
EXPLANATION
Major active faults
Other major faults, some active
Base of continental slope
Boundary of geomorphic province
Refer to caption for names of faults (numbered) and geologic regions (lettered)
KILOMETRES
0 100 200 300
0 50 100 150 200
MILES
OREGON
NEVADA
CALIFORNIA
ARIZONA
U.S.A.
MEXICO
BAJA
N
GV
SN
CM
SC
PA
BA
SJ
SB
SM
CI
VB
BB
TR
LA
MB
GB
ET
ST
PR
CO
BP
BO
GC
SO

The transform boundary at the western margin of North America was established in late Oligocene time, about 30 Ma, and then began to lengthen (Fig. 1D). Miocene and Pliocene events were dominated by the transform boundary after the passage southward of the diverging ridge, and the margins of both the Pacific and North American plates were sliced and fragmented. Beginning about 4.5 Ma (Pliocene) plate boundary fragmentation resulted in the Mesozoic arc belt of Baja California and the Peninsular Ranges breaking off along a tectonic zone to the E of the belt. This long slice of continental lithosphere is now joined tenuously to the Pacific Plate and is moving relatively northwestward with it. As it does so, the Gulf of California is widening through sea-floor spreading.

Throughout this long interval the rates and directions of relative plate motions have changed and some onshore and borderland tectonic events can perhaps be related to these changes. Even after the Pacific Plate encountered the North American Plate, the rates of transform movement were variable: about 4 cm/yr between 29 and 21 Ma, 1.3 cm/yr between 21 and 10 Ma, 4 cm/yr from 10 to 4.5 Ma, and 5.5 cm/yr since then (Atwater & Molnar 1973). Altogether about 1000 km of slip probably took place, during and since Oligocene time, but across a broad distributive belt. Some of this displacement might have taken place along a fault at the base of the continental slope. Hein (1973), for example, suggested on the basis of provenance studies that a submarine fan now lying upon oceanic floor has been displaced 300 to 550 km from its source area since about 16 Ma. Much of this proposed slip could have taken place on a fault bordering the continent.

This review in plate tectonic terms brings out problems in viewing the San Andreas Fault and system. Today the fault and its zone of closely associated faults and crushed and shattered rocks can be traced from the Cape Mendocino region in the NW 1200 km to the Salton Trough in the SE. Over this distance it is a continuous active fault and today forms the principal boundary between Pacific and North American plates. Other right slip faults, closely associated with it and subparallel to it are commonly assigned to the San Andreas system. The Hayward, Calaveras, San Jacinto, and Elsinore Faults, for example, are in this class and are also active. Some others, such as the San Gabriel Fault, are related genetically to the San Andreas Fault, but are not now active. Several faults of regional extent and the same general characteristics are now recognized but are long dead, locally deformed, and here and there overlapped by younger strata. Many faults of this type no older than early Miocene can be conceptually related to the broad transform boundary between the Pacific and North American plates and therefore assigned to the San Andreas system. Others, including some that are older than Miocene, may or may not be logically placed in the system.

In central California, however, the total right slip on the San Andreas Fault proper is usually given as about 600 km based mainly upon the total offset of the western edge of Sierran basement from the southern Great Valley (Fig. 2, GV) to the Point Arena region (Fig. 2, PA) and in addition, upon the interpretation of some Palaeocene–Cretaceous sedimentary facies. Miocene facies and structures are displaced a total of only about 300 km. With these displacements restored, however, a mis-match in the arrangement of older rocks suggests that the fault then existed. Sedimentary facies of Eocene and Oligocene age, however, show no indication of the fault's presence. For these and other reasons, most Californian geologists credit the San Andreas Fault with two stages of development: the first episode of late Cretaceous and Palaeocene age contributed about 300 km of right slip and the second, beginning in early Miocene time, another 300 km to make a total of 600 km. The interval between was a time of quiescence when the San Andreas was inactive. This two-stage history apparently fits expectations arising from plate tectonic reconstructions: an early episode of right slip may correspond to conjectural transform interaction between the Kula and North American plates (Fig. 1A, B) and a later episode to transform interaction between the Pacific and North American plates (Fig. 1D–H).

The San Andreas fault system today

An attractive approach to understanding the history of the San Andreas Fault system is to work back from the present to the past. In California the Cenozoic and Mesozoic records, although complex, are unusually complete. By peeling off the effects of later events, such as the movement of slices caught up within the junction between the Pacific and North American plates, a reconstructed kaleidoscope of successive panoramas should result, but Californian geologists have only just begun to use this palinspastic approach.

The active San Andreas Fault today extends from the Mendocino triple-junction southeastward into a series of side-stepped spreading centres within the Gulf of California. The fault trace is not straight but is characterized by straight stretches broken by several bends. Only three reaches: (*a*) bordering the central Salinian Block (Fig. 2, SB); (*b*) SE of Point Arena (Fig. 2, PA); and (*c*) along the San Jacinto Fault in the SE (its principal active strand) are parallel to the lateral motion between the Pacific and North American plates. Elsewhere the fault trace is curved.

Data from seismology and geodesy are not reviewed here although they provide the best information on present behaviour of the plate boundary. These data show that the straight reaches of the faults are indeed parallel to the direction of plate motion and lie along small circles with poles in accord with recent plate tectonic evaluation (Minster *et al.* 1974). Slippage along the straight stretches is shown both by creep and by many small earthquakes (Savage & Burford 1973; Nason 1973; Scholz 1977). The curved segments, in contrast, appear to be 'locked' at the moment (Allen 1968) and may be sites of major earthquakes such as the 1857 Fort Tejon Earthquake in the Big Bend Region (Fig. 2, BB) It is an attractive hypothesis that, as the two lithospheric plates glide by each other at about 5.5 cm/yr (Atwater & Molnar 1973), strain is easily released through creep and small earthquakes on the reaches parallel to the motion, but is stored and released only intermittently in strong quakes where slip is inhibited by constraining bends. Southern Californians are therefore waiting timorously for the Big Bang in the Big Bend.

The pattern of bends and splays along the San Andreas system suggests a tendency for splays to originate as bends grow, so that the fault becomes straighter. Northward from the San Juan Bautista region (Fig. 2, SJ) the Calaveras and Hayward Faults are marked by creep and frequent small quakes, whereas the San Andreas proper to the W is not. The Hayward Fault is directly in line with the straight stretch of the San Andreas SE of the San Juan Bautista Bend and may become the major transform. Perhaps in time it will extend northward to join up with the San Andreas Fault near Shelter Cove (Fig. 2, SC) and isolate the long slice to the W. As these jostlings for dominance among faults go on, crustal slices are apparently squeezed, stretched and rotated. San Francisco Bay is probably the result of crustal sagging where the slice between the Hayward and San Andreas is widest, and where it is squeezed to the SE mountains have risen. Such an interpretation suggests the way that this splintered transform boundary has deformed through time.

To the SE the San Gorgonio Bend (Fig. 2, GB) is apparently being bypassed as major movements are largely taken up by the San Jacinto Fault. Across this region the plate boundary zone is especially broad and includes the offshore borderland. Farther SE, the San Jacinto and Elsinore Faults enter into the pattern of side-stepped spreading centres characteristic of the Gulf of California. Each fault in turn ends its prime activity at the margin of a pull-apart (Elders *et al.* 1972; Crowell 1974*c*). This pattern results in oblique rifting to open the Gulf with the associated oblique movement of the Baja California Peninsula northwestward from the mainland. Major faults within and marginal to the Gulf are of two types: (*a*) transform faults with straight NW trends parallel to plate movements, and (*b*) irregular dip-slip faults, many with NNE trend, that have formed through marginal collapse and sagging at the pull-apart edges of growing basins or where thick continental rocks have lost their support along the edges of both the peninsula and the mainland (Crowell 1974*c*). Geophysical data at depth permit spreading centres to be shown with pull-apart margins of almost any trend, including trends only slightly oblique to the straight transform margins and almost in line with them. In the southern Gulf of California rhomboid depressions are forming by this pull-apart mechanism and show up in the bathymetry (King 1969). In fact, the same topographic pattern in the offshore borderlands of both California and Baja California (Fig. 2, BO, CO) suggest similar tectonic causes.

The Neogene San Andreas

The San Andreas Fault, as a continuous system including the San Gabriel Fault, has a right slip of about 300 km that began in Miocene time. In central California (Fig. 2, SB) Eocene and Oligocene facies show the same offset (Addicott 1968) and so does a transected volcanic pile 22 Ma old (Huffman 1972; Matthews 1976). Therefore the Neogene episode of slip began after the beginning of the middle Miocene, as shown also by folding dynamically related to the plate motions which developed with rapid growth of the San Andreas Fault (Harding 1976). Surface and subsurface data from several oil fields along the southwestern margin of the Great Valley (Fig. 2, GV) show folds with orientations indicative of a right couple. Upper Miocene beds nearby display coarse submarine debris flows shed eastward from a steep scarp along the San Andreas Fault (Fletcher 1967; Huffman 1972). These sedimentary tongues have been displaced right-laterally from their sources for about 240 km. The difference between their displacement (240 km) and that of the 22 Ma markers (300 km slip) shows that about 60 km of right slip took place during the late middle Miocene and early late Miocene. The system was therefore active before the opening of the Gulf of California, an event that began about 4.5 Ma (Fig. 1, GBH) (Larson *et al.* 1968; Elders *et al.* 1972).

During the late Miocene and Pliocene the San Gabriel Fault connected directly with the San Andreas proper NW of the Big Bend region and with the fault system in the Salton Trough. Basement terrains are offset and sets of sedimentary breccias are displaced as much as 60 km from their source areas along the San Gabriel Fault, successively younger units are offset progressively less, showing that the fault was active during their deposition (Crowell 1975*b*). Within Pliocene time, *c.* 4 Ma ago the San Gabriel strand of the system was abandoned and the strand lying to the NE and extending from the Big Bend region into the San Gorgonio Bend took over. Unusual basement rocks overlain by unusual stratal and volcanic sequences are quite certainly displaced from the San Gabriel Mountain region to the Salton Sea vicinity, a distance of about 240 km. Other coarse sedimentary prisms piled against scarps from the Big Bend to the Salton Trough show that the system has been active, at least intermittently, since 10 or 12 Ma.

In southern California the recognized strands of the San Andreas system are no older than late Miocene, since all older rocks are displaced the same amount (Crowell 1975*a*; Ehlig *et al.* 1975). The total right slip across the set of subparallel faults is more than 360 km if the 300 km on the combined San Andreas–San Gabriel (Crowell 1975*a*), 24 km on the San Jacinto (Sharp *et al.* 1972), and about 40 km on the Elsinore (Sage 1975) are added up. To this must be added the slip on faults within the system to the W, such as those in the Borderland (Howell 1976*b*). The northern and central sector, as already seen, may have as much as 600 km of slip.

In summary, the Neogene San Andreas system has a complex history with three aspects: (*a*) a slow beginning in the Miocene, (*b*) rapid displacement in late Miocene and culminating in Pliocene time, and (*c*) displacements associated with the opening of the Gulf of California. During later movements the fault system apparently was a transform fault, transecting continental crust. The Mio–Pliocene fault, however, cut obliquely across the belt of arc rocks traceable southward from the Sierra Nevada into the Peninsular Ranges and Baja California. To the SE this fault system apparently followed a proto-Gulf and lay to the E of Andean-type arc rocks that were emplaced during late Mesozoic and early Tertiary plate convergence (Karig & Jensky 1972). The long and narrow proto-Gulf was occupied by a sea at the beginning of Pliocene time but with continental crust probably beneath marine sediments.

The Transverse Ranges

The NW tectonic grain of California is conspicuously broken by the E–W trending Transverse Ranges which lie athwart the arc belt of the Sierra Nevada–Peninsular Ranges and include the Big Bend of the San Andreas. The region is active tectonically and undergoing uplift (Castle *et al.* 1976; Grantz *et al.* 1971; Hadley & Kanamori 1977) to form the present mountains, but the rocks and structures within the range have had a long history. E-trending faults were active during the late Oligocene–early Miocene, as shown by the age of overlapping strata. Some of these old faults have been tentatively identified as displaced about 300 km on the San Andreas system to the SE where there is no vestige of the mountains themselves but only of the bedrock structure (Crowell 1975*a*). The transverse structure therefore may have existed in Oligocene time, but may be considerably older and involve Mesozoic tectonic arrangements as yet unsatisfactorily deciphered. Jurassic and Cretaceous strata around the Los Angeles region, for example, now disjunct and sparsely exposed, display strikingly variable facies.

The Oligocene structure, however, may owe its origin to the meeting of the Pacific and North American plates (Fig. 1C). The Mendocino fracture zone at the spreading centre between the Pacific and Farallon plates first made contact at about this time, and has since moved northwestward to its present position. During the Miocene the Transverse Ranges region was fragmented and deep adjoining basins such as the Los Angeles, Santa Maria, and parts of the Ventura were formed. Significant left slip on several major faults (e.g. Santa Monica system and Santa Ynez) took place, beginning in the Miocene, resulting in complex tectonics and geography (Campbell & Yerkes 1976). Fragmentation continued locally in the Pliocene with some blocks elevated and intervening terrain stretched and depressed (Crowell 1976; Yeats 1976). Later and probably associated with the opening of the Gulf of California, the region was compressed and elevated. At this time the Big Bend in the San Andreas Fault became more marked and the San Gabriel Fault was abandoned as the major strand of the San Andreas system. During this sequence of strong deformations, blocks have not only been squeezed and uplifted, but some may have been strongly rotated horizontally in a clockwise direction as suggested by palaeomagnetic studies (Kamerling & Luyendyk 1977). The borderland offshore of southern California presumbly has a similar complex Neogene history (Vedder *et al.* 1974; Howell 1976*b*).

The basement terrains of the Transverse Ranges and adjacent regions to the NW, E and SE, contain exposed remnants of an extensive overthrust system: the Vincent–Orocopia–Chocolate Mountain system (Ehlig 1968; Haxel & Dillon 1973). The thrust has brought Precambrian through upper Cretaceous granitic and metamorphic rocks along a thick mylonite zone upon greenschists of late Mesozoic or Palaeocene age. Sparse local minor structures indicate probable emplacement of the huge thrust sheet from the SW toward the NE, probably in Palaeocene time. Dillon and Haxel (1975) suggested that a back-arc basin with oceanic floor formed during late Mesozoic time, and in which thick sedimentary and some volcanic rocks accumulated. In the Palaeocene the basin was closed and the beds within it metamorphosed during thrusting. Much of southeastern California may therefore be allochthonous, including the Transverse Ranges (Crowell 1974*b*). The W–NW trend and original position of this thrust system, now folded and dismembered by later movements, show that it either originated the Transverse Ranges structure or accentuated it at an early date. Moreover, its trend broke across some arc-belt rocks and may therefore have aided the positioning of the later San Andreas system.

The Cretaceous–Palaeocene San Andreas

In central California a long block, underlain by granitic terrain, of rocks somewhat similar to those of the

Sierra Nevada and Peninsular Ranges lies to the W of the present San Andreas Fault. This, the Salinian Block, is apparently out of alignment by nearly 600 km with the arc belt, by displacement on the San Andreas system (Hill & Dibblee 1953). This offset, together with the interpretation of facies relations and correlations of Cretaceous–Palaeocene strata, argues for a pre-Miocene displacement episode of about 300 km of right slip. The Cretaceous–Palaeocene facies (the Gualala beds) are interpreted as deposited on the W side of the San Andreas at the base of a scarp between 80 and 60 Ma (Wentworth 1968; Ross *et al.* 1973). These beds, now in the Point Arena region apparently came from the Big Bend region before offset and contain debris derived from the E. The time of deposition was apparently the interval when the Kula Plate and the spreading centre between the Kula and Farallon plates were moving relatively northward along the coastal region (Fig. 1 A, B) so strike slip is to be expected.

Studies of Palaeocene and Eocene strata in western California show that an irregular coast with islands lay between a linear arc on the NE and ocean to the SW (Nilsen & Clarke 1975; Nilsen 1977*b*). Deep submarine valleys, now filled with sediments, notched a narrow borderland on the N (Almgren 1978) and a broader and embayed borderland lay farther S (Howell 1975). Although the role of a presumed proto-San Andreas in the development of this geography is not yet clear, most reconstructions of California involve it in order to account for local linear coastlines as in the Point Arena region (Wentworth 1968) and of upper Cretaceous red beds in the southern Salinian Block (Vedder *et al.* 1977).

A significant problem lies in finding the original position of the basement rocks within the anomalous Salinian Block (Gastil *et al.* 1972; Ross 1977). Judging from their petrology, isotopic ages, and inferred history they may have originally lain along the eastern margin of the Mesozoic arc, but their source has not yet been identified. The basement block had arrived near the continental margin and W of the continuing arc during late Cretaceous time, however, because it is overlain unconformably by the coastal prism of upper Cretaceous strata. These sedimentary rocks seem to consist of forearc facies laid down in a linear belt that can be traced roughly along the western flanks of both the Sierra Nevada and the Peninsular Ranges. In summary, data in hand suggest that by Cretaceous time a slice of basement had somehow been displaced from behind the arc belt to its front, that a linear coast was developed along the eastern margin of an elongate forearc basin, and that this belt was then further displaced.

Identifying the role of a possible proto-San Andreas in these events poses other problems. First, the inferred strike-slip between the Kula and North American plates may have taken place at the boundary between deep ocean and continental crust and so not now identifiable on the continent. Secondly, the late Cretaceous, Palaeocene, and Eocene topography reconstructed from facies studies may have been associated with an arc–trench gap, and in particular, with a forearc basin which would imply continuing subduction. Oblique convergence presumably prevailed during earliest Tertiary time as the Farallon Plate began to meet the North American Plate and so some strike slip of this age can be expected. Thirdly, the displacements observed, such as the nearly 600 km offset of basement terrain, may be entirely the result of post-early Miocene slip, accumulated on several converging faults of the Neogene San Andreas system (Johnson & Normark 1974). This explanation would limit the movements credited to the San Andreas entirely to mid-Tertiary and later events associated with the meeting of the Pacific and North American plates.

This latter hypothesis acquires credibility since several major faults have recently been delineated within the Salinian Block and in the offshore area N of the Transverse Ranges and some may possess considerable strike slip displacements and join with the San Andreas Fault on the N. The Hosgri–San Gregorio Fault may have from 60 to 100 km of right slip (Hall 1975; Graham & Dickinson 1978) and the Rinconada at least 60 km (Dibblee 1976). If these and others (Silver *et al.* 1971) are eventually demarcated and shown to have right slip measured in tens of kilometres, the Salinian Block may be sliced into many elongate segments. The Red Hills–San Juan–Chimeneas Fault (Smith 1977; Ross 1977) may be another example, across which there is a strong contrast in basement rock types. If, in turn, these segments are strewn out, the total terrain underlain by basement may be elongated so that basement slices are brought adjacent to the San Andreas zone, to increase the apparent displacement. Critical to this hypothesis is proof that the faulting is no older than about 30 Ma (late Oligocene). In addition, it requires changes in the present interpretation of the Gualala beds near Point Arena: perhaps these beds were laid down at the base of a steep slope near or along the margin of the continent during plate convergence rather than along a transform boundary.

Review

The total Neogene right slip across the wide plate boundary between the Pacific and North American plates should equal about 1000 km (Atwater & Molnar 1973) and the post-early Miocene slip now recognized is a significant proportion of this total. Major right-slip faults are but a part of this belt of deformed crust, however, in which pull-apart basins, fault divergence and convergence, and tipping wedges with rising sectors all play a role in a complex scheme (Crowell 1974*a*; Dibblee 1977). In such a moving system some faults are discontinuous and some bring wedges of different terrains next to each other. Blocks between faults are squeezed and rotated, or stretched and dismembered; a pattern that prevails at one time is soon modified as plate motion continues. Such a mobilistic view is visualized for California since Oligocene times when the Pacific and North American plates impinged and the transform boundary formed. The hallmark of this tectonic style is complexity in an irregularly moving system along a broad and yieldable plate boundary.

Working back in time, the San Andreas has the following episodes in its history. At present it is performing as a transform system connecting the spreading Gulf of California to the Mendocino triple junction. Four bends along the fault are apparently unstable; three are in stages of being by-passed by active right slip splays whereas the Big Bend is largely constrained so that tectonic relief appears to be mainly through uplift, folding, thrusting and occasional large earthquakes. The southern Sierra Nevada and Mojave Block are pressing westward, with discontinuities on the White Wolf and Garlock Faults, to reinforce the Big Bend. During Pliocene time, the San Andreas Fault was straighter and through-going from the Point Arena region (and perhaps even on the NW to the Mendocino fracture zone) all the way SE to the Salton Trough. In the region of the yet-to-rise Transverse Ranges, the San Gabriel Fault was the active strand and it connected directly through and beyond the San Gorgonio Bend. The San Jacinto Fault originated as this bend developed, so that the San Jacinto Fault now performs as a principal element of the San Andreas system.

In late Miocene and Pliocene time, 300 km of right slip began to accumulate on the through-going San Andreas, both in central and southern California. To the W the Los Angeles and Ventura basins were reshaped during the Pliocene by extension, and some faults both here and offshore contributed to transform movements. The Santa Maria, Ventura, and Los Angeles basins, along with several other basins in both the onshore and offshore regions, originated through complicated extension and crustal thinning in Miocene time, during both transform and divergent events. Volcanism accompanied these events in mid-Miocene time and may have been associated with the passage southward along the plate boundary of the spreading ridge between the Pacific and Farallon plates. Throughout California major tectonic changes took place in Miocene time, including the delineation of the modern San Andreas fault system. The sum of all of these late Cenozoic movements may have so juxtaposed and arranged slices of the Salinian Block that the basement offset of the San Andreas Fault today in central California is about 600 km, twice that recognized on this single fault to the SE.

Whether the San Andreas system existed prior to the Miocene is problematic. Sparse evidence, such as the arrangement of Cretaceous and Palaeocene facies along a steep slope is suggestive but not strong. Strike slip may be inferred if the Kula Plate moved laterally with respect to the North American Plate during this time interval, but the geological record supporting this possible early episode of displacement is still disputable.

ACKNOWLEDGMENTS: My recent studies of the San Andreas fault system and California tectonics have been supported by the US National Science Foundation (Grant DES71-00498), the US Geological Survey (Grants G-395 and G-473), and the University of California, Santa Barbara. I am indebted to many colleagues and students for fruitful discussions and field trips, and to T. L. Davis, T. W. Dibblee, Jr., W. B. Hamilton, D. G. Howell, R. M. Norris, B. M. Page, and A. G. Sylvester for critical comments on a draft of this paper.

References

ADDICOTT, W. O. 1968. Mid-Tertiary zoogeographic and paleogeographic discontinuities across the San Andreas Fault, California. (Conference-proceedings). *Stanford Univ. Publs. geol. Sci.* **11,** 144–65.

ALLEN, C. R. 1968. The tectonic environments of seismically active and inactive areas along the San Andreas Fault. *Stanford Univ. Publs. geol. Sci.* **11,** 70–83.

ALMGREN, A. A. 1978. Timing of Teriary submarine canyons and marine cycles of deposition in the southern Sacramento Valley, California. In: STANLEY, D. J. & KELLING, G. H. (eds). *Sedimentation in submarine canyons, fans, and trenches*, Dowden, Hutchinson, and Ross, Inc., Stroudsburg, Pennsylvania, 276–91.

ATWATER, T. 1970. Implications of plate tectonics for the Cenozoic tectonic evolution of western North America. *Bull. geol. Soc. Am.* **81,** 3513–35.

—— & MOLNAR, P. 1973. Relative motion of the Pacific and North American plates deduced from sea-floor spreading in the Atlantic, Indian, and South Pacific oceans. *Stanford Univ. Publs. geol. Sci.* **13,** 136–48.

BLAKE, M. C., JR., CAMPBELL, R. H., DIBBLEE, T. W., JR., HOWELL, D. G., NILSEN, T. H., NORMARK, W. R., VEDDER, J. G., & SILVER, E. A. 1978. Neogene basin formation in relation to plate-tectonic evolution of the San Andreas fault system, California. *Bull. Am. Ass. Petrol. Geol.* **62,** 344–72.

CAMPBELL, R. H. & YERKES, R. F. 1976. Cenozoic evolution of the Los Angeles basin area—relation to plate tectonics. *Pacif. Sect. Am. Ass. Petrol. Geol. Misc. Publ.* 24, 541–58.

CASTLE, R. O., CHURCH, J. P. & ELLIOT, M. R. 1976. Aseismic uplift in southern California. *Science* **192,** 251–3.

CROWELL, J. C. 1974*a*. Origin of late Cenozoic basins in

southern California. *Spec. Publ. Soc. econ. Palenot. Miner., Tulsa*, **22,** 190–204.
—— 1974*b*. The Orocopia Thrust, southeastern California. *Abstr. with progs. geol. Soc. Am.* **6,** 159.
—— 1974*c*. Sedimentation along the San Andreas Fault, California. *Spec. Publ. Soc. econ. Paleont. Miner. Tulsa*, **19,** 292–303.
——1975*a*. The San Andreas Fault in southern California. *Spec. Rep. Calif. St. Min. Bur.* **118,** 7–27.
—— 1975*b*. The San Gabriel Fault and Ridge Basin, southern California. *Spec. Rep. Calif. St. Min. Bur.* **118,** 208–19.
—— 1976. Implications of crustal stretching and shortening of coastal Ventura basin, California. *Pacif. Sect. Am. Ass. Petrol. Geol. Misc. Publ.* **24,** 365–82.
DIBBLEE, T. W., Jr. 1976. The Rinconada and related faults in the southern Coast Ranges, California, and their tectonic significance. *Prof. Pap. U.S. geol. Surv.* **981**.
—— 1977. Strike-slip tectonics of the San Andreas Fault and its role in Cenozoic basin evolvement. *In*: *Late Mesozoic and Cenozoic sedimentation and tectonics in California.* Short Course: San Joaquin geol. Soc., Bakersfield, 26–38.
DICKINSON, W. R., COWAN, D. S., & SCHWEICKERT, R. A. 1972. Test of new global tectonics [discussion]. *Bull. Am. Ass. Petrol. Geol.* **56,** 375–84.
DILLON, J. & HAXEL, G. 1975. The Chocolate Mountain–Orocopia–Vincent thrust system as a tectonic element of late Mesozoic California. *Abstr. with progs. geol. Soc. Am.* **7,** 311–12.
EHLIG, P. L. 1968. Causes of distribution of Pelona, Rand, and Orocopia schists along the San Andreas and Garlock faults. *Stanford Univ. Publs. geol. Sci.* **11,** 294–306.
—— EHLERT, K. W., & CROWE, B. M. 1975. Offset of Upper Miocene Caliente and Mint Canyon Formations along the San Gabriel and San Andreas faults. *Spec. Rep. Calif. St. Min. Bur.* **118,** 83–92.
ELDERS, W. A., REX, R. W., MEIDAV, T., ROBINSON, P. T., & BIEHLER, S. 1972. Crustal spreading in southern California. *Science* **178,** 15–24.
FLETCHER, G. L. 1967. Post Late Miocene displacement along the San Andreas fault zone, central California: *In*: *Guidebook, Gabilan Range and adjacent San Andreas Fault*, Pacific Section Am. Ass. Petrol. Geol. 74–80.
GASTIL, G., PHILLIPS, R. P. & RODRIGUEZ-TORRES, 1972. The reconstruction of Mesozoic California. Section 3, *24th Int. Geol. Congr.* 217–29.
GRAHAM, S. A. & DICKINSON, W. R. 1978. Evidence for 115 km of right slip on the San Gregorio–Hosgri fault trend. *Science* **199,** 179–81.
GRANTZ, A. *et al.* 1971. The San Fernando, California, earthquake of February 9, 1971. *Prof. Pap. U.S. geol. Surv.* **733.**
HADLEY, D. & KANAMORI, H. 1977. Seismic structure of the Transverse Ranges, California. *Bull. geol. Soc. Am.* **88,** 1469–78.
HALL, C. A., Jr. 1975. San Simeon–Hosgri fault system, coastal California: economic and environmental implications. *Science* **190,** 1291–94.
HARDING, T. P. 1976. Tectonic significance and hydrocarbon trapping consequences of sequential folding synchronous with San Andreas faulting, San Joaquin Valley, California. *Bull. Am. Ass. Petrol. Geol.* **60,** 356–78.
HAXEL, G. & DILLON, J. 1973. The San Andreas fault system in southeasternmost California. *Stanford Univ. Publs. geol. Sci.* **13,** 322–33.
HEIN, J. R. 1973. Deep-sea sediment source areas: Implications of variable rates of movement between California and the Pacific Plate. *Nature, Phys. Sci.* **241,** 40–1.
HILL, M. L. & DIBBLEE, T. W., JR. 1953. San Andreas, Garlock and Big Pine faults, California: *Bull. geol. Soc. Am.* **64,** 443–58.
HOWELL, D. G. 1975. Eocene paleogeography of southern California. *In*: *Paleogene symposium and selected technical papers, conference on Future Energy Horizons of the Pacific Coast*, Annual Meeting., Pacif. Sect., Am. Ass. Petrol. Geol. Soc. econ. Paleont. Miner. Tulsa, Soc. Expl. Geophys. 272–93.
—— 1976*a*. A model to accommodate 1000 kilometers of right-slip, Neogene, displacement in the southern California area. *Pacif. Sect., Am. Ass. Petrol. Geol. Misc. Publ.* **24,** 330–40.
—— 1976*b*. (ed.) Aspects of the geologic history of the California continental borderland. *Pacif. Sect., Am. Ass. Petrol. Geol. misc. Publ.* **24,** 561.
HUFFMAN, O. F. 1972. Lateral displacement of Upper Miocene rocks and the Neogene history of offset along the San Andreas Fault in central California. *Bull. geol. Soc. Am.* **83,** 2913–46.
JOHNSTON, J. D. & NORMARK, W. R. 1974. Neogene tectonic evolution of the Salinian Block, west-central California. *Geology* **2,** 11–14.
KAMERLING, M. J. & LUYENDYK, B. P. 1977. Tectonic rotation of the Santa Monica Mountains in southern California (Abstract). *Eos*, **58,** 1126.
KARIG, D. E. & JENSKY, W. 1972. The proto-Gulf of California. *Earth Planet. Sci. Lett.* **17,** 169–74.
KING, P. B. 1969. Tectonic map of North America. *US Geol. Survey*, Scale: 1:5,000,000.
LARSON, R. L., MENARD, H. W. & SMITH, S. M. 1968. Gulf of California: a result of ocean floor spreading and transform faulting. *Science* **161,** 781–4.
MATTHEWS, V. III. 1976. Correlation of Pinnacles and Neenach volcanic formation and their bearing on the San Andreas Fault problem. *Bull. Am. Ass. Petrol. Geol.* **60,** 2128–41.
MINSTER, J. B., JORDAN, T. H., MOLNAR, P. & HAINES, E. 1974. Numerical modeling of instantaneous plate tectonics. *Geophys. J. R. astr. Soc.* **36,** 541–76.
NASON, R. D. 1973. Fault creep and earthquakes on the San Andreas fault. *Stanford Univ. Publs. geol. Sci.* **13,** 275–85.
NILSON, T. H. 1977*a*. Introduction to late Mesozoic and Cenozoic sedimentation and tectonics in California. *In*: *Late Mesozoic and Cenozoic sedimentation and tectonics in California*. Short Course: San Joaquin geol. Soc., Bakersfield, 7–19.
—— 1977*b*. Early Tertiary tectonics and sedimentation in California. *In*: *Late Mesozoic and Cenozoic sedimentation and tectonics in California.* Short Course: San Joaquin geol. Soc., Bakersfield, 86–98.
—— & CLARKE, S. H., Jr. 1975. Sedimentation and tectonics in the early continental borderland of central California. *Prof. Pap. U.S. geol. Surv.* **925**.
ROSS, D. C. 1977. Pre-intrusive metasedimentary rocks of the Salinian Block, California—a paleotectonic dilemma. *In*: *Paleozoic paleogeography of the western United States*. Pacif. Sect., Soc. econ. Paleont. Miner. 371–80.
——, WENTWORTH, C. M. & MCKEE, E. H. 1973. Cretaceous mafic conglomerate near Gualala offset 350 miles by San Andreas Fault from oceanic crustal source near Eagle Rest Peak, California. *J. Res. U.S. geol. Surv.* **1,** 45–52.

SAGE, O., JR. 1975. Sedimentological and tectonic implications of the Paleocene San Francisquito Formation, Los Angeles County, California. *Spec. Rep. Calif. St. Min. Bur.* **118,** 162–9.

SAVAGE, J. C. & BURFORD, R. O. 1973. Geodetic determination of relative plate motion in central California. *Jl. geophys. Res.* **78,** 832–45.

SHARP, R. V. *et al.* 1972. The Borrego Mountain Earthquake of April 9, 1968. *Prof. Pap. U.S. geol. Surv.* **787**.

SCHOLZ, C. H. 1977. Transform fault systems of California and New Zealand: similarities in their tectonic and seismic styles. *Jl geol. Soc. Lond.* **133,** 215–29.

SILVER, E. A., CURRAY, J. R. & COOPER, A. K. 1971. Tectonic development of the continental margin off central California. *Field trip guidebook*, geol. Soc. Sacramento, 1–10.

SMITH, D. P. 1977. San Juan–St. Francis fault—hypothesized major middle Tertiary right-lateral fault in central and southern California. *Spec. Rep. Calif. St. Min. Bur.* **129,** 41–50.

VEDDER, J. G., BEYER, L. A., JUNGER, A., MOORE, G. W., ROBERTS, A. E., TAYLOR, J. C., & WAGNER, J. C. 1974. Preliminary report on the geology of the continental borderland of southern California: *U.S. geol. Surv. map MF*–624.

——, HOWELL, D. G., & MCLEAN, H. 1977. Upper Cretaceous redbeds in the Sierra Madre–San Rafael Mountains, California. *Pacific Coast paleogeography field guide* 2, Pacif. Sect., Soc. econ. Paleont. Miner. 71–8.

WENTWORTH, C. M. 1968. Upper Cretaceous and lower Tertiary strata near Gualala, California, and inferred large right slip on the San Andreas Fault. *Stanford Univ. publs. geol. Sci.* **11,** 130–43.

YEATS, R. S. 1976. Neogene tectonics of the central Ventura Basin, California. *Proc. Neogene Symposium*, Pacif. Sect., Soc. econ. Paleont. Miner. San Francisco, 19–32.

20

Reprinted from pages 15–16, 22, 23, and 24 of *Science* **178:**15–24 (1972)

Crustal Spreading in Southern California

The Imperial Valley and the Gulf of California formed by the rifting apart of a continental plate.

Wilfred A. Elders, Robert W. Rex, Tsvi Meidav, Paul T. Robinson, and Shawn Biehler

More than 45 years ago Wegener, as part of his comprehensive scheme of continental drift, proposed that the Gulf of California formed by separation of the peninsula of Baja California from mainland Mexico (*1*). This proposal, ignored for several decades during which the concept of continental drift was found unacceptable by many geologists, has now been revived (*2*). The idea is particularly appealing in the light of later ideas on sea floor spreading (*3*), which put continental drift in a wider context and gave birth to the newer, unifying theory of plate tectonics that is currently revitalizing geology (*4*).

The northeast Pacific Ocean appears to be a flank of the East Pacific Rise modified by the westward boundary of the North American plate (Fig. 1) (*5*). Marine geophysical studies both in the Gulf of California and in the adjacent ocean strongly support the interpretation that the gulf originated in the spreading apart of the continental crust. It appears that the Gulf of California is part of the active boundary between the North American and Pacific plates (*6*).

In this article we discuss geological and geophysical evidence for crustal spreading for the Salton trough, the landward extension of the gulf into North America (*7*). Patterns of magnetic, gravitational, and thermal anomalies, together with geodetic measurements, strike-slip faulting, and young volcanism suggest that the continental crust is rifting apart, as part of the dynamic system that includes the southern part of the San Andreas and related faults in the Salton trough. In the Imperial Valley of California, at the northern end of the trough, where our work has been concentrated (*8*), the high heat flow associated with this active tectonism is potentially a major energy resource (*9*).

Structure of the Valley

The Mexicali-Imperial Valley, the northern extension of the Gulf of California, is a broad, structural trough partly filled with lacustrine and deltaic silts, sands and gravels of late Tertiary age, and by great thicknesses of Quaternary alluvium and lake sediments. It is a complex rift valley bordered by mountains consisting of Mesozoic, and older, granitic and metasedimentary basement rocks, with subordinate Tertiary volcanic rocks (*10*). The oldest sedimentary rocks cropping out in the valley may be as old as Miocene, but most of the sediments were probably deposited by the Colorado River during the last few million years (*11*).

A preliminary investigation of the basement configuration under the valley north of the Mexican border shows that the Imperial Valley has steep, step-faulted margins and a broad, relatively flat basement floor 6 to 7 kilometers deep (*12*). Numerous subsidiary blocks and basins are aligned along the major faults. This trough is comparable in shape and size with the deeper submarine basins of the southern part of the Gulf of California, but it is partly filled here by the vast accumulation of sediments of the Colorado River Delta (*13*).

The major faults in the region are shown in Fig. 2. The map also shows three newly inferred faults, named the Brawley, Calipatria, and Holtville faults, which apparently transect the valley south of the Salton Sea (*14*). Together with the Sand Hills fault they may be part of the San Andreas fault system.

We believe that the Salton trough formed by a combination of tensional and right-lateral strike-slip movements associated with the opening of the gulf as Baja California was transferred from the North American to the Pacific lithospheric plates (*13, 15*).

A complete Bouguer anomaly map of the Salton trough (*16*) indicates that the earth's crust is isostatically compensated south of the Salton Sea, where the valley, although underlain by 6 km of low-density sediments, is characterized by a broad gravity maximum. Immediately north of the Salton Sea, where the thickness of sediments in the Coachella Valley is less than 3 km, the residual gravity anomaly is markedly negative, reaching a minimum of −44 milligals. A regional Bouguer anomaly map of southern California based on 28,000 gravity measurements was compiled by averaging Bouguer anomalies within 20-km squares and producing a smoothly contoured map (Fig. 3). Because features like the Imperial Valley have widths considerably in excess of 20 km, this does not remove the effect of the low-density sediments in the southern portion of the Salton trough. On the basis of basement seismic refraction, borehole, and geological data

Drs. Elders, Rex, Robinson, and Biehler are members of the faculty of the Department of Geological Sciences, University of California, Riverside 92502; Dr. Meidav is now with the Resources and Transport Division of the United Nations, New York 10017.

(*16*), a density model of the sediments of the Imperial Valley was constructed and used for the gravity calculation (dashed lines in Fig. 3A). If the effect of the low-density sediments in the Imperial Valley is removed by gravity stripping, crust-mantle models can be computed. The offshore gravity data and crustal seismic refraction data have been used for control (*17*).

These models indicate that the crust beneath the axis of the trough is either (i) about 8 km thinner (thickness 20 to 22 km) or (ii) of higher density (density excess 0.10 gram per cubic centimeter) than the normal continental thickness and density for the surrounding area (*18*). Combinations of these two models will also fit the gravity anomaly. These rapid lateral changes are attributed to processes that cause the crust to undergo ductile thinning and basification, becoming more oceanic in character to the south.

[*Editor's Note:* Material has been omitted at this point.]

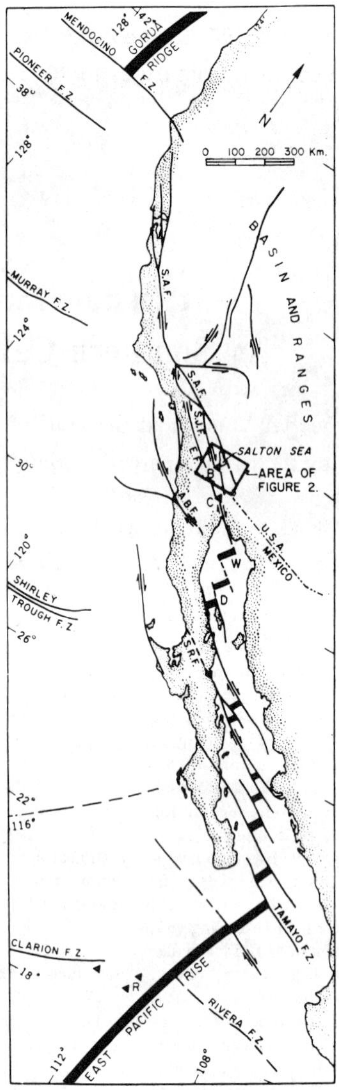

Fig. 1. Fractures and spreading centers along part of the Pacific Coast of North America between the East Pacific Rise and the Gorda Ridge (*6, 41, 43, 50*). Oceanic fracture zones (*F.Z.*) and continental faults (*F.*) are solid black lines, dashed where uncertain. *S.A.F.*, San Andreas fault; *E.F.*, Elsinore fault; *S.J.F.*, San Jacinto fault; *A.B.F.*, Agua Blanca fault; *S.R.F.*, Santa Rosalia fault. Postulated spreading centers in the Gulf of California are shown in black: *W*, Wagner Basin; *D*, Delfin Basin. The black triangles are Holocene or Recent volcanoes: *B*, Buttes; *C*, Cerro Prieto; *R*, Revillagigedo.

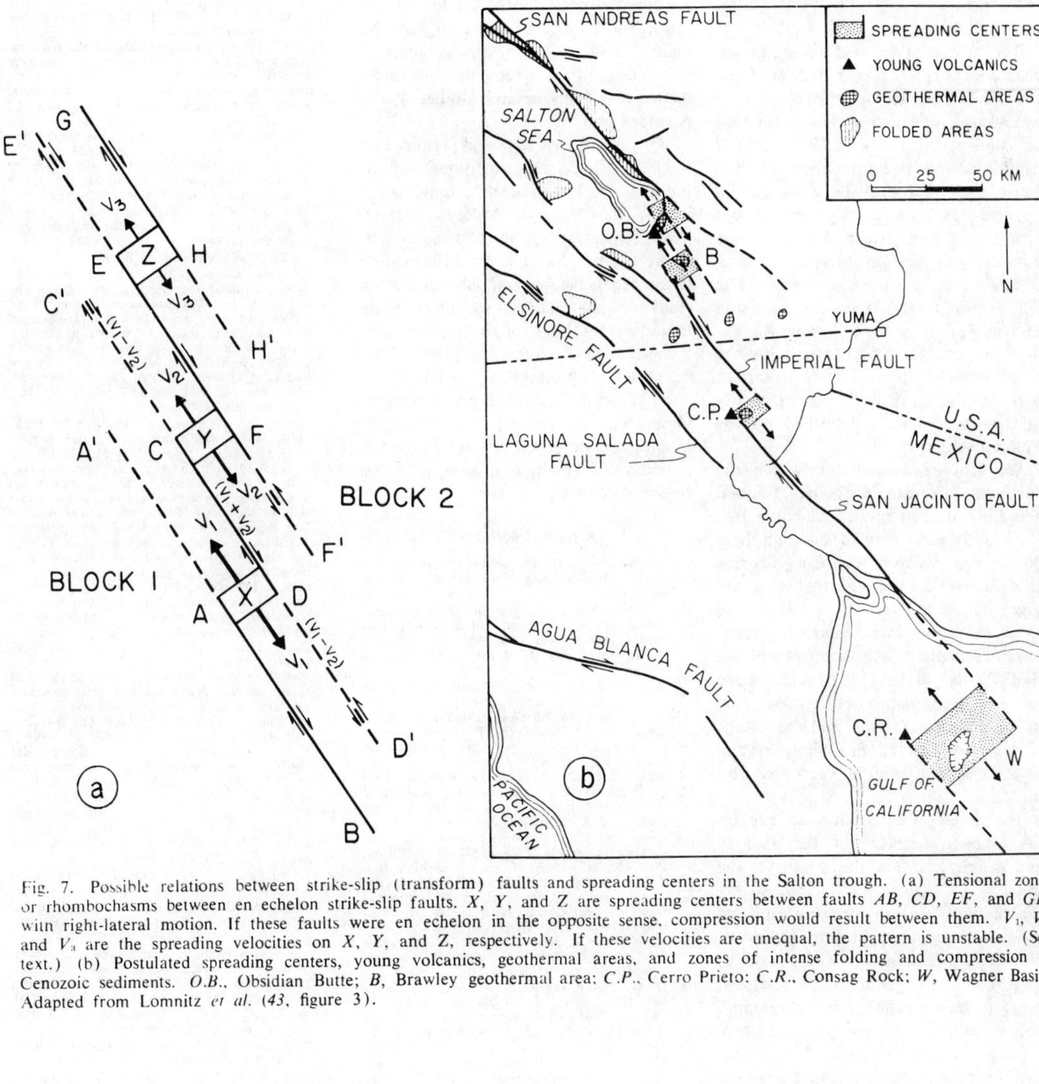

Fig. 7. Possible relations between strike-slip (transform) faults and spreading centers in the Salton trough. (a) Tensional zones or rhombochasms between en echelon strike-slip faults. *X*, *Y*, and *Z* are spreading centers between faults *AB*, *CD*, *EF*, and *GH*, with right-lateral motion. If these faults were en echelon in the opposite sense, compression would result between them. V_1, V_2, and V_3 are the spreading velocities on *X*, *Y*, and *Z*, respectively. If these velocities are unequal, the pattern is unstable. (See text.) (b) Postulated spreading centers, young volcanics, geothermal areas, and zones of intense folding and compression in Cenozoic sediments. *O.B.*, Obsidian Butte; *B*, Brawley geothermal area; *C.P.*, Cerro Prieto; *C.R.*, Consag Rock; *W*, Wagner Basin. Adapted from Lomnitz *et al.* (*43*, figure 3).

Summary

The current excitement among geologists and geophysicists stemming from the "new global tectonics" has led to a widespread, speculative reinterpretation of continental geology. The Gulf of California and its continuation into the Imperial Valley provide an excellent opportunity for studying the border zone between the North American and Pacific plates, and an interface of continental and oceanic tectonics. The Salton trough, the landward extension of the gulf, is a broad structural depression, comparable in size with the deeper marine basins of the southern part of the gulf, but here partially filled with sediments deposited by the Colorado River.

We propose a model in which the continental crust is being thinned beneath a deepening and widening rift. Dilation is accompanied by high heat flow and magmatism. The trough forms as successive sections of the crust are sliced off along strike-slip faults. These slices move northwest and are transferred from the North American to the East Pacific plate. Our studies relate to the present nature of a small part of the plate boundaries, and the details of the evolution of the whole system in the last few million years are still unclear.

The model is an attempt to explain the following aspects of the geology and geophysics: (i) the youth of the trough, (ii) active right-lateral, strike-slip faulting, (iii) many large areas of anomalous high-temperature gradients, (iv) volcanic episodes suggestive of localized melting in the basement beneath the sedimentary pile, (v) rapid lateral change in the geophysical properties of the crust from north to south, (vi) gravity models suggestive of thinning of the crust from 32 km in the north to a maximum of 22 km in the south with apparent upwelling of the mantle under some of the geothermal anomalies, (vii) geodetic measurements, and (viii) seismic studies in the northern gulf.

We intend to test and refine our model by further investigations, including drilling. The pace of tectonic activity in the Imperial Valley is rapid enough to make studies of time dependence meaningful. The Imperial Valley is an accessible natural laboratory for studying sea floor spreading, plate boundary tectonics, ore formation, and a wide range of geothermal phenomena. A series of geophysical, geochemical, and geological studies are presently under way as part of a program to investigate the scientific and economic potential of geothermal energy in the Imperial Valley.

References and Notes

1. A. Wegener, *The Origin of Continents and Oceans* (Dutton, New York, ed. 3, 1924), pp. 184–185.
2. S. W. Carey, in *Continental Drift* (Geology Department Symposium, University of Tasmania, Hobart, Australia, 1958), pp. 177–355; W. Hamilton, *Bull. Geol. Soc. Amer.* **72**, 1307 (1961).
3. A. Holmes, *Principles of Physical Geology* (Nelson, London, ed. 1, 1944), pp. 505–509; H. H. Hess, in *Petrologic Studies*, A. E. J. Engel, H. L. James, B. F. Leonard, Eds. (Geological Society of America, Boulder, Colo., 1962), p. 599.
4. W. J. Morgan, *J. Geophys. Res.* **73**, 1959 (1968); B. Isacks, J. Oliver, L. R. Sykes, *ibid.*, p. 5855 (1968); X. Le Pichon, *ibid.*, p. 3661 (1968); J. R. Heirtzler, *Sci. Amer.* **219**, 60 (December 1968); D. P. McKenzie and R. L. Parker, *Nature* **216**, 1276 (1967); J. T. Wilson, *ibid.* **207**, 343 (1965). Wilson's suggestion, that the San Andreas fault is a transform fault connecting ocean ridges, provided an intellectual framework for further advances [W. R. Dickinson, *Science* **174**, 107 (1971)].
5. H. W. Menard, *Marine Geology of the Pacific* (McGraw-Hill, New York, 1964); F. J. Vine, *Science* **154**, 1405 (1966).
6. D. G. Moore and E. C. Buffington, *Science* **161**, 1238 (1968); W. R. Normark and J. R. Curray, *Bull. Geol. Soc. Amer.* **79**, 1589 (1968); C. G. Chase, H. W. Menard, R. L. Larson, G. F. Sharman III, S. M. Smith, *ibid.* **81**, 491 (1970); L. R. Sykes, in *History of the Earth's Crust*, R. A. Phinney, Ed. (Princeton Univ. Press, Princeton, N.J., 1968), pp. 140–143.
7. Diverse aspects of this problem were discussed at a recent symposium on the Salton trough. See *Geol. Soc. Amer. Abstr. Programs* **3** (No. 2), xiii, xvi (1971).
8. Rex is principally responsible for the temperature gradient studies, Meidav for the electrical resistivity studies, Robinson and Elders for petrology, Biehler for gravity and seismic work, and Elders for the overall synthesis. All the authors have contributed to this synthesis and our continuous debate has made the writing of this article a lengthy task.
9. R. W. Rex, *California Water: A Study in Resource Management*, D. Seckler, Ed. (Univ. of California Press, Berkeley, 1971), pp. 190–205; *Bull. At. Sci.* **27**, 52 (1971).
10. T. W. Dibblee, Jr., *Calif. Div. Mines Geol. Bull.* **170** (1954), chap. 2, pp. 21–28.
11. Coarse clastic sedimentary formations, derived from the erosion of the bordering mountains, crop out along the sides of the valley. The oldest is the Split Mountain Formation of possible Miocene age. The oldest fine-grained Colorado Delta sediments belong to the Imperial Formation, thought to be Pliocene or possibly partly late Miocene. R. Merriam and O. L. Bandy, *J. Sediment. Petrology* **35**, 911 (1965); L. J. P. Muffler and B. R. Doe, *ibid.* **38**, 384 (1968).
12. An isopach map has been prepared by W. R. Hail, W. Randall, and R. W. Rex [*Trans. Amer. Geophys. Union Abstr.* **51**, 421 (1970)]. It is based on (i) gravimetric, magnetometric, and electrical resistivity data; (ii) five oil test wells in the valley, together with a steam well in the Cerro Prieto geothermal field in Mexico, all of which reached basement; (iii) 12 seismic refraction profiles; (iv) surface mapping of basement outcrops compiled from published state geologic maps; (v) the maximum depth reported for seismic hypocenters; and (vi) the degree of isostatic compensation (low-density sedimentary fill compensated by high-density subcrustal material).
13. D. G. Moore, *Geol. Soc. Amer. Abstr. Programs* **3** (No. 2), 166 (1971).
14. The Brawley, Calipatria, and Sand Hills faults are inferred from electrical soundings by Meidav during the past 4 years and unpublished aeromagnetic data [T. Meidav, *Geothermics, Special Issue 2*, **2**, 303 (1970); *Trans. Amer. Geophys. Union Abstr.* **51**, 421 (1970)]. The Holtville fault was discovered by infrared aerial photography [E. A. Babcock, *Bull. Geol. Soc. Amer.* **82**, 3189 (1971)].
15. W. A. Elders, T. Meidav, R. W. Rex, P. T. Robinson, *Geol. Soc. Amer. Abstr. Programs* **2** (No. 7), 545 (1970); W. R. Normark, *ibid.* **3** (No. 2), 172 (1971).
16. S. Biehler, thesis, California Institute of Technology, Pasadena (1964); *Geol. Soc. Amer. Abstr. Programs* **3** (No. 2), 82 (1971).
17. G. Shor and R. W. Raitt, *Proc. 20th Int. Geol. Congr. Mex.* (Sect. 9), 253 (1958).
18. The gravity models are in accord with models derived from seismic studies by W. Thatcher, J. N. Brune, and D. N. Clay [*Geol. Soc. Amer. Abstr. Programs* **3** (No. 2), 208 (1971)].
41. W. Thatcher and J. Brune, *Geophys. J. Roy. Astron. Soc.* **22**, 473 (1971)
43. C. Lomnitz, F. Mooser, C. Allen, J. Brune, W. Thatcher, *Geofis. Int.* **10**, 37 (1970).
50. P. B. King, *Tectonic Map of North America* (U.S. Geological Survey, Washington, D.C., 1969).

[*Editor's Note:* Only the references cited in the preceding excerpts are reprinted here.]

21

Reprinted from pages 1-5 of *Rio Grande Rift: Tectonics and Magmatism,* R. E. Riecher, ed., American Geophysical Union, Washington, D.C., 1979, 445p.

EVOLUTION OF THE RIO GRANDE RIFT — A SUMMARY

Charles E. Chapin

The purpose of this summary is to provide a concise overview of the Rio Grande rift. For details and other interpretations see the individual papers in this volume, regional syntheses by Chapin and Seager (1975) and Cordell (1978), and Guidebook to the Rio Grande Rift in New Mexico and Colorado compiled by Hawley (1978).

Rifting began between 32 and 27 m.y. ago when regional extension reactivated the southern Rocky Mountains, a major north-trending zone of weakness that had developed during late Paleozoic and late Cretaceous-early Tertiary orogenies. By 26 m.y. ago, the crust along the developing rift had sagged sufficiently to form broad, shallow basins in which mafic flows and volcanic ash beds were intercalated with alluvial fill. Preservation within rift basins of complete pre-rift volcanic sections and early-rift volcanic-alluvial sections eliminates regional doming as an early rift process. As the rift opened, it broke en echelon across a series of northeast- and west-northwest-trending flaws which appear as lineaments in the basement terrane of the southern Rocky Mountains. Basins forming on opposite sides of lineaments tended to develop opposing symmetries -- hinged on the west versus hinged on the east. Consequently, those portions of lineaments connecting the ends of staggered basins were subjected to a scissors-like torque in the brittle near-surface rocks and to a transverse shear at depth. Near-surface strain along these transverse structures has been relieved mainly by a complex intermeshing of normal faults of opposing sense rather than by strike-slip faulting. The deeply penetrating transverse structures have tended to leak magmas and to be zones of high heat flow and geothermal activity. See Chapin et al. (1978) for a map showing major lineaments transected by the Rio Grande rift and for geological and geophysical documentation of a transverse shear zone at Socorro.

The main portion of the Rio Grande rift can be divided into three segments, each with its own structural style and history but with a common thread of major events and timing. The northern segment, Leadville to Alamosa (Figure 1), began to open about 27 m.y. ago and is characterized by: a north-northwest trend paralleling the late Paleozoic and Laramide structural grain; a near absence of synrift volcanism in the axial basins; and a shift in extension away from the axial grabens into a broad belt along the east shoulder of the rift (Tweto, this volume). The axial graben system tapers northward and pinches out about 20 km north of Leadville. Extension continues northward to near the Wyoming border as a broad zone of block faulting (Tweto, this volume).

The central segment, Alamosa to Socorro (Figure 1), is characterized by a north-northeast-trending series of en echelon basins separated by complex transverse structures. This segment traverses an area of northeast-trending basement lineaments and crosses 1.4-1.8 b.y.-old Precambrian rocks (Hedge et al., 1968) in the north and 1.3-1.5 b.y.-old Precambrian rocks (White, 1978) in the south. Except near Taos,

L
N
UTAH
KANSAS
AI
COLORADO
NEW MEXICO
OKLA.
T
J
A
ARIZONA
SA
S
DM
LC
TEXAS
EP
MEXICO
0
100
200 Kilometers

early-rift magmatism was relatively sparse within this segment, but late-rift bimodal volcanism was voluminous. The Jemez lineament and the Taos Plateau have been especially active during the past 5 m.y.

The southern segment, Socorro to El Paso (Figure 1), has undergone the most extension as evidenced by the widening of the rift near Socorro into a north-trending series of parallel basins and ranges with a total width about 2.5 times that of the large single basins to the north. The rift also bifurcates near Socorro with a weaker arm (the San Augustin rift) extending southwestward along the Morenci lineament into Arizona. Between the San Augustin arm and the southern segment of the main rift is a triangular area about 80 km on a side which has undergone moderate to extreme (> 50 percent) crustal attenuation and voluminous early-rift and late-rift magmatism. "Domino-style" normal faulting (Chamberlin, 1978) and 30 to 60° of rotation of early Miocene and older strata is common within this triangle, wherever intrarift horsts expose the floor of an early rift basin. A similar structural style is present in the Rincon-Radium Springs area north of Las Cruces. The Socorro triangle is bisected by a northeast-trending transverse shear zone across which the sense of rotation of strata and normal faulting changes 180°. A large, present-day, sill-like magma body at mid-crustal depths of 19 km (Rinehart et al., this volume) ends against this transverse structure and is apparently leaking magmas along it to form shallow reservoirs.

Rifting began about 5 m.y. earlier (32 m.y.) in the southern segment of the Rio Grande rift. The change to bimodal volcanism occurred near the beginning of ash-flow magmatism in the Datil-Mogollon field (Elston and Bornhorst, this volume) instead of at the end as in the San Juan field (Lipman et al., 1978). Large volumes of basaltic-andesite lavas with high initial Sr^{87}/Sr^{86} ratios (0.7055-0.7090, Stinnett and Stueber, 1976) are interbedded with high-silica ash-flow sheets dated at 32 to 26 m.y. in age. Basaltic andesite volcanism continued to about 20 m.y. and was followed by a lull in magmatism between about 20 and 13 m.y. ago. A similar hiatus occurred throughout the Intermountain West but was shorter in the Basin and Range province(s) (20-17 m.y., McKee et al., 1970). Following the middle Miocene lull, bimodal magmatism changed to a basalt-rhyolite suite with low initial Sr^{87}/Sr^{86} ratios (basalt, 0.7034-0.7051; rhyolite, 0.7048-0.7070). Everson and Silver (this volume) report highly radiogenic rock leads similar to those of oceanic islands in Pliocene and Pleistocene basalts of the southern segment and suggest a deep mantle upwelling. The structural characteristics of the southern segment and its magmatic evolution support their conclusion. In contrast, magma source regions in the central and northern segments remained in the lithosphere throughout rifting as evidenced by the lack of systematic chemical changes between early-rift and late-rift rocks (Lipman et al., 1978; Baldridge, this volume; Williams and Murthy, 1978; Dungan et al., 1978). The more silicic pre-rift and early-rift rocks of the southern

---

Fig. 1. Shaded relief map of the Rio Grande rift. L = Leadville, Al = Alamosa, SJ = San Juan volcanic field, T = Taos, J = Jemez volcanic field, A = Albuquerque, S = Socorro, SA = San Augustin rift, DM = Datil-Mogollon volcanic field, LC = Las Cruces, EP = El Paso. Modified from cover of New Mexico Bureau of Mines and Mineral Resources Circular 163, 1978.

segment and their higher initial Sr^{87}/Sr^{86} ratios probably reflect differences in age and composition of the Precambrian lithosphere between the two areas.

Volcanism along the Rio Grande rift increased slowly after the middle Miocene lull (20-13 m.y.) with activity concentrated initially in the Socorro triangle and the Jemez Mountains where the rift transects major northeast-trending lineaments. The southern Rocky Mountains and adjacent areas were strongly uplifted between about 7 and 4 m.y. ago. Total uplift since middle Miocene has been about 1100 m (Axelrod and Bailey, 1976), much of which probably occurred during this brief interval. The uplift is probably due to mantle upwelling like that documented for the Great Basin and adjacent areas by Eaton et al. (in press). Block faulting accompanying epeirogenic uplift disrupted the early broad basins of the Rio Grande rift and formed the modern topography, with narrower, more sharply defined basins. Preliminary calculations based on deflection of lineaments indicate a total extension across the Rio Grande rift of 100-150%. These figures are an order of magnitude larger than previous estimates, but they are consistent with the quantities of basaltic magma that must be injected into the crust to maintain the heat flow anomaly (A. Lachenbruch, 1978, oral commun.) and to explain the residual positive gravity anomaly along the rift axis (Decker and Smithson, 1975; Cordell, 1978).

Increased runoff from newly elevated alpine regions integrated the drainage to form the ancestral Rio Grande prior to 4 m.y. ago. Widespread geomorphic surfaces, graded to the ancestral Rio Grande and its tributaries, became mantled with basalt flows following sharp acceleration of basaltic volcanism about 5 m.y. ago. Lowering of base levels following capture of the Rio Grande at El Paso during middle Pleistocene time (Kottlowski, 1958) resulted in dissection of basins along the Rio Grande drainage and isolation of basalt-capped mesas. Epeirogenic uplift has continued since 4 m.y. ago but at a reduced rate. Rifting continues today as evidenced by abundant fault scarps cutting Pleistocene deposits, high heat flow, modern elevation changes, modern magma bodies, and geophysical evidence for anomalous crust and upper mantle beneath the rift.

References

Axelrod, D.I., and H.P. Bailey, Tertiary vegetation, climate, and altitude of the Rio Grande depression, New Mexico—Colorado, <u>Paleobiology</u>, <u>2</u>, 235-254, 1976.

Baldridge, W.S., Petrology and petrogenesis of Plio-Pleistocene basaltic rocks from the central Rio Grande rift, New Mexico, and their relation to rift structure, <u>this volume</u>, 1979.

Chamberlin, R.M., Structural development of the Lemitar Mountains, an intrarift tilted fault-block uplift, central New Mexico, <u>Los Alamos Sci. Lab. Conference Proceedings</u>, <u>LA-7487-C</u>, 22-24, 1978.

Chapin, C.E., R.M. Chamberlin, et al., Exploration framework of the Socorro geothermal area, New Mexico, <u>N. Mex. Geol. Soc. Spec. Publ. 7</u>, 114-129, 1978.

Chapin, C.E., and W.R. Seager, Evolution of the Rio Grande rift in the Socorro and Las Cruces areas, <u>N. Mex. Geol. Soc. Guidebook of Las Cruces Country</u>, <u>26th Field Conf.</u>, 297-322, 1975.

Cordell, L., Regional geophysical setting of the Rio Grande rift, Geol. Soc. Amer. Bull., 89, 1073-1090, 1978.

Decker, E.R., and S.B. Smithson, Heat flow and gravity interpretation in southern New Mexico and West Texas, J. Geophys. Res., 80, 2542-2552, 1975.

Dungan, M.A., P.W. Lipman, and T.G. Pronold, A geochemical reconnaissance of the Taos Plateau volcanic field, Los Alamos Sci. Lab. Conference Proceedings, LA-7487-C, 30-31, 1978.

Eaton, G.P., R.R. Wahl, et al., Regional gravity and tectonic patterns: their relation to late Cenozoic epeirogeny and lateral spreading in the western Cordillera, in Smith, R.B., and Eaton, G.P., eds., Cenozoic tectonics and regional geophysics of the western Cordillera, Geol. Soc. Amer. Mem. 152, in press.

Elston, W.E., and Bornhorst, T.J., The Rio Grande rift in context of regional post-40 m.y. volcanic and tectonic events, this volume, 1979.

Everson, J.E., and L.T. Silver, Lead systematics of late Cenozoic basalts from the Rio Grande rift, Los Alamos Sci. Lab. Conference Proceedings, LA-7487-C, 36-37, 1978.

Hawley, J.W., ed., Guidebook to Rio Grande rift in New Mexico and Colorado, N.M. Bur. Mines Min. Res. Circ. 163, 241 p., 2 maps, 1978.

Hedge, C.E., Z.E. Peterman, et al., Precambrian geochronology of the northwestern Uncompahgre Plateau, Utah and Colorado, in Geological Survey research 1968, U.S. Geol. Surv. Prof. Paper 600-C, C91-C96, 1968.

Kottlowski, F.E., Geologic history of the Rio Grande near El Paso, W. Texas Geol. Soc. Guidebook of the Franklin and Hueco Mountains, 46-54, 1958.

Lipman, P.W., B.R. Doe, et al., Petrologic evolution of the San Juan volcanic field, southwestern Colorado: Pb and Sr isotope evidence, Geol. Soc. Amer. Bull., 89, 59-82, 1978.

Lipman, P.W., and H.H. Mehnert, The Taos Plateau volcanic field, northern Rio Grande rift, New Mexico, this volume, 1979.

McKee, E.H., D.C. Noble, and M.L. Silberman, Middle Miocene hiatus in volcanic activity in the Great Basin area of the western United States, Earth Planet. Sci. Letters, 8, 93-96, 1970.

Rinehart, E.J., A.R. Sanford, and R.M. Ward, Geographic extent and shape of an extensive magma body at mid-crustal depths in the Rio Grande rift near Socorro, New Mexico, this volume, 1979.

Stinnett, J.W., Jr., and A.M. Stueber, A strontium isotopic and geochemical study of volcanic rocks from the Datil-Mogollon field, southwestern New Mexico (abstract), Geol. Soc. Amer. Abstr. w. Programs, 8, 636-637, 1976.

Tweto, O., The Rio Grande rift system in Colorado, this volume, 1979.

White, D.L., Rb-Sr isochron ages of some Precambrian plutons in south-central New Mexico, Isochron/West, 21, 8-14, 1978.

Williams, S., and V.R. Murthy, Genetic relationships of volcanic rocks from the Taos Plateau, Rio Grande rift: Rb-Sr and Sm-Nd isotopic systematics, Los Alamos Sci. Lab. Conference Proceedings, LA-7487-C, 100, 1978.

Part VI

SOUTH AMERICA

Editor's Comments on Papers 22 and 23

22 ZEIL
Excerpts from *The Andes: A Geological Review*

23 KATZ
Continental Margin in Chile—Is Tectonic Style Compressional or Extensional?

Paper 22 is made up of excerpts from Zeil (1979), the most fruitful single publication for study of rifting within the Andes. Professor Werner Zeil is with the Institut für Geologie und Paläontologie, Technische Universität, Berlin. Unexcerpted chapters should be referred to for the full background. This is an exhaustive and valuable assembly of information from many sources, supplemented with fieldwork by the author.

Gansser (1973) also furnishes a useful overview, but the information on rift structures is dispersed. He is concerned essentially with geodynamic development since the start of the Triassic and discusses the formerly dominant view that geosynclinal development was uniform along the entire Andean system.

Correlation of volcanism with recent extension processes in three regions is described by Zeil on pp. 27 and 72-76 (see also Jenks, 1956, 274). The tectonics are then considered, from Colombia southward. Maps (on differing scales) are included covering each country in turn. In Colombia (p. 151 and Fig. 65) the Magdalena valley is of particular interest (see also Harrison, 1930; Julivert, 1970; Jenks, 1956, 324). The Cauca depression continues into Ecuador (Fig. 70) as the Quito depression, which ends near the Peru border (p. 155). Near Guayaquil (Fig. 70) are the Guayas and Esmeralda basins (Garner, 1959, 1974). In Peru (Fig. 70) block movements are important. In northern Chile (Fig. 79) and Bolivia (Fig. 72) are the most northerly of the *salars* (closed, saline basins), described as tectonic depressions. The Altiplano extends from Lake Titicaca into Bolivia (Fig. 72) (Jenks, 1956, 239; Newell, 1949). The Chilean cordilleras (Fig. 80), with the young longitudinal valley (the Vallee Central, Jenks, 1956), introduce a change in expression that is to continue southward with modifications. The high Cordillera de Los Angeles in the east are paralleled in the west by the Pampa de Tamaragal (see also Jenks, 1956, 191). The elongate depressions (Fig. 82) are generally seen as grabens or tectonic depressions. The descriptions (following p. 173) that continue to the south are of lessening interest. The Atacama dextral strike-slip fault zone in northern Chile is described by St. Amand and Allen (1960).

Paper 23 by Katz (New Zealand Geological Survey), which contains a structural synthesis of central Chile, is included because it indicates the dominance of extensional forces across a wide belt. This ties in with the "taphrogenic breakdown" resulting in the tectonic subsidence and rift features. This type of synthesis can no doubt be applied also to other segments of the Andes.

[*Editor's Note:* The reader is referred to the list following the comments on Papers 24 and 25 (page **271**) for the references cited in these comments.]

22

Reprinted from pages 1–4, 27, 72, 75, 98, 151–160, 163, 166–173 and Figures 65, 70, 72, 79, 80, and 82 of *The Andes: A Geological Review,* Gebrüder Borntraeger, Berlin, 1979, 260p.

THE ANDES: A GEOLOGICAL REVIEW

Werner Zeil

Introduction

Extending as they do over about 7,500 km, the Andes form the Pacific boundary of South America and represent, as it were, the backbone of this southern continent. They are part of the circumpacific mountain system, some features of which differ from those of the W-E trending orogenic zones of the earth:

There are more igneous rocks present than in many parts of the Tethyan geosynclines.

The tectonic structure is characterized in particular by significant normal faulting while there are fewer overthrusts or deformations caused by folding, especially in the overlying rock.

The majority of the deep-focus earthquakes and megaseisms, and also most of the recent volcanic activity, are connected with this region.

Prominent deep-sea trenches run parallel to the continents and to the island arcs which encircle the Pacific.

The actual Andean orogen is distinguished by further dominant geological aspects:

From the morphology one would think that the orogenic events were contemporaneous whereas in fact the mountain range has been formed, since the Precambrian, by temporally and spatially superimposed crustal sections.

There is no such thing as a uniform Andean geosyncline, but instead there are several separate fully developed sections which were welded together during different geodynamic phases.

The Andes are an asymmetrical range of mountains with a foreland which is developed only on the eastern side.

Large sections of the range are intersected by deep fractures extending presumably down to the mantle. These fractures can be regarded as tectonic grabens.

A Coastal Cordillera, with a basement which was consolidated during the Paleozoic, runs between Tierra del Fuego and Peru. The overlying strata have undergone little folding since the Triassic.

Since the Cretaceous, the western sections of the northern and southern parts of the Andes have developed differently. In Colombia and Ecuador there are indications of the existence of an oceanic crust which is missing from Peru on southwards.

The Andes contain a particularly large number of ore deposits and in this respect are amongst the richest ranges in the world.

The range is arc-shaped. The deviation from the S-N direction in Peru and Bolivia is caused by the Precambrian shields of South America which extend here in the bedrock to the boundary of the young mountain range. The orogen can be subdivided into individual structures, some of which run S to N, for the most part following the trend of the mor-

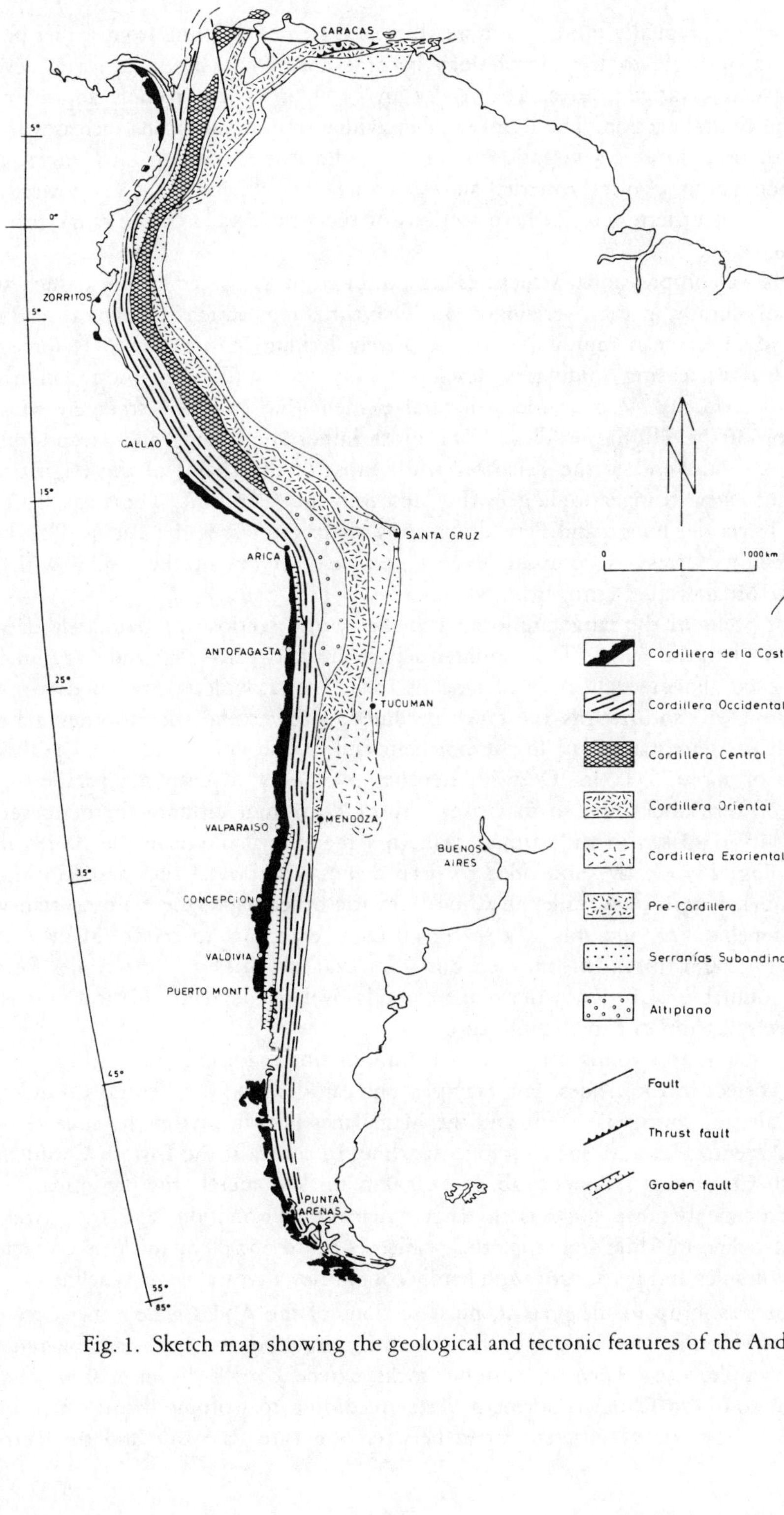

Fig. 1. Sketch map showing the geological and tectonic features of the Andes.

phology. However, gradually more and more structural elements dating from earlier periods and striking at an angle to the morphology become apparent. The geodynamic development has pursued a varied course. The Andes are at their widest, namely approximately 900 km, in the central section. The number of individual mountain chains increases from S to N. In northern Colombia a virgation occurs, i. e. the Andes split up into single ranges which continue on to Central America and Venezuela or which drop away towards the Caribbean. In the southern and northern sections of the Andes parts of the range turn in a W-E direction.

While intensive compressional structures are exhibited in the basement rock, such structures occur only locally in the overlying rocks. Overthrusting, which is a typical feature of some folded Mediterranean mountains, is completely lacking in the Andes. A foreland – foredeep – with decreasing folding is developed only in the E. The lower and narrow Coastal Cordillera in the W is an old structural element that contrasts strangely with the younger chains of the High Cordilleras. The most important folding and metamorphism took place here at the end of the Paleozoic while most of the chains of the High Cordilleras underwent even stronger folding in the Mesozoic and Cainozoic. The Coastal Cordillera between Tierra del Fuego and Peru drops away sharply towards the Pacific. The intensity of the tectonic stress is comparable only in a few sectors of the Andes with that observed in the Mediterranean orogens.

The highest peaks in the range are located in the central section approximately between the equator and latitude 35° S. The mountains here are between 5,000 and 7,000 m high and are composed almost exclusively of igneous rocks – stratovolcanoes or bodies of plutonic rock. From 35° southwards the range becomes narrower and the altitudes diminish rapidly. North of the equator, up to the northern end of the range, individual peaks still attain heights of about 6,000 m. Deep-sea trenches over 7,000 m deep run parallel to the Pacific coast of Peru and Chile so that over a short horizontal distance there are vertical differences of up to 15 km in the earth's crust. In a meridional direction the Andes are in places morphologically clearly subdivided by deep depressions (Rio Cauca and Rio Magdalena in Colombia, longitudinal valley in Chile). On the other hand, there are no transverse valleys of reasonable size, and this is a severe obstacle, especially to traffic. Many Andean passes are over 4,000 m high. In some sections – for example in Peru – the Andes form an almost insurmountable wall. Since the range extends over 65 degrees of latitude, it is exposed to a wide spectrum of climatic influences.

The classic view that a young range of folded mountains is formed from geosynclines is only partially correct in the Andes. For example, one can describe the Central Cordillera of Bolivia as a Paleozoic miogeosyncline and the Magallanes geosyncline in the southern part of Chile and Argentina as a Mesozoic miogeosyncline. In contrast, the Eastern Cordillera of Colombia with Cretaceous sediments about 14,000 m thick is merely the preliminary stage of a miogeosyncline because there is no later extensive deformation, and the "Andean" Mesozoic geosyncline in Chile and Argentina, which is up to 35,000 m thick, is characterized mainly by volcanics and pyroclastics and hardly constitutes a typical eugeosyncline.

From the Mesozoic up to the present, most sections of the Andes were either characterized by violent volcanic production, or they were interspersed and underlain by plutonic bodies. For example, the "Andean" plutonic rocks extend over a distance of more than 4,000 km from southern Chile to northern Peru. According to isotopic dating these intrusions took place however at different times between the Late Paleozoic and the Tertiary.

Fig. 2. Sharp contrast between the Andes with Aconcagua (6,958 m) in the background and the Argentinian pampas in the foreground.

During the Mesozoic period volcanic sequences up to 20 km thick were formed in some Cordilleran ranges. These series are interbedded with thick terrestrial sediment complexes. From the start of the Tertiary onwards to the present, three large sections of the range are characterized by enormous volcanic activity. The deep faults within the Andes developed approximately at the same time as and parallel to these young volcanic zones. Most of these faults can be described as tectonic grabens.

In statistical terms many regions of the Andes are shaken almost daily by earthquakes. Modern research shows that, in particular in the central sections of the range, the hypocentres of the quakes occur from the Pacific boundary eastwards along a zone plunging deep below the continent ("Benioff zone", "subduction zone"). On the other hand, the spatial distribution of the hypocentres is irregular in Colombia. In the countries through which the Andes pass, the violent seismic activity frequently gives rise to severe catastrophes and proves that the mountain range is not yet consolidated with its substructure.

[*Editor's Note:* Material has been omitted at this point.]

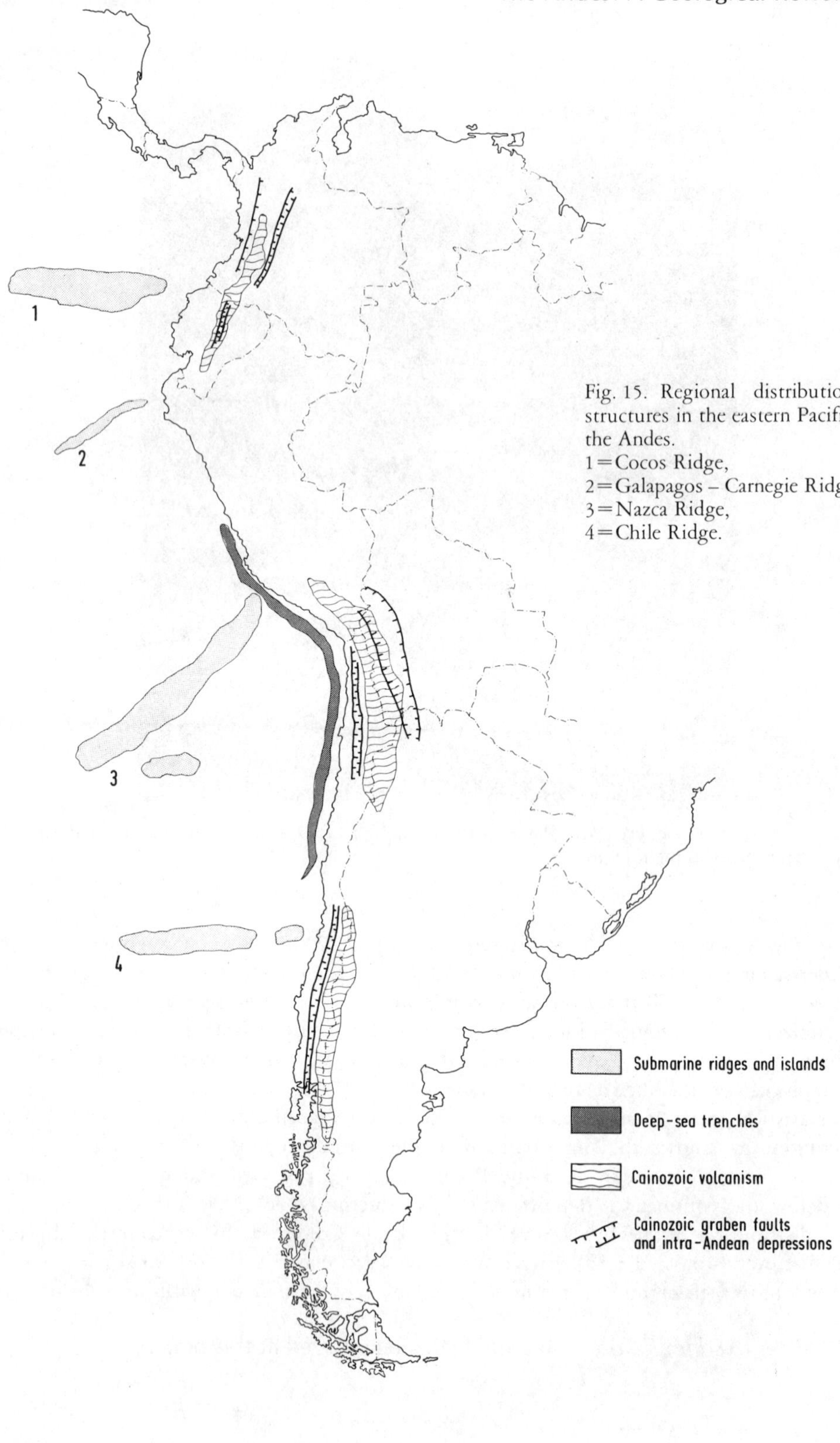

Fig. 15. Regional distribution of structures in the eastern Pacific and the Andes.
1 = Cocos Ridge,
2 = Galapagos – Carnegie Ridge.
3 = Nazca Ridge,
4 = Chile Ridge.

[*Editor's Note:* Material has been omitted at this point and the maps appearing on pages 73 and 74 have also been omitted.]

Correlations with the tectonics

It has already been indicated further above that there is a link between the centres of recent volcanism and the recent extension processes in the range. The volcanic regions are located at precisely the points where the Andes are intersected by graben-like depressions running parallel to the strike of the range. Unfortunately there is no exact information available on the age of the fracture tectonics and little is also known about the internal structure of the grabens. According to the morphology and by analogy with better investigated grabens, abrupt vertical variations in the relief amounting to several thousands of metres can be expected to occur. In fact, for example, the localities in the "valle longitudinal" in Chile are situated at an altitude of about 500 – 700 m above sea level while the High Cordillera east of Santiago rises abruptly to 5,500 m. The graben-like depression of the salar of Atacama is 2,500 m above sea level and the volcanoes in the immediate vicinity rise to over 6,000 m. The salar of Uyuni in Bolivia and the Altiplano are between 3,000 and 4,000 m above sea level, and the Sajama volcano is 6,500 m high. The longitudinal valley of Quito runs at altitudes between 2,500 and 3,000 m while Chimborazo on the western margin of the valley is 6,300 m high. The localities in the upper Cauca graben and in the Magdalena graben in Colombia are situated between 1,800 and 2,500 m while the volcanoes in the Central Cordillera between the two grabens attain altitudes of over 5,400 m.

The age of the inner-Andean longitudinal fracture structures can be approximately determined on the basis of the material with which they are filled. They are certainly recent and their main development took place in the Cainozoic. The thickness of the sediments in the depressions and the tectonic inventory within the structures are still largely unknown. Only a few boreholes or geophysical measurements have been carried out.

VAN HOUTEN & TRAVIS (1968) for example state that the upper Magdalena valley between the Central and Eastern Cordilleras of Colombia contains non-marine, molasse-like Cainozoic sediments 12,000 m thick. EVERNDEN, KRIZ & CHERRONI (1966) describe Tertiary molasse facies up to 14,000 m thick which they discovered in several profiles run in the Altiplano of Bolivia between the Western Cordillera and the Cordillera Real. These enormous thicknesses indicate considerable vertical fracture tectonics during the Tertiary.

In view of the inadequate data available it is not possible to say whether, in addition to the widening movements, strong narrowing movements also took place in these depressions. However, the former type of movement is probably the more important of the two. By analogy with the graben zone of East Africa where there is likewise a close link between recent volcanism and recent fracture structures, the faults in the Andes probably also extend very deep into the crust.

When considering the mass of recent volcanics in the Andes, there are several potential flow paths along which the material may have travelled, namely not only deep-reaching expansion joints **within** the range but also structures **outside** the orogen, i. e. the deep-sea trenches which parallel the Pacific coast of South America and which form strips of highly negative gravitational anomalies. The bottoms of the trenches are mostly smooth and the trenches themselves contain only a thin filling of sediment with an often relatively unfaulted stratification. Although the formation of these probably geologically young structures is still largely unexplained, their deepening – which was associated with a loss of mass – is causally linked with the production of volcanics. Consideration of the regional distribution of the deep-sea trenches and of recent volcanics however reveals great discrepancies. The deep-sea trench running parallel to the Andes is deepest in the Peru trench, with a depth of 6,282 m, and in the Atacama trench, with a depth of 8,066 m, i. e. in the area of the bend in the Andes. No clearly identifiable deep-sea trenches have formed to the north and south of the Andes.

[*Editor's Note:* Material has been omitted at this point.]

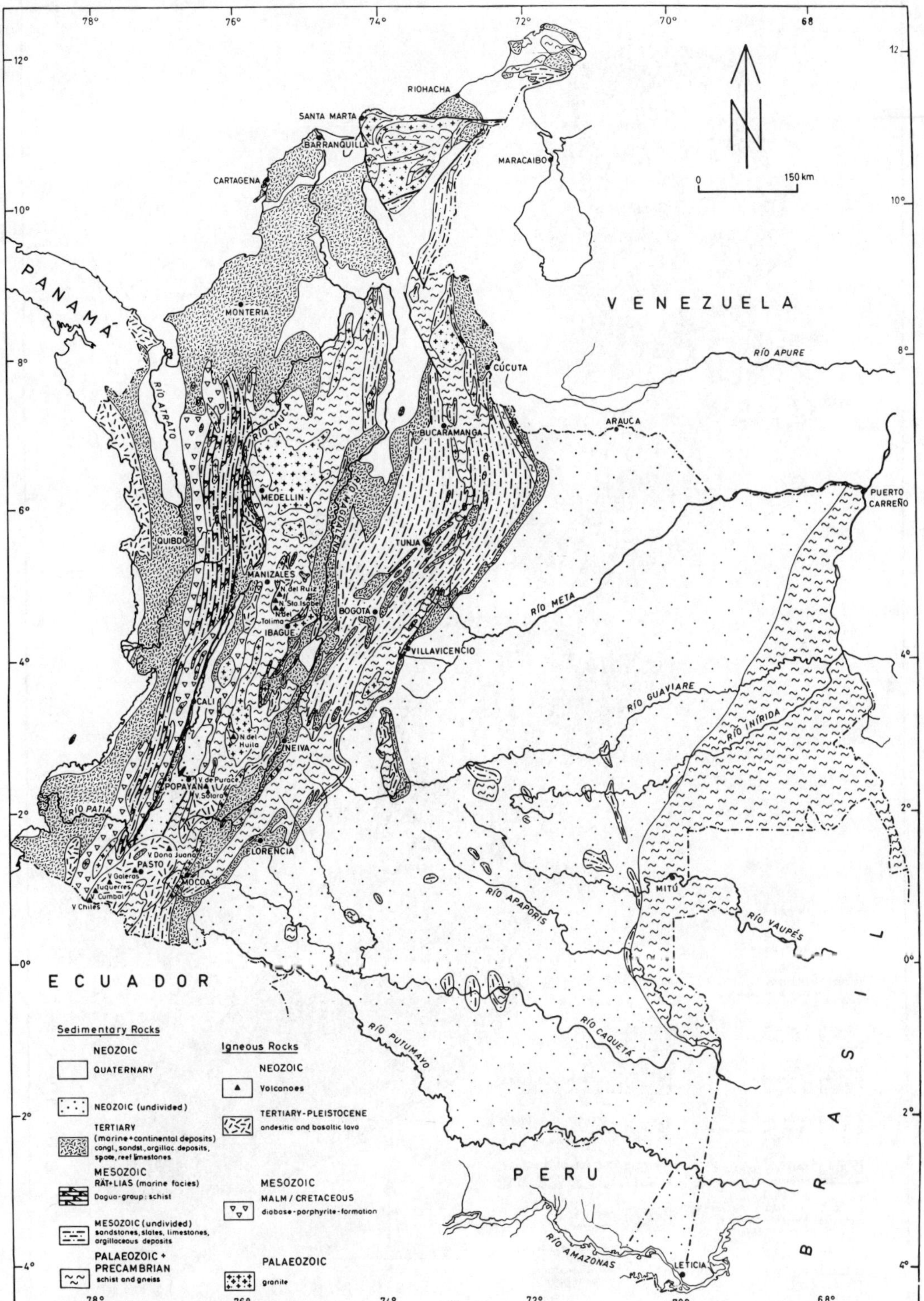

Fig. 65. General geological sketch map of Colombia.

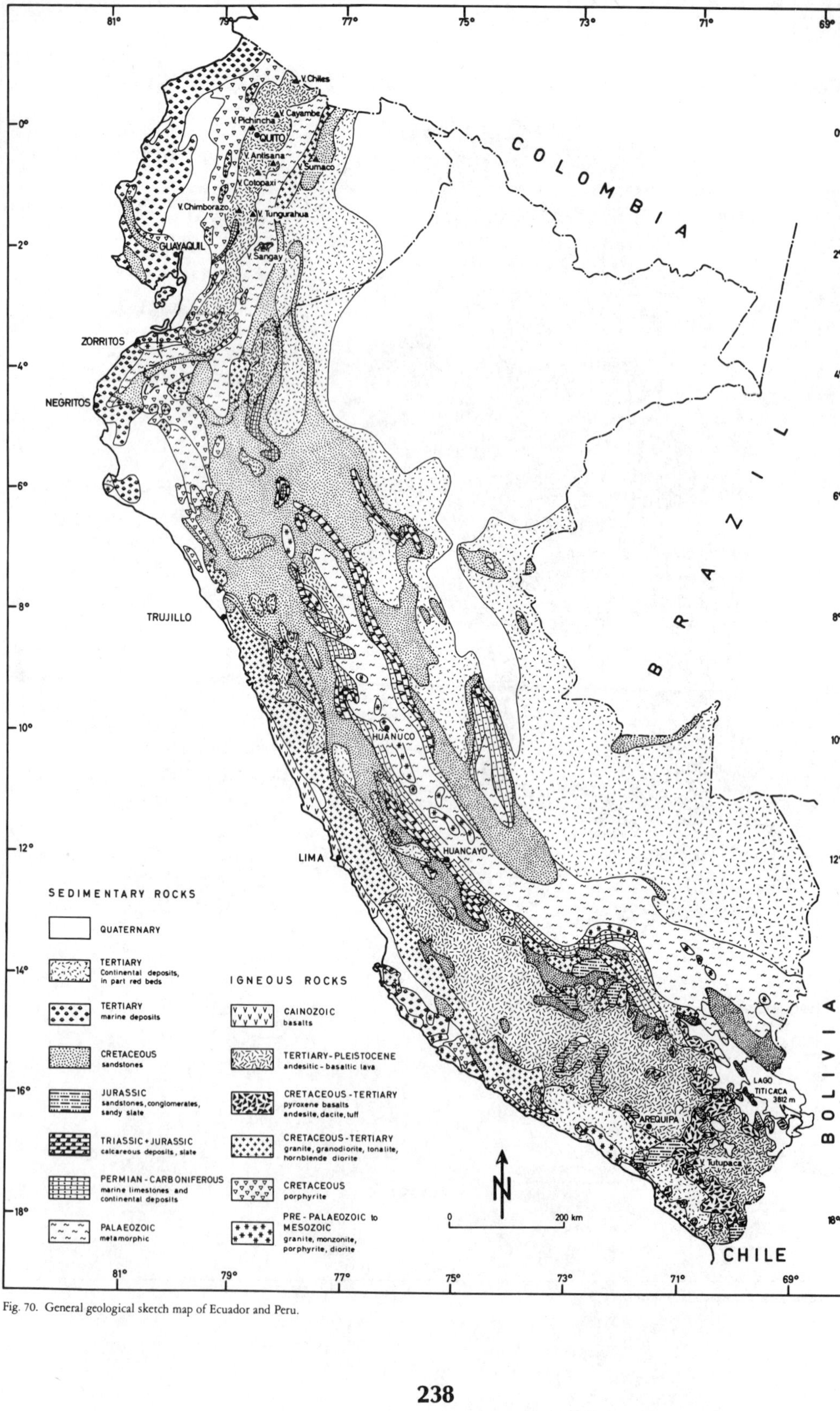

Fig. 70. General geological sketch map of Ecuador and Peru.

ad p. 96

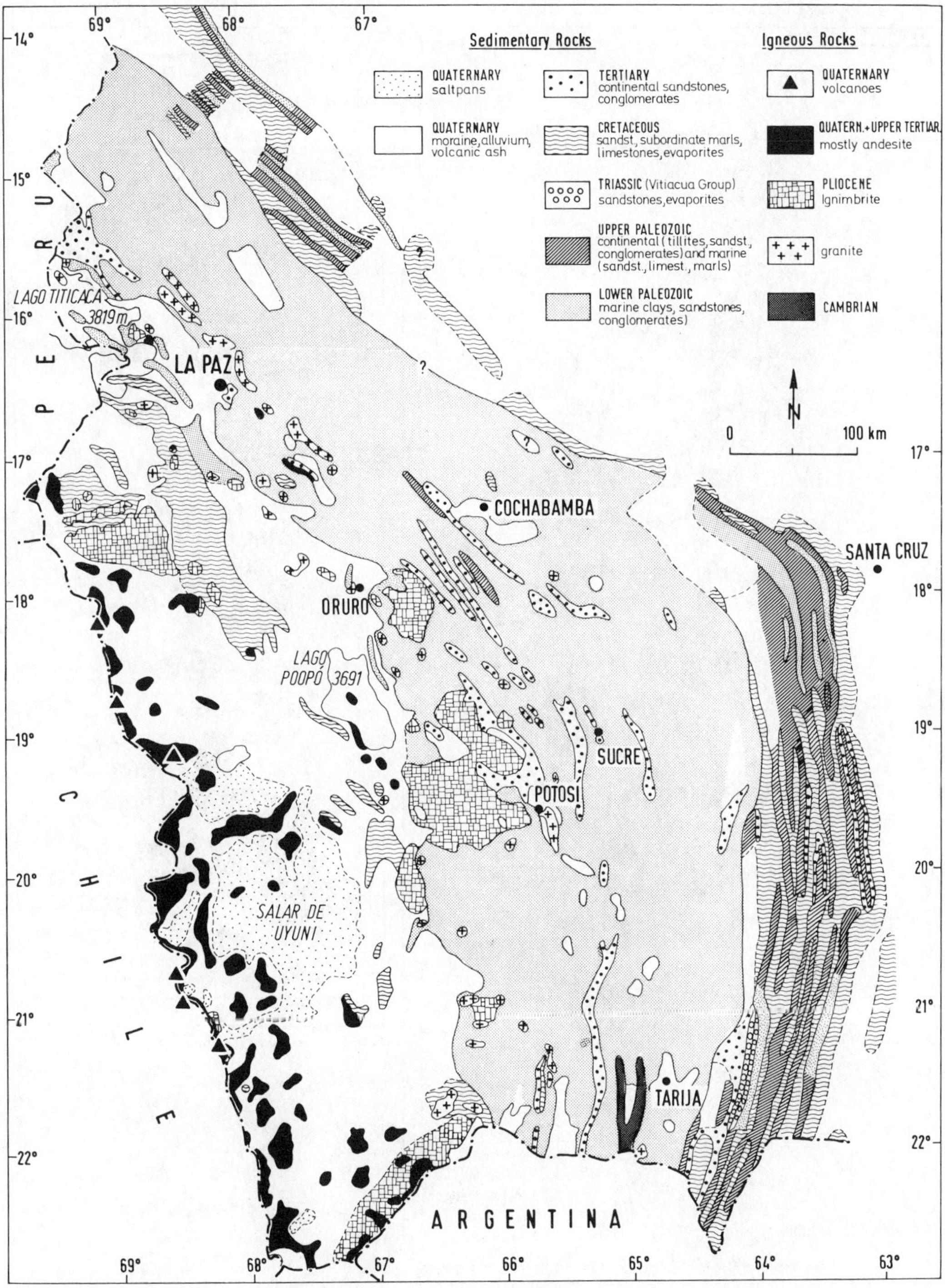

Fig. 72. General geological sketch map of Bolivia.

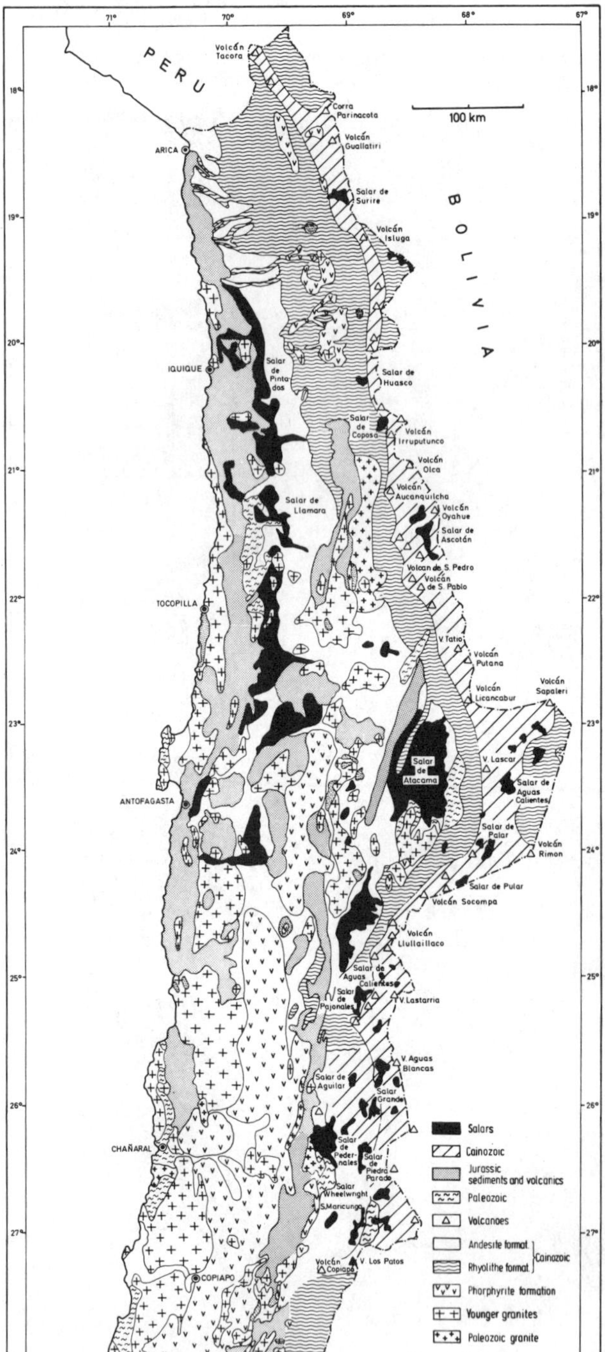

Fig. 79. General geological sketch map of northern Chile.

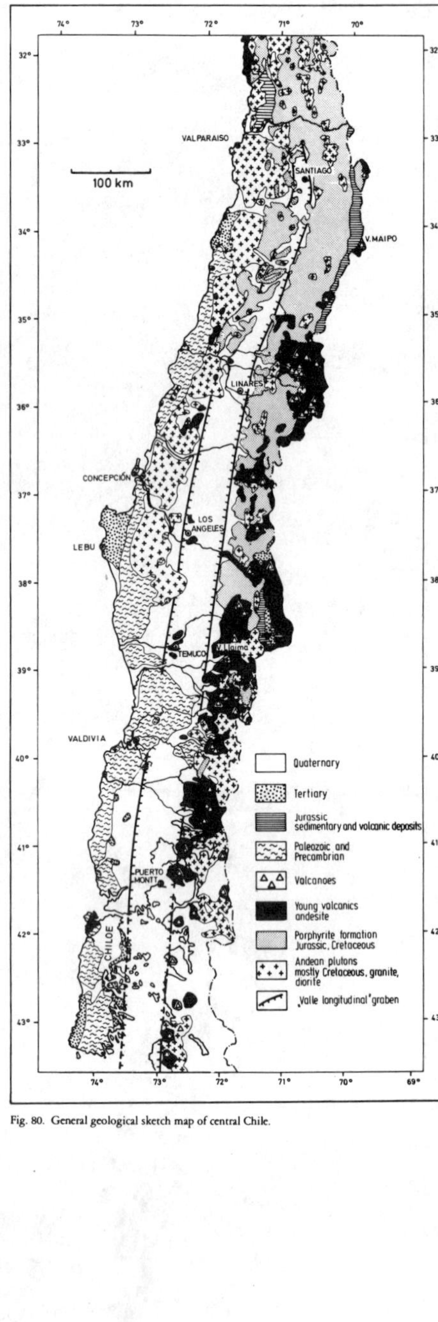

Fig. 80. General geological sketch map of central Chile.

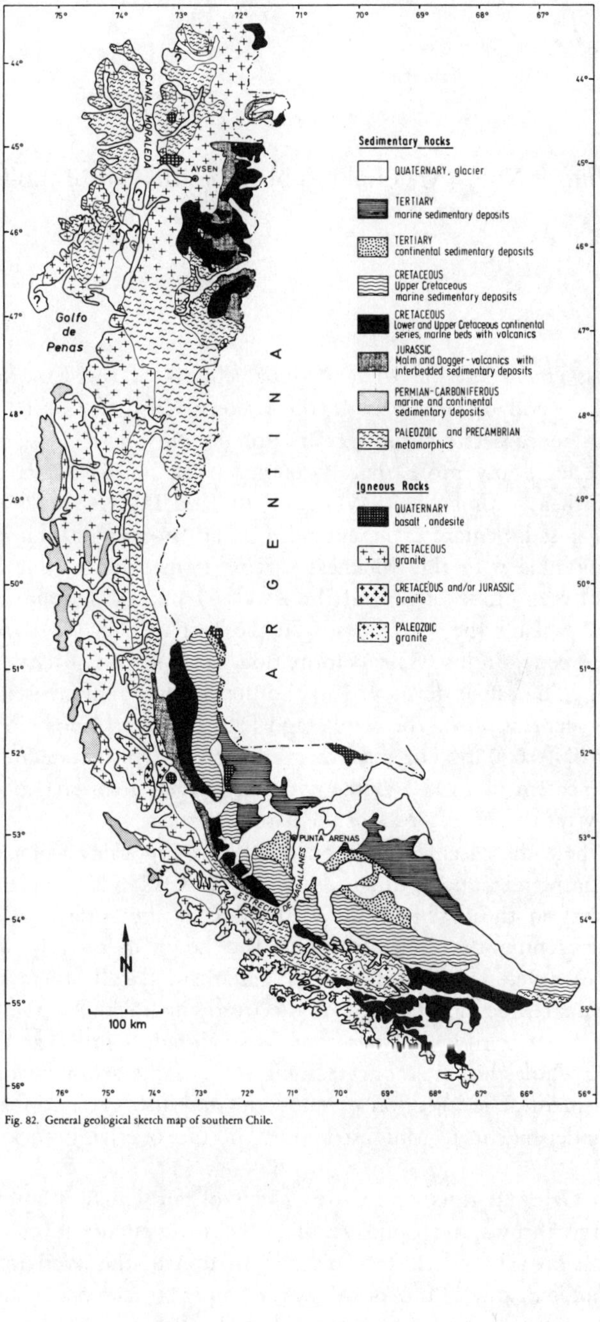

Fig. 82. General geological sketch map of southern Chile.

[*Editor's Note:* Material has been omitted at this point.]

During the post-Triassic period the individual **chains of the Colombian Andes** were subjected to varying geodynamic stress. In the **Eastern Cordillera** JULIVERT (1970) clearly established the difference between the tectonics of the pre-Triassic basement and the recent overlying strata. The locally more than 11,000 m thick marine Cretaceous exhibits only gentle folding which according to JULIVERT (1970) and IRVING (1975) took place in the Tertiary. This thick sedimentary layer extended as an orographic unit through the entire eastern part of Colombia with the thickness varying from time to time but always remaining in the order of magnitude of several thousands of metres. Despite the great thickness of the sedimentary package the "orogenesis" in the Tertiary did not cause a metamorphism nor any well developed jointing. The deformation close to the surface caused only gentle flexure-like folding. The range developed in the upper edge of the basement as an uplifted block between the depression of the Rio Magdalena and the Llanos along faults with differences in height of 10,000 m. The structures of the overlying strata are dominated by the deformation of the basement rock, and the overlying rock – according to JULIVERT (1970) – reacted in various ways to the tectonics of the basement rock.

In the north, where the basement is subdivided into a mosaic of blocks, the sedimentary cover remained completely undeformed and did not develop any independent structures. Inclinations and tilts in the basement show through at the surface. Where the basement experienced major arching the cover was also laid down in broadly sweeping anticlines. Monoclinal flexures develop over faults in the basement. Locally there are incipient occurrences of gravitational slides in the area of the overlying rock. At points where the basement tectonics are more complex, disharmonious small-radius folds can form in the overlying strata. On the whole, however, the tectonic style of the sedimentary cover is matched to that of the basement. The injection of salt series and also gravitational slipping can lead locally to small independent tectonic structures in the overlying rock in the Sabana of Bogotá.

The **Cordillera Oriental** thus constitutes a type of fan which is bounded on both sides by faults with large throws and opposite dip. The sedimentary cover has not undergone any major compression. The gentle tectonic deformation of the overlying strata is dictated by the basement and is mainly a "blueprint" of the deeper structure.

The structures were presumably first formed in the Paleocene and continued to develop right throughout the whole of the Tertiary. It is not possible to detect a typical orogenic compression phase. There is no evidence in the Eastern Cordillera of any folding taking place during the Cretaceous. The synclines consisting of Tertiary filling in the Eastern Cor-

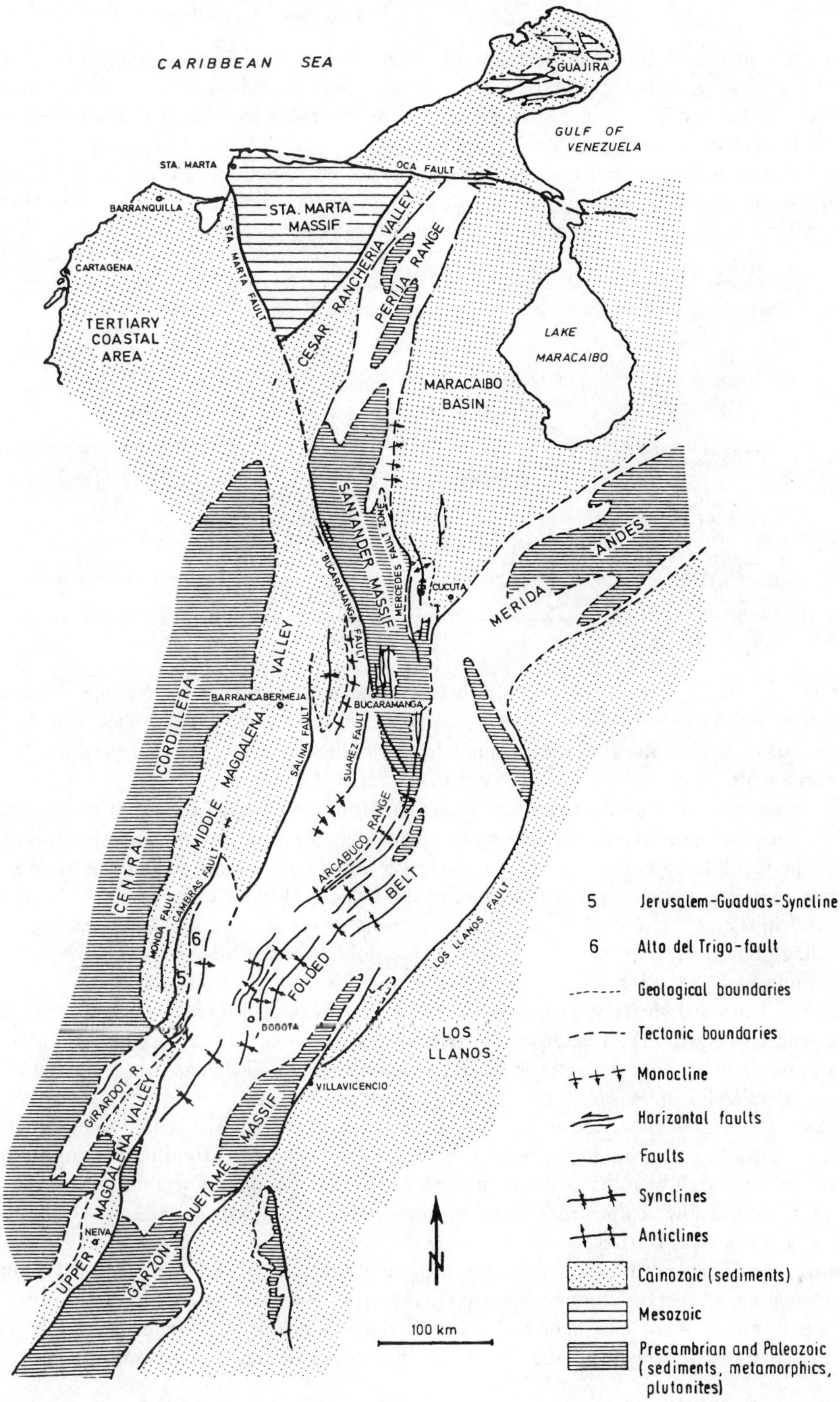

Fig. 108. Structural sketch of the eastern part of the northern Andes. According to JULIVERT (1970).

dillera were deformed together with the older complexes. The Tertiary sediments in the Magdalena valley and in the Llanos, which are several thousands of metres thick, formed between the Paleocene and the Pliocene. They are separated by several discordances and their molasse facies dips steeply.

To the west of the Santander massif runs the important Bucaramanga fault in the northern part of the Eastern Cordillera. This fault can be seen as extending into the Santa Marta fault. All researchers are in agreement here in postulating large horizontal faults which

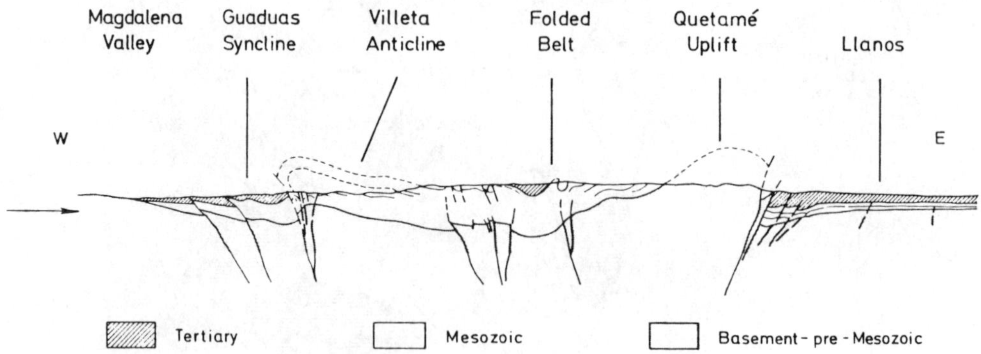

Fig. 109. Structural sketch of the Cordillera Oriental of Colombia. According to JULIVERT (1970).

together with the uplifting of the Cordillera Central and of the Sierra Nevada de Santa Marta have given rise to a complicated mosaic of blocks in the northern Andes, namely to Tertiary basins with unusually thick sediments and a block-like uplifting of crystalline parts of the basement.

The tectonic development of the **Cordillera Central** and of the **Cordillera Occidental** of Colombia followed an entirely different course from the start of the Triassic onwards. While the Cordillera Central functioned chiefly as a highland region during the Mesozoic and was uplifted as a positive block soon after the Upper Paleozoic orogenesis, the "eugeosynclinal" filling of the Western Cordillera at the end of the Cretaceous underwent regional bending. This Late Cretaceous to Early Tertiary folding phase affected the entire Western Cordillera of Colombia and its continuation into Ecuador. The Dagua group with its fine-grained slates and cherts several thousands of metres thick is described by many researchers as "eugeosynclinal" facies because it is overlain by the thick "Basic Igneous Complex". The sediments are slightly metamorphic and in places sharply folded. The facies and tectonic style are completely different from those of the Cordillera Oriental.

IRVING (1975) and other researchers are of the opinion that the Mesozoic filling of the Western Cordillera in the uppermost Cretaceous experienced a significant orogenesis. Because of the flysch-like character of the sediments and the slight metamorphism under high pressure and low temperature some researchers postulate that a subduction occurred in this case during the Tertiary.

During the Mesozoic and Cainozoic the three main branches of the Cordillera of Colombia experienced a separate tectonic history. Recent fractures were often predetermined by the older tectonics of the basement and up to the Late Cainozoic they intensified the block faulting. This can be seen from the enormous thicknesses of the Tertiary and Quaternary in the Cauca and Magdalena depressions between the mountain chains and the subsidence basin in northern Colombia.

The geodynamic development of the **Andes in Ecuador** followed a similar course. As in Colombia, the Western Cordillera is characterized during the Jurassic and Cretaceous by thick volcanic series (Formación Piñon) and by volcanic-sedimentary overlying strata (Formación Cayo) – nowadays known as the "Basic Igneous Complex". In the west a broad coastal lowland lies in front of the Cordillera; its basement has been relatively well studied in the course of oil exploration work. Drilling has shown that the base of thick Cainozoic sedimentary basins in this coastal plain is formed by the Cretaceous Cayo Formation. Cainozoic fracturing produced a block mosaic here as well.

The most striking tectonic element in western Ecuador is a WNW-ESE trending highland region between 1° and 2° 10′ S in the area of Guayaquil where the Jurassic-Cretaceous series of the "Basic Igneous Complex" come to the surface. This Cordillera of Chongon-Colonche runs perpendicular to the strike of the Andes and has therefore given rise to various speculations. KEHRER (written communication 1977) interprets the Cordillera and the associated Daule ridge, which extends north into the Manta-Portoviejo region, as the result of positive and negative fracturing. The ridge corresponds to a positive gravitational anomaly of +200 mgl. Along the Cordillera of Chongon-Colonche there are other transverse elements in Ecuador which also appear in the Oriente. According to KEHRER the Daule ridge tilted in the post-Miocene period with the inclination greatest towards the edge of the present western Cordillera. This formed the N-S trending Guayaquil depression which is filled with Quaternary sediments. The Daule ridge, which presumably formed in the Upper Cretaceous as a broad highland region, may extend to the east as far as the Oriente in Ecuador. This is where the transverse element of the Cononaco highland region is located which can be detected in the Middle Cretaceous from the thinner sediments. If the Daule ridge and the Cononaco ridge belong to the same transverse element then, in KEHRER's view, no northerly directed lateral shift occurred along the so-called Andean Guayaquil fault.

While the rock series of the coastal plain are chiefly fractured by recent fracture faults and no clear bending is detectable, the **Cordillera Occidental (Sierra)** was hit by episodes of sharp folding in the Mesozoic and Cainozoic. According to KEHRER & KEHRER (1969) and GOOSSENS (1970) an important folding phase took place at the end of the Cretaceous. FAUCHER & SAVOYAT (1973) in addition mention folding phases in the Western Cordillera dating from the Lower Eocene, the end of the Oligocene and the boundary between the Miocene and Pliocene. These Tertiary folding processes acted to a large extent like fracture faults in the rock series of the coastal plain and in the Eastern Cordillera (Cordillera Real). From my own observations I can say that flysch-like sediments of the Upper Cretaceous and molasse-like Tertiary series are sharply folded and separated by discordances in the Sierra.

HERBERT (1977) states that from the Paleozoic to the Lower Cretaceous the **Cordillera Oriental of Ecuador (Cordillera Real)** was a continental region of low elevation on the eastern margin of which the marine series of the Margajitas-Formación were deposited. The flat truncated upland of the Eastern Cordillera was uplifted and exposed to erosion in the Upper Cretaceous (Senonian). In the west the eroded material formed the more than 2,000 m thick flysch-like San Juan Formación (KEHRER & KEHRER 1969) of the deep Maestrichtian/Campanian and in the Oriente the approximately 700 m thick Tena For-

Fig. 110. Recent fracture tectonics in Cainozoic pyroclastics near Guayllabamba, north of Quito/Ecuador.

mación of the Maestrichtian. Whether the most important phase of folding in the Cordillera Oriental occurred as early as in the Albian, as KEHRER & KEHRER (1969) assume, is a question that cannot be positively answered because of the lack of corresponding sediments in the area of the actual Eastern Cordillera. FAUCHER & SAVOYAT (1973) do not mention any folding phases but only uplift processes in the region of the Cordillera Real in the Mesozoic and Cainozoic. HERBERT (1977) cites K/Ar datings on micas, crystalline schists, granites and granodiorites from the Eastern Cordillera which yielded ages of between 60 ± 2.2 and 30.0 ± 1.3 m. y. thus indicating different episodes of heating during the Tertiary. Major uplift episodes in the Cordillera Real took place according to FAUCHER & SAVOYAT (1973) in the Lower Eocene at the Miocene/Pliocene boundary and at the end of the Pliocene. The graben-like intra-Andean trench of Ecuador formed between the two Andean chains of the Sierra and the Cordillera Real as a result of this strong uplifting which also affected the Sierra. The internal tectonic structure of this depression is still to a large extent unknown. However, deep-reaching fracture zones probably exist, some of which were definitely triggered by expansion, as the magnificent recent volcanoes in this region show.

The **sub-Andean zone of Ecuador** (Oriente) corresponds to a sweeping basin in which marine and continental sediments over 10,000 m thick were deposited between the Devonian and the Quaternary. Between the Cordillera Real and the Brazilian shield the region is divided into individual subsidence zones which underwent epirogenic movement at various periods in time. For example, the Cretaceous Napo facies was definitely laid down in a largely closed basin. According to FAUCHER & SAVOYAT (1973) during the Campanian and at the boundary of the Cretaceous and Tertiary this region was affected by weak undulations while an imbricate structure inclined towards the east was laid down at the Miocene/Pliocene boundary.

Southern section (Peru, Bolivia, Argentina, Chile)

The **Cordilleras of Peru,** which morphologically act as a wall, unlike the northern and southern sections of the Andes have undergone violent tectonic deformation since the start of the Mesozoic but in particular in the Cainozoic. STEINMANN (1929) in his monograph, which is still the only review on the subject, noted that in the Peruvian Andes the fold structure is more impressive and more clearly developed than in other sections of the range even if the compression is less intense than in alpinotype folded mountains. He drew attention to flat-to-steep overthrusts which resulted in imbricate structures, for example in the region of Cerro de Pasco, but he stressed the limited horizontal extent of these overthrusts. With a few exceptions the inclination of the folds and overthrusts is to the northeast.

The greatest amount of folding occurs in a zone about 150 km wide between the coast and the Marañon line in the east. To the east of this the folding becomes more gentle and fades away in the foreland of the Amazon. He distinguished between three fold belts running from NE to SW in Peru:

- an outer zone with moderate folding without any great overthrusting and without an imbricate structure (to the east of the Rio Marañon)
- a middle zone with maximum folding, frequent overthrusts and imbricate structure
- an inner zone with weak undulations

In addition to this folding STEINMANN frequently mentions the great importance of block-like fracture faults. He identifies three "Andean" orogenetic episodes; the Peruvian folding in the Upper Cretaceous, the Incaic folding in the Eocene and the Quechuan folding in the Lower Pliocene. Forty years later BELLIDO (1969) summarized the result of many detailed studies which had in the meantime produced great regional progress in geological knowledge. While he by and large confirms the temporal classification of STEINMANN's orogenic phases, he assumes two large basins separated by the geanticline in the east and a highland region in the west which was more extensive than the present coastal zone. According to BELLIDO thick volcanics and interbedding of marine and continental facies occurred in this subsidence zone during the Jurassic and Lower Cretaceous. As the geanticline is approached in the east the continental facies increases. On the other side of the Marañon-Mantano geanticline a miogeosyncline stretching as far as the Brazilian shield developed. Various marine ingressions poured into this miogeosyncline during the Lower Cretaceous. The region of the eugeosyncline in the west and parts of the miogeosyncline were filled mainly with red terrestrial sediments and volcanics in the period from the Upper Cretaceous to the Pliocene.

In the last ten years research into the Mesozoic and Cainozoic geodynamic history of the Peruvian Andes has regionally made great progress. At the same time the picture became much more complex. Marine facies receded in the sediments during that period, also there are very few good index fossils available, therefore radiometric dating was used

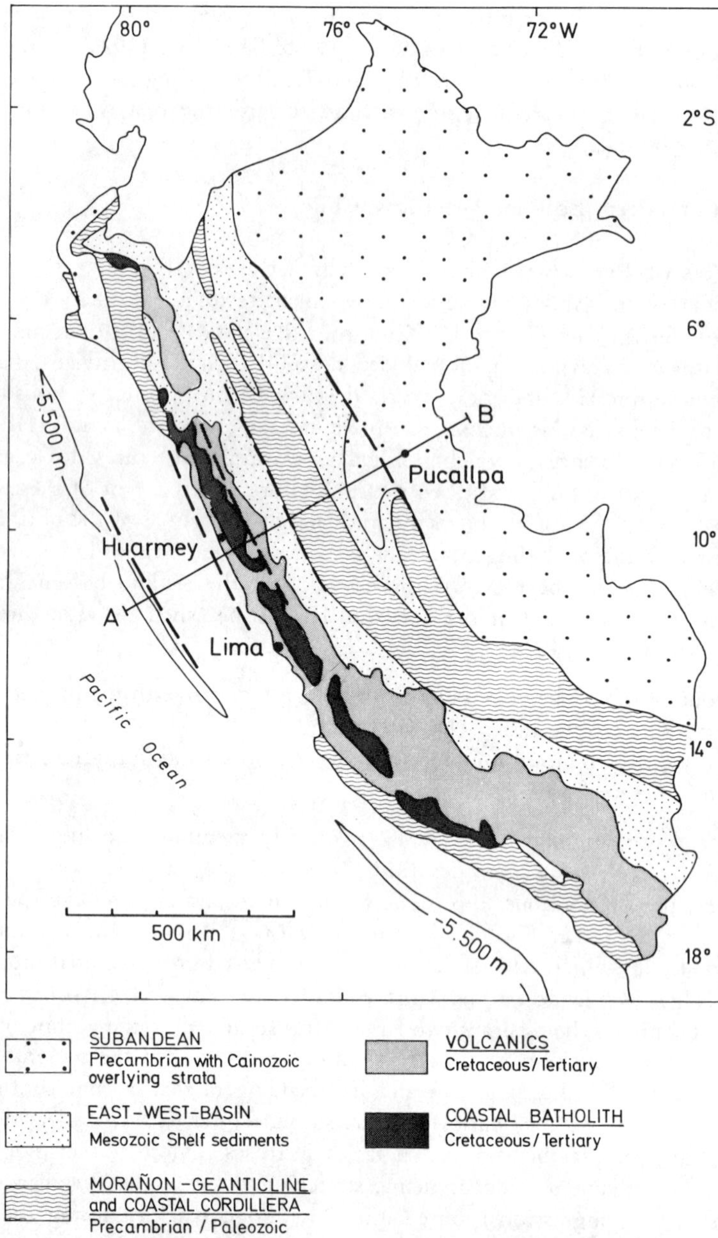

Fig. 111. Geological-tectonic sketch of Peru during the Cretaceous and Lower Tertiary. In Figs. 112 and 113 an attempt is made along profile line A – B to demonstrate tectonic and magmatic events occurring 100 m.y. ago (112) and about 95 m.y. ago (113). According to MYERS (1975 b).

also to study the tectonic processes. The tectonic structures of the Cordilleras were more carefully catalogued. For example, AUDEBAUD, LAUBACHER & MAROCCO (1976) list the following elements for a cross section of the Andes in southern Peru between the Pacific and the Brazilian shield during the Mesozoic:

volcanic belt (coast) with sialic basement
neritic marine basin (Western Cordillera)
mainly continental basin (Altiplano)
positive zone (Eastern Cordillera)
continental basin (sub-Andean)

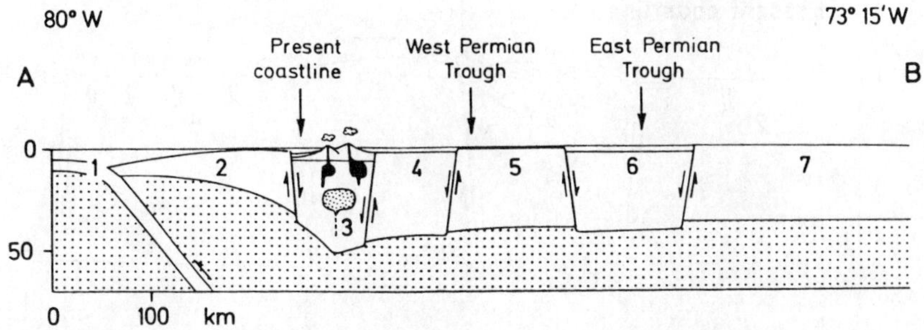

Fig. 112. Profile A – B in Fig. 111: Subduction of oceanic crust (1) and movement of large crustal blocks: 2 = Paracas block; 3 = Paramonga block; 4 = Chavin block; 5 = Marañon block; 6 = East Peruvian Trough block; 7 = Brazilian Shield. The rise of the coastal batholith is indicated (closely spaced dots) in the Paramonga block; the Casma volcanics flow from this batholith. Continental crust = white; mantle = widely spaced dots. Diagram according to MYERS (1975 b).

Three NW-SE trending folds formed these elements, namely in the Upper Cretaceous, the Upper Eocene (marine folding) and at the Miocene/Pliocene boundary. The zonal structure of the Peruvian Andes, in these authors' opinion, is evidence of the existence of a subduction zone dating from the start of the Mesozoic, although only the magmatic development has been clearly elucidated.

Other researchers have devoted themselves to studying in particular the extensive plutonite bodies in Peru and they therefore reject a simple plate tectonics model as a means of explaining the development of the Peruvian Andes. COBBING & PITCHER (1972) did not find any rock material (e. g. ophiolites) in Peru that clearly derives from an oceanic crust. STEWART, EVERNDEN & SNELLING (1974) came to the conclusion that the plutons of the coastal batholith are with few exceptions posttectonically intruded so that no parallel exists between folding and intrusions. MYERS (1975 b) stresses that the tectonic style of the Peruvian Andes corresponds to an extensive range of fault mountains made up of various crustal blocks and that this cannot be reconciled with a process of oceanic subduction. Even earlier than MYERS two other researchers, DEBELMAS & TROTTEREAU (1964) discovered that uplift events predominated over compression forces in the development of the Peruvian Andes. NOBLE, MCKEE, FARRAR & PETERSEN (1974) studied the relationships between volcanism and tectonic events in the Cainozoic of Peru and discovered one or more volcanic episodes and intensive deformation during the Paleocene and Eocene, a volcanically and tectonically quiet period in the Oligocene, a reactivation of volcanism and tectonic events at the start of the Miocene with a tectonic paroxysm in the Middle Miocene and intensive activity in the Pliocene. The orogenesis in the Middle Miocene and the Late Tertiary volcanism is seen by these authors as being directly correlated with the start of a very rapid spread of the Eastern Pacific Ridge.

According to SZEKELY (1967, 1969), CHANOVE, MATTAUER & MEGARD (1969), AUDEBAUD (1970), DALMAYRAC (1973) or MEGARD (1973) the tectonic analysis of individual sections of the Peruvian Andes showed that the Mesozoic/Cainozoic overlying rock was locally exposed to violent folding, foliation and overthrusts. From the region of Lago

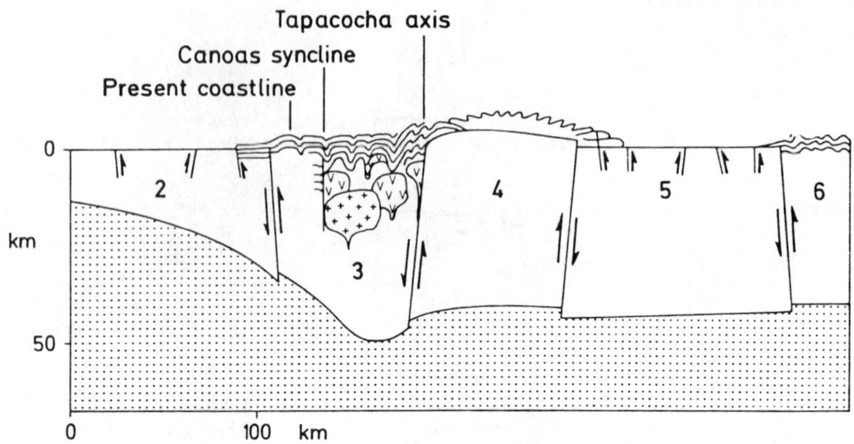

Fig. 113. Part of the section shown in Fig. 111: The diagram indicates various steps in the deformation of the sediments of the Mesozoic and of the volcanics as a result of the rise of various coastal blocks about 95 m. y. ago. Numbers explained in Fig.112. According to MYERS (1975 b).

Huallacocha and between Cocha and Yauricocha in the Western Cordillera ESE of Lima, SZEKELY described what is for Andean conditions a rather complicated fold structure with tight folds, inverse throws, upthrusts and tectonic imbrication extending as far as overthrusts in the northeast. The most important deformation period is situated in the Early to Middle Tertiary.

In the massif of Pirin southwest of Lago Titicaca CHANOVE, MATTAUER & MÉGARD (1969) found intensive folding and westward-directed overthrusting to the west of the Eastern Cordillera in terrestrial red sediments of the Cretaceous. They were created by two different tectonic phases, each differing in style: The older phase – at the end of the Cretaceous in the Eocene – created recumbent folds and overlapping imbrication; the Tertiary of the Formación Puno is superjacent upon this discordance. The more recent phase – after the Oligocene and probably before the Miocene/Pliocene boundary – created simple structures with partially reactivated imbrication. AUDEBAUD (1970) observed a multi-phase complex tectonic development with significant tangential movement and NW-SE structures in Cretaceous/Tertiary red sediments of the Eastern Cordillera of southeast Peru near Sicuani and Nunoa. DALMAYRAC (1973) drew a line through the Andes of Peru between 9° and 11° S. He describes the Andean chains as intra-cratonic mountains that have developed symmetrically on a Precambrian/Paleozoic continental basement. The two epicontinental basins in the west and in the east have undergone multi-phase tectonic stressing which produced tight folds and fracture cleavage. The intensity of the deformation increases from the coastal zone to the central Andes and rapidly fades away in the sub-Andean zone.

The Andean geodynamics are marked by three tectonic phases: one at the end of the Cretaceous – synchronous with the end of the marine sedimentation; a violent phase between the Upper Eocene and Lower Oligocene – both with the onset of NW-SE trending axes; and a phase in the Pliocene which gently deformed the young andesitic volcanics. It is often difficult to distinguish between the first two phases. All three tectonic episodes however created structures with the same NNW-SSE trend. MEGARD (1973) drew up a geological map between 10° and 13° S down to the sub-Andean region and in this region he detects four folding phases: Upper Cretaceous, Lower Oligocene, Middle Miocene and Middle Pliocene which are interrupted by phases of expansion with volcanism. Individual Andean elements were subjected to various stresses in the process. For example, the phase in the Middle Oligocene produced strong deformation in the Western Cordillera but weaker deformation in the Eastern Cordillera while the Pliocene phase on the other hand produced weaker deformation in the Western than in the Eastern Cordillera.

With the aid of radiometric dating FARRAR & NOBLE (1976) and MÉGARD & PHILIP (1976) tried to unravel the Late Tertiary and Early Quaternary tectonic-magmatic development of the Peruvian Andes. According to these authors the folding started in Central Peru between 17 and 15 m. y. ago and the subsequent uplift and erosion 10.5 m. y. ago. The authors correlate this with a worldwide variation in the movement of the lithosphere plates. MEGARD & PHILIP contrast the regions to the north and south of 15° S. In the north – northern and central Peru – the hypocentres of the quakes dip beneath the continent at an angle of 10 – 15° to the northeast. There is no recent volcanic activity, and compression tectonics prevail. South of 15° S – in southern Peru and northern Chile – the hypocentres dip to the east at an angle of 30°. Calc-alkaline volcanism is still active and recent expansion tectonics are still having an effect.

A critical evaluation of this brief review of the geodynamic history of the Peruvian Andes since the Mesozoic reveals that on the one hand the fundamental points of STEINMANN's description are confirmed by more recent research. The intensity of the "Andean" deformations is stronger in Peru than in the northern and southern Andes. The range clearly rests on top of continental crust. Oceanic rock elements are lacking at the surface and the calc-alkaline chemistry of the enormous masses of young plutonites and volcanics does not indicate that they originated directly from a submerged oceanic plate.

Some signs point toward the continent having expanded extensively towards the west during the Mesozoic-Cainozoic period. The term "eugeosyncline" for the Western Cordillera is imprecise. Fracture movements of thick crustal blocks may be a major tectonic principle. Episodes of expansion in the crust promoted the rise of granodiorites and of corresponding volcanic rocks, particularly in western Peru. The geodynamic history of the **Bolivian Andes** since the start of the Mesozoic is characterized primarily by fracture and block tectonics. From west to east the following morphological and tectonic macroelements of the range emerged in the Cainozoic (Fig. 105)

Cordillera Occidental
Altiplano
Cordillera Oriental with Cordillera Real
Mountainous region of eastern Bolivia
Sub-Andean zone

[*Editor's Note:* Material has been omitted at this point.]

This tectonic zone of weakness is said to have been formed as early as in the Precambrian and some researchers extend it to the west right up to the Pacific coast at Arica.

All researchers agree that the Andean region of Bolivia was not formed until the Tertiary. Older, Mesozoic deformation processes did not have any decisive effect, even if this seems possible from the general statements made by BOTELLO, MARTINEZ, TOMASI & ZUBIETA (1973). The first decisive uplifts of the Western and Eastern Cordilleras took place in the Middle Tertiary and simultaneously with them the graben-like subsidence of the Altiplano which lasted up to the Late Tertiary. The broad subsidence zone of the Altiplano was filled from the west and from the east with more than 10,000 m thick, chiefly continental sediments which in places developed as an intra-Andean molasse facies. According to AHLFELD (1970) the most important uplift phase in the Bolivian Andes took place in the Pliocene. The Altiplano was also affected by fracture tectonics and broken up into blocks, and gentle folding occurred only locally.

In contrast to the Peruvian Andes no reports of recent deformation processes leading to intensive folding, foliation or overthrusts, have so far been received from the individual regions of the Bolivian Andes. The clearly dominant tectonic inventory was instead generated by fracture and block tectonics.

No intensive studies have been made of tectonic events occurring in the Andean part of Bolivia since the start of the Mesozoic. AHLFELD (1970) comes to the conclusion that the entire Andean block was subjected to mountain-forming forces in the Late Tertiary with an important folding process taking place in the Early Pliocene, while the most important uplift phase occurred in the Late Pliocene. According to DOBROVOLNY (1962) Pleistocene sediments also contain evidence of uplift processes. BOTELLO, MARTINEZ, TOMASI & ZUBIETA (1973) in the explanation to the "Mapa Tectónico de Bolivia" found that as a result of the Andean orogenetic cycle the intensity and geometry of the recent tectonic inventory was controlled and marked by older Variscan structures. A folding phase occurred presumably in the Pliocene in the sub-Andean zone. The intramontane basins of the Variscan orogen were affected by two tectonic phases. Cretaceous to Upper Eocene red sediments were folded between the Pre-Oligocene to Lower Oligocene and the volcanics and pyroclastics of the Lower Oligocene to Pliocene were displaced by two deformation phases during the Pliocene. Only a phase of recent fracture tectonics can be detected in the Late Cainozoic sediments of the Altiplano and of the Cordillera Occidental.

However, no special detailed studies of the most recent tectonics have so far been carried out. KUSSMAUL, HÖRMANN, PLOSKONKA & SUBIETA (1977) distinguish between four different tectonic episodes in the Altiplano of SW Bolivia:

Lower Oligocene:	open folding, upthrusts and fissuration
Lower Miocene:	folding
Upper Miocene:	folding and upthrusting in ignimbrites and volcanoclastic material, reactivation of older structures
Pliocene (3 m. y.):	faults and displacements

The intensity of the gentle folding decreases from the Oligocene to the Pliocene; in the Pleistocene there was only compression and no more folding.

Further to the north in the Altiplano only three deformation episodes can be detected in the Tertiary, namely one in the Oligocene, one in the Middle Pliocene and one in the uppermost part of the Pliocene.

[*Editor's Note:* Material has been omitted at this point.]

In the **Chilean Cordilleras,** which are more than 4,000 km long, the Mesozoic-Cainozoic geodynamics created three major morpho-tectonic structures: the **Coastal Cordillera**

Fig. 117. Map showing the position of Chile within South America.

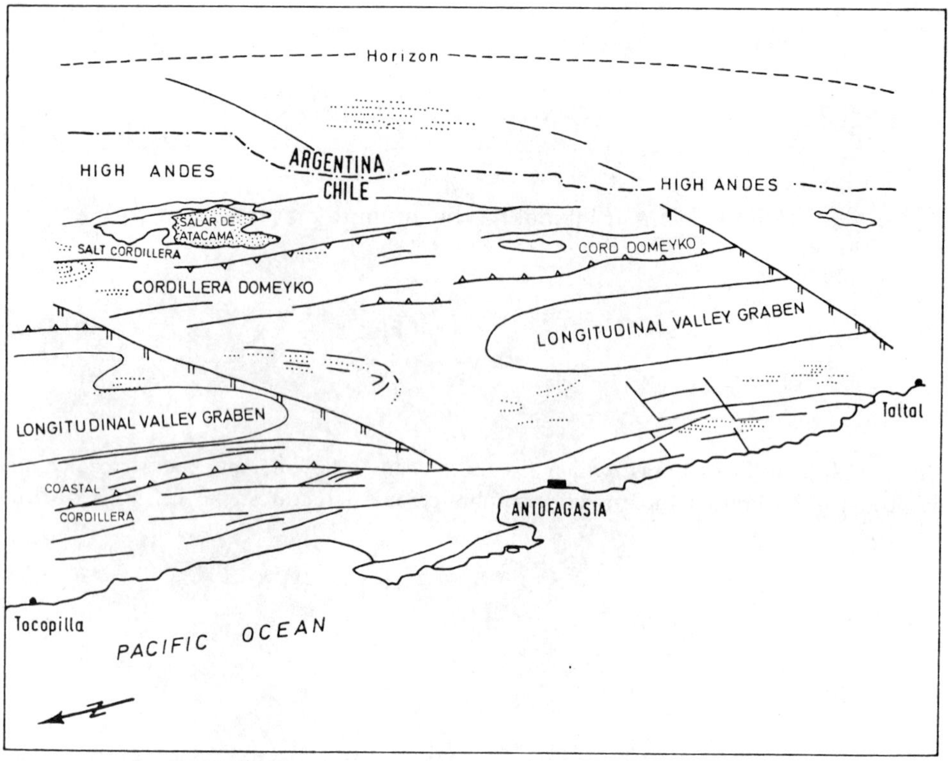

Fig. 118. Structural sketch of the Cordilleras of northern Chile. Drawn from satellite photographs by THOMAS (1970).

(Cordillera de la Costa), **longitudinal valley troughs** (valle longitudinal) and the **High Cordillera** (Cordillera de los Andes). These structures are not sharply delineated everywhere but they do have a different internal structure and they underwent non-synchronous development. Alongside this clear macro-division the tectonic development is characterized by two contrasting basin fillings dating from the start of the Mesozoic, namely the **Andean "geosyncline"** and the **Magallanes geosyncline.** Reference to the facies differences between these two sedimentation basins has already been made in an earlier chapter (p. 114). The deformation of these two regions took place neither synchronously nor with equal intensity. The Chilean Cordilleras thus did not emerge from a uniform "Andean geosyncline" as is frequently claimed in modern reviews of global tectonics. All researchers who have carried out geological-tectonic analyses in the region of the Chilean Cordilleras agree that the Andes rest on continental crust and that the inclusion of mantle material in the structural plan of the range since the start of the Mesozoic is a process of only secondary importance. The consequences arising from these facts must be kept in mind when fitting the Chilean Cordilleras into theoretical global-tectonic models. We will return to this point in one of the concluding chapters of this review.

The three large tectonic units of Chile were examined in detail in the last 15 years. There is only scope in this review to give a very simplified account of the results of these

Fig. 119. Lateral faults in the Devonian (dark-coloured area) of the Coastal Cordillera of northern Chile. Section from aerial photograph 331 – V – 147. Central dark block 3 km long, 1.6 km wide. Deformation presumably occurred in the Upper Mesozoic.

studies. Reducing the events to one simple formula we can say that the overlying strata of the **Coastal Cordillera** underwent only gentle folding but also significant block-like faulting that extends right up to the Holocene and is permeated by important intrusions. As in other parts of the Andes the strongly deformed and stable Precambrian/Paleozoic basement of the Coastal Cordillera, supported by enormous plutonic bodies, to a large extent prevented the effects of compression tectonics in the overlying strata. The considerable fracture faults in the overlying strata follow older zones of weakness in the orogen which were clearly traced in the upper strata. Nowadays the highest mountains in the Coastal Cordillera rise to 3,000 m in the Sierra Vicuna Mackenna south of Antofagasta, otherwise the average height attained in the northern part of the country is between 2,000 and 2,500 m, in central Chile between 1,000 and 2,000 m and south of 42° a height of 1,000 m is only seldom reached. The much less pronounced uplift that has taken place since the start of the Tertiary, in contrast to the High Cordillera, also the lack of Cainozoic volcanism indicate that the tectonic history of this early consolidated coastal belt followed an independent course from the start of the Mesozoic onwards.

The Coastal Cordillera is a narrow range. In the extreme northern part where the Andes bend in the NW direction near Arica, the structure wedges out almost completely at the surface. The range increases in width towards the south but seldom exceeds 80 km. The morphological and tectonic delineation from the longitudinal valley troughs – Pampa de Tamarugal in the north and "valle longitudinal" in the south – is very distinct in some places but fluid in others. In the arid northern part of the country this question can only be exactly answered by geophysical measurements because of the fanglomerate cover which is thick in places. No symmetrical geological structure can be detected within the Coastal Cordillera. Anabatic Cainozoic fracturing and block faulting occurred in the northern part of the country in individual isostatic events (STIEFEL 1974). In the Pleistocene and Holocene this gave rise to an extremely steep coastline at the edge of the Pacific which stopped the flow of all rivers with the exception of the Rio Loa and at other points generated marine terraces at various heights.

The internal structure of the Coastal Cordillera is characterized above all by chequerboard fragmentation (KATZ 1970). ZEIL (1964) summarized the tectonic analyses that had been made up to that time. The morphological N-S trend of the range gives a false impression. The overlying rock contains very noticeable NW-trending structures, E-W trending graben-like structures and important lateral faults which in certain areas are more important than the likewise present N-S structures (THOMAS 1970). Considerable vertical faults are reported in some areas. MILLER (1971) describes examples of transverse folding in northern and central Chile and SEGERSTROM (1970) interprets E-W trending lineaments on satellite images of northern Chile. Vertical movements at the coast were still occurring right up until very recently. Uplifting and subsidence over distances of several metres resulted from the severe earthquakes in southern Chile in 1960. Thus the region of the present Coastal Cordillera was a self-contained unit during the recent geodynamic events and it contrasts with the development that took place in the High Cordillera. The western boundary of the graben-like longitudinal valley troughs thus also forms a boundary in the internal tectonic history.

The geologically young **intra-Andean morpho-tectonic longitudinal valley** troughs in the South American Cordilleras are striking structures. In Chile the "valle longitudinal" that stretches over 1,100 km between Santiago and the southern end of Chiloe stands out in particular. Although all the researchers who have studied the tectonic development of the longitudinal valley argue that it is a geologically recent feature, there are few exact data available on the internal structure. KATZ (1963) summarizes the data gathered up to that time from petroleum prospecting between 35° 30′ and 43° 30′. According to these data the graben-like depression in this region is a very young structure (Pliocene/Pleistocene to Recent). Seismic measurements show that the sediments of the subsided Cainozoic are 2,000 – 3,000 m thick.

Fig. 120. Recent uplift at the Pacific coast near Antofagasta, northern Chile. Plio-Pleistocene littoral sediments with rich fauna, discordantly superjacent on the Jurassic "Porphyrite Formation".

The few scattered data available on the thickness of the Quaternary in the "valle longitudinal" indicate that the graben subsided by varying amounts and in individual sections from north to south in the most recent Cainozoic. According to KAUSEL & LOMNITZ (1969) the Cainozoic filling at 33° S (Santiago) is 500 m thick; at 36.5° S (Chillán) it is 1,000 m, at 40.5° S (south of Valdivia) it is 2,500 m on the eastern flank and 1,500 m on

Fig. 121. Northern end of the "valle longitudinal" near Santiago/Chile. Graben fault. In the centre, the city of Santiago, 520 m. In the background, the High Cordillera which quickly rises to 5,200 m.

Fig. 122. View from the Coastal Cordillera over the depression of the longitudinal valley to the N–S trending chains of the High Cordillera. Llay-Llay, Valparaiso province, central Chile.

the western flank, and finally at 41.5° S (Puerto Montt) it is about 4,000 m thick. Since these figures are based chiefly on seismic and gravimetric measurements the reasons for these significant differences are unknown.

Tectonic measurements are available only for small sections of the graben. AGUIRRE (1960), for example, describes at 33° S (Los Andes) on the western slope of the High Cordillera a sharp N-S fault with a vertical displacement of 2,000 m and ILLIES (1960) at 40° S on the western margin of the longitudinal valley towards the Coastal Cordillera describes a system of step faults.

The present morphology of the longitudinal valley makes it clear that this is a block which has subsided in relation to the Coastal and High Cordilleras. Thus, the depression – in which Santiago is located – is on average 500 m above sea level while the High Cordillera to the east is sharply offset and rises rapidly in several stages to a height of over 5,000 m. To the north of Santiago the graben-like structure is no longer clearly morphologically developed but broken up into individual basins which are separated from each other by island-like chains of hills. Nor is there any clear graben structure at the surface in the south down to 35° S (Talca). Between 36° S and 39° S (Linares – Temuco) the longitudinal valley is clearly and distinctly developed. Between 39° and 40° S comes the 70 km wide section of mountain range from Valdivia where an intermediate massif up to 800 m high and consisting of metamorphic rocks interrupts the longitudinal valley (Fig. 80). South of 40° S a distinct graben trough is developed again and it can be traced across the straits between the island of Chiloe and the mainland up to the Taitao peninsula (46° S).

The internal structure of the intra-Andean graben-like depressions which run parallel to the trend of the recent range is still largely unknown. The depressions were, however, caused by expansion and distortion processes in the crust and they formed during the Cainozoic. The amount of subsidence varies greatly and it can be deduced from the thickness of the filling of Cainozoic sediments: in Colombia – Rio Magdalena – the subsidence amounts in places to more than 10,000 m, in Ecuador – Quito depression – 1,000 – 2,000 m (estimated), in Bolivia – Altiplano – more than 10,000 m, in Chile – "valle longitudinal" – between 500 and 4,000 m. Simultaneous with the tectonic formation of these depressions the neighbouring Cordilleran arcs began to rise up; parallel to the grabens and half-grabens violent volcanism with a strong explosive component erupted; finally the deep-sea trenches off the Pacific coast formed synchronously with the movements in the continental crust. As a result of the foregoing the morphogenetic development of the Andes has been given some decisive impetus in the last 20 – 30 million years.

The morphology of the individual depressions in the northern part of the country is less easy to define than in the "valle longitudinal" between 33° S and 42° S. The course of the **Pampa de Tamarugal** can be traced in the western part of the country between 18° S and 25° S where it separates the narrow Coastal Cordillera from the western slope of the High Cordillera. Only the western slope can be detected from aerial photography in the "Atacama fault" and in the Cañon of the Rio Loa as a major fault extending over a length of 250 km. The eastern flank is hidden under thick Cainozoic detrital formations. Part of the intra-Andean graben system follows in the east between a Precordillera (Cerro Juan de Morales, Cordillera Domeyko, Cordillera Claudio Gay) (Fig. 118) and the actual High Cordillera. Nowadays this system contains a large number of salars between 18° S and 27° S, mostly trending N-S. Their tectonic formation is assumed by many researchers (VILAS 1975) and it has only been verified in detail at the Punta Negra salar between 24°

and 25° S by VAN OVERMEEREN & STAAL (1976) and in the Laguna del Negro Franzisco at 27° 30′ S by SEGERSTROM (1968). DINGMAN (1965) also assumes a tectonic formation for the salar de Atacama between 22° 45′ and 23° S. DIAZ DEL RIO, BONILLA & PERALTA (1972) discovered Cainozoic material 575 m thick in this area. Geological surveys and gravimetric measurements have shown that the basement of the Punta Negra salar was formed as a graben by N-S and NNE-SSW trending faults between the Upper Tertiary and the Pleistocene. The graben was filled by episodic detrital fans. Drilling down to a depth of 430 m revealed fanglomerates similar to those formed recently. To the east of the present salar a further gravimetric system with 550 – 750 m thick Quaternary filling was gravimetrically explored in the basement rock.

The salt pans in the region of the Chilean-Argentinian-Bolivian border in the High Cordillera are on average located at an altitude of between 2,000 and 4,000 m and extend between 18° 30′ and 27° S. They lie between chains of Cordilleras which rise in the west to heights up to 4,500 m and the High Cordilleras in the east, which are dominated by a large number of volcanic structures up to 6,880 m high. The most recent research, in particular that carried out by HOLLINGWORTH (1964), HOLLINGWORTH & RUTLAND (1968), GALLI (1967), MORTIMER (1973), MORTIMER & SARIC (1975) and CHONG (1977), attempted to reconstruct the sequence of uplift in the Andes during the Cainozoic by making use of geological and radiometric data. The block-like rise of the Andes in northern Chile began, according to them, in the Upper Miocene, built up in individual phases until the Middle Pliocene and, with the aid of new radiometric data from the widely distributed ignimbrite superstructure, can be traced up to the turn of the Pliocene and Pleistocene.

Strong epirogenic block movements are at the same time reported – as described above – from the High Cordilleras of Bolivia and Argentina. The Andes thus acquired their striking relief within a relatively short and recent period of the earth's history. At the same time that individual elongated blocks were uplifted the narrow intra-Andean grabens subsided. There is no doubt about their tectonic nature, and a detailed study of these features promises to yield important knowledge of the recent tectonic history of the range.

The tectonic development of the **High Cordillera of Chile** (Cordillera de los Andes) has become more accurately known since 1960 through a series of special geological studies and maps. Although the tectonic inventory and the structural style vary frequently from north to south, some general characteristics can be detected:

1. The general strike of almost all the structures is N-S in contrast to the Coastal Cordillera. Locally N-W striking lineaments show through the surface, in particular in the northern part of the country.

2. All the series up to and including the Lower Cretaceous are mostly gently folded although sharp folding occurs in places. The intensity of the folding increases from west to east and gives rise locally to complex structures in the area of the Chilean-Argentinian frontier.

3. It is not possible to fix exactly and over great distances the folding events that occurred between the start of the Triassic and the end of the Jurassic. The most important folding episodes occurred in the Middle and Upper Cretaceous. Gentle folding occurred locally in the Tertiary before the Miocene. Older ignimbrite strata (14 – 10 m. y.) in the northern part of the country are discordantly superjacent on the folded basement (Fig. 125).

4. Violent block movements in the Pliocene and Pleistocene created fracturing and faulting. The vertical faults are more important than the horizontal displacements.

[*Editor's Note:* Material has been omitted at this point and only the references cited in the preceding excerpts have been reproduced here. In the original, Evernden, Kriz, and Cherroni (1966), Katz (1963), Aguirre (1960), Illies (1960), Vilas (1975), Dingman (1965), and S. Diaz del Rio et al. (1972) were omitted.]

REFERENCES

AHLFELD, F., (1970): Zur Tektonik des andinen Bolivien.—Geol. Rdsch., **59**, 1124-1140.

AUDEBAUD, E. (1970): Premières observations sur la tectonique tangentielle polyphasée des terrains secondaires, de la Cordillère orientale du sud-est péruvien.—Acad. Sci., C. R. Sér. D, **270.** 26, 3190-3193, Paris.

AUDEBAUD, E., LAUBACHER, G. & MAROCCO, R. (1976): Coupe géologique des Andes du Sud du Pérou de l'Océan Pacifique au Bouclier Brésilien.—Geol. Rdsch., **65,** 223-264.

BELLIDO, B. E. (1969): Sinopsis de la Geologia del Peru.—Bol. Serv. Geol. Miner. (Peru), **22,** 54p., Lima.

BOTELLO, R., SUBIETA, T., MARTINEZ, R. & TOMASI, P. (1973): Mapa Tectónico de Bolivia.—4 p., Serv. Geol. Bolivia, La Paz.

CHANOVE, G., MATTAUER, M. & MÉGARD, F. (1969): Précisions sur la tectonique tangentielle des terrains secondaires du massif de Pirin (Nord-Ouest du lac Titicaca, Pérou).—Acad. Sci., C. R., Sér. D, **268,** 13, 1698-1701, Paris.

CHONG, D. G. (1977): Contribution to the knowledge of the Domeyko Range in the Andes of Northern Chile.—Geol. Rdsch., **66,** 2, 374-404.

COBBIN, E. J. & PITCHER, W. S. (1972a): The coastal batholith of Central Peru.—J. geol. Soc. London, **128,** 421-460.

COBBING, E. J. & PITCHER, W. S. (1972b): Plate tectonics and the Peruvian Andes.—Nature, **240,** 99, 51-53.

DALMAYRAC, B. (1973): Coupe Générale des Andes péruviennes.—Bull. Inst. Fr. Et. Andes, **2,** 1, 1-15.

DEBELMAS, J. & TROTTEREAU, G. (1964): Essai sur les grands traits structuraux et l'évolution des Andes du Pérou.—Rev. Géogr. phys. Geol. dyn., **6,** 4, 259-268.

DOBROVOLNY, E., (1962): Geologia del valle de La Paz, Bolivien.—Bol. Dept. Nac. Geol. (Bolivia) **3,** 152p.

FARRAR, E. & NOBLE, D. C. (1976): Timing of Late Tertiary deformation in the Andes of Peru.—Bull. geol. Soc. Amer., **87,** 1247-1250.

FAUCHER, B. & SAVOYAT, E. (1973): Esquisse géologique des Andes de l'Equateur. —Rev. Géogr. phys. Géol. dyn., (2), **15,** 1-2, 115-142.

GALLI, C. (1967): Pediplain in northern Chile and the Andean Uplift.—Science, **158,** 653-655.

GOOSSENS, P. J. (1970): The geology of Ecuador. Explanatory note for the Geological Map of the Republic of Ecuador.—Ann. Soc. Géol. Belgique, **93,** 255-263.

HERBERT, H.-J. (1977): Die Grünschiefer der Ost-Kordillere Ecuadors und ihr metamorpher Rahmen.—183 p., Diss Univ. Tübingen.

HOLLINGWORTH, S. E. (1964): Dating the uplift of the Andes of northern Chile. —Nature, **201,** 4914, 17-20.

HOLLINGWORTH, S. E. & RUTLAND, R. W. R. (1968): Studies of Andean uplift; part I, Post-Cretaceous evolution of the San Bartolo area, North Chile.—geol. **J.,** 6, 1, 49-62.

IRVING, E. M. (1975): Structural evolution of the northernmost Andes, Colombia.—U.S. geol. surv., prof. Pap., **846,** 47p.

JULIVERT, M. (1970): Cover and basement tectonics in the Cordillera Oriental of Colombia, South America, and a comparison with some other folded chains.—Bull. geol. Soc. Amer., **81,** 12, 3623-3646.

KATZ, H. R. (1970): Randpazifische Bruchtektonik am, Beispiel Chiles und Neuseelands.—Geol. Rdsch., **59,** 3 898-926.

KAUSEL, E. & LOMNITZ, C. (1969): Tectonics of Chile.—Simp. Panamer. Manto Superior Mexico 1968, **2,** 47-67.

KEHRER, W., & KEHRER, P. (1969): Die oberkretazische San Juan-Formation der Westkordillere Ecuadors. -N. Jb. Geol. Paläont.,Abh, **133,** 1-22.

KUSSMAUL, S., HÖRMANN, P. K., PLOSKONKA, E. & SUBIETA, T. (1977): Volcanism and structure of southwestern Bolivia.—J. Volc. Geotherm. Res., **2,** 73-111.

MÉGARD, F. (1973): Etude géologique d'une transversale des Andes au niveau du Pérou Central.—Thèse, 282p., Montpellier.

MÉGARD, F. & PHILLIP, H. (1976): Plio-Quaternary tectono-magmatic zonation and plate tectonics in the central Andes.—Earth planet. Sci. Lett., **33,** 231-238.

MILLER, H. (1971): Querfaltung in mesozoischen Schichten Nord -und Zentral-Chiles.—Münster. Forsch. Geol. Pälaont., **20/21,** 199-214.

MORTIMER, C. (1973): The Cenozoic history of the southern Atacama Desert, Chile.—J. geol. Soc. London, **129,** 505-526.

MORTIMER, C. & SARIC, N. (1975): Cenozoic studies in northernmost Chile.—Geol. Rdsch., **64,** 2, 395-420.

MYERS, J. S. (1975b): Vertical crustal movements of the Andes in Peru.—Nature, **254,** 672-675.

NOBLE, D. C., MCKEE, E. H., FARRAR, E. & PETERSEN, U. (1974): Episodic Cenozoic volcanism and tectonism in the Andes of Peru.—Earth planet. Sci. Lett., **21,** 213-220.

SEGERSTROM, K. (1968): Geologiá de las Hojas Copiapó y Ojos del Salado, Provincia de Atacama.—Chile Inst. Invest. Geol., **24,** 58p., Santiago.

SEGERSTROM, K. (1970): Apollo 7 photography in Antofagasta province, Chile —an interpretation.—Prof. Pap., U.S. geol. Surv., **700 D,** D 10-D 17.

STEINMANN, G. (1929): Geologie von Peru.—448 p., Heidelberg.

STEWART, J., EVERNDEN, J. F. & SNELLING, N. J. (1974): Age determinations from Andean Peru. A reconnaissance survey.—Bull. geol. Soc. Amer., **85,** 1107-1116.

STIEFEL, J. (1974): Zur tektonischen Interpretation jungkänozoischer Sedimente und Landformen in der Küstenzone Mittelchiles.—Geotekt. Forsch., **46,** 70-194.

SZEKELY, T. S. (1967): Geology near Huallacocha lakes, Central High Andes, Perù.—Bull. Amer. Assoc. Petrol. Geol., **51,** 7, 1346-1353.

SZEKELY, T. S. (1969): Structural Geology, Cochas to Yauricocha, Central High Andes, Peru.—Bull. AAPG, **53,** 3, 553-567.

THOMAS-NOVOA, A. (1970): Beiträge zur Tektonik Nordchiles.—Geol. Rdsch., **59**, 3, 1013-1027.

VAN HOUTEN, F. B. & TRAVIS, R. B. (1968): Cenozoic deposits, upper Magdelena Valley, Colombia.—Bull. AAPB, 52, 4, 675-702.

VAN OVERMEEREN, R. A. & STAAL, J. H. (1976): Floodfan sedimentation and gravitational anomalies in the Salar de Punta Negra, Northern Chile.—Geol. Rdsch., **65**, 1, 195-211.

ZEIL, W. (1964): Geologie von Chile. 233 p.—Berlin (Gebr. Borntraeger).

23

Reprinted from *Am. Assoc. Petroleum Geologists Bull.* **55:**1753-1758 (1971)

Continental Margin in Chile—Is Tectonic Style Compressional or Extensional?[1]

H. R. KATZ[2]
Lower Hutt, New Zealand

Abstract Sea-floor spreading and continental drift seem to imply, for the Pacific segment of South America, upthrust of the continent and compression along its margin. Geologic evidence, however, indicates extension in this area since at least the Miocene. Andean folding has migrated east, away from the Pacific and toward the continent, whereas the tectonic development near the continental margin, which is backland of Andean orogeny, is retrograde and has resulted in a breakdown of the crust mainly along old lineations. There extensional stress dominates the upper crust over an area 300–400 km wide.

The observed extensional deformation is not necessarily incompatible with the forces and stress conditions as postulated for an active continental margin. Compression farther inland away from this margin, and continentward migration of fold belts appear to be related directly to the destructive processes that occur in the continental border area.

Introduction

The hypothesis of sea-floor spreading and recent geophysical investigations have contributed to the discussion and understanding of deep-seated crustal structure and evolution along active continental borders. Extension of the Pacific crust from the East Pacific Rise, underthrust of the continent along an eastward-dipping plane marked by the foci of shallow to intermediate and deep-seated earthquakes, and/or westward drift of South America and upthrust over the Pacific plate along the same plane—are basic concepts which are generally accepted and not easily disputed. Implied is the existence of a belt between continent and ocean that is under compressional stress. This should be borne out by the tectonic style and evolution of the crustal segment adjacent to the continental border.

Off the coast, along the deep-sea trench bordering the continental margin, geophysical evidence has shown a thinning of the crust which was interpreted as an indication of extension (Worzel, 1965; Hayes, 1966; Le Pichon, 1968); sediments in the Chile Trench were found undisturbed (Scholl *et al.*, 1968). This evidence has been considered anomalous and incompatible with an active spreading of the ocean floor. In an attempt to reconcile the issues, Lomnitz (1969), on the basis of a local study in southern Chile, proposed a mechanism involving sea-floor spreading and compression in a form that would agree with his tectonic interpretation of the particular area. Also, new interpretations of profiler sections indicate that some disturbed sediments may in fact be present near the continental border and could represent buckled trench sediments, thus suggesting compression (Ewing *et al.*, 1969).

But the question whether compression or extension is the prevailing tectonic condition, as shown by the regional geology, has never been discussed. Surface and subsurface investigations on land now suggest that, during its Neogene history, the continental border area has been mainly under extensional stress; individual fault block movements and the depositional and facies pattern of upper Tertiary and Quaternary sediments indicate an area of disruption and breakdown with a marked absence of compressional features. Recent surface tension in an east-west direction—*i.e.*, across the trend of the continental margin—also is deduced from an analysis of deformation and mechanism of the great 1960 earthquake (Plafker and Savage, 1970, p. 1018). In seismic-reflection records of the sediments filling the trench, no structures produced by compressional forces were observed by von Huene and Scholl (1969); the only deformation reported at the foot of the continental slope is from subsidence and slumping (*cf.* Scholl *et al.*, 1970, Fig. 5).

Thus, extension rather than compression is the tectonic condition of the upper crust at the juncture of continental and oceanic plates. This condition and the observed extensional deformation along the continental margin, however, are not necessarily incompatible with an underthrust of the Pacific plate beneath the continent. Quite the contrary, as interpreted here,

[1] Manuscript received, June 23, 1970; revised and accepted, December 31, 1970.

[2] New Zealand Geological Survey; formerly Empresa Nacional del Petróleo, Santiago, Chile.

I am indebted to C. A. Fleming for critically reading an early draft of the manuscript, and to C. A. Burk for continued help and encouragement.

extension appears to be closely related to the underthrust.

Geologic Situation in Central Chile

The characteristic deformation pattern during the Neogene is one of block-faulting with local drag folds; folding due to crustal shortening, of an age younger than about Oligocene, nowhere is of importance west of the Andes. Although the regional morphotectonic arrangement is roughly north-south—*i.e.*, parallel with the coast (deep sea trench, Coast Range, Central Valley, Andes)—the trend of faults, fractures, and lineaments is mainly transverse (west-northwest to northwest). This orientation can be shown to be inherited from the pre-Andean (Gondwanan) architecture which, to a great extent, has been reactivated during Neogene and Holocene times (Katz, 1970). A second trend which is oriented northeast is particularly well developed near Valdivia (Fig. 1), where apart from widespread faulting Miocene volcanism and a more recent breakdown of the coastal segment are aligned in this direction (Illies, 1967).

Such lineaments oblique to the regional strike, which are of pre-Andean as well as post-Andean origin, include basement magnetic anomalies, *en échelon* arrangement of granitic intrusions, fold axes of Triassic rocks, fracture trends in the Coast Range and the Andes and tectolinears in general as revealed by photogeologic analysis, and Plio-Pleistocene andesitic volcanism in the Andes in the east and also within the Central Valley (Fig. 1 and unpublished material on file with Empresa Nacional del Petróleo, Santiago). These lineaments are also expressed by the parallel alignment of rivers, which generally run west-northwest in more or less straight courses across both the alluvial fill of the Central Valley and the uplifted, hard-rock core of the Coast Range.

Mainly bounded by northwest- and northeast-trending fault systems, crustal blocks throughout the late Tertiary and Quaternary have been subject to repeated, differential up-and-down movements. This is indicated by the thickness and facies distribution of terrestrial and marine deposits that fill the various local sedimentary basins (Katz, 1970). At the well-known locality of La Cueva within the Coast Range (34° 13′; Brüggen, 1950), 30 km inland and 300 m high, a regionally peneplaned and tilted granitic basement is capped by fluviatile gravels, which are overlain by fossiliferous marine sands; the transgression is late Pliocene, according to recent investigations by Herm (1969). Many promontories and peninsulas along the present coastline have moved differently and independently from the rest of the coast, commonly with a greater net upward displacement. From Punta Morro near Caldera in northern Chile (27°) Herm (1969, p. 20) mentioned such individual movements going on from the Pliocene to the youngest Pleistocene; fault line and fracture trends are north-northeast, northwest, and west-northwest, and thus correspond to the pattern shown previously.

During the Miocene in southern Chile, the Coast Range of Osorno was a high area of rugged topography with fault-controlled coastal cliffs. In the Pliocene or later a subdued peneplain formed and is tilted and partly uplifted to 700-800 m. Not only in the Coast Range, however, but also below the plains of the Central Valley on the east, vertical displacements have occurred widely throughout the later history, with various blocks moving independently of each other and partly in opposite directions. Conditions are greatly varied in the Osorno basin. In the central basin, basement is only 150 m below the surface. Within a few tens of kilometers, considerable thicknesses of sedimentary rocks overlie basement. In one area, marine-bathyal upper Tertiary beds are as thick as 1,100 m, but very little Quaternary is present. In another area of the basin, thick Tertiary beds are overlain by almost 1,300 m of Quaternary beds; the Tertiary beds are nonmarine, and the Quaternary deposits are terrestrial, glacio-fluviatile, and marine (Katz, 1970, Fig. 4). There is evidence of repeated transgressions and regressions, probably of tectonic origin, during the late Quaternary near Valdivia (Illies, 1960) and between Puerto Montt and Chiloé. Recent dislocations after the 1960 earthquake follow the same pattern (Lomnitz, 1969; Plafker and Savage, 1970). Also in northern Chile (Paskoff, 1966, 1970) and southern Peru (Rüegg, 1956, 1960) there is clear evidence of comparable later movements of coastal blocks.

Although the Central Valley between the Coast Range and Andes constitutes a generally depressed, elongate area, its complex internal structure and development were controlled by the same transverse pattern that is typical for the various uplifted blocks which form the Coast Range. Detailed mapping, in fact, nowhere has revealed longitudinal boundary faults as a characteristic and controlling feature (Fig. 1; Katz, 1970). Even farther east into the

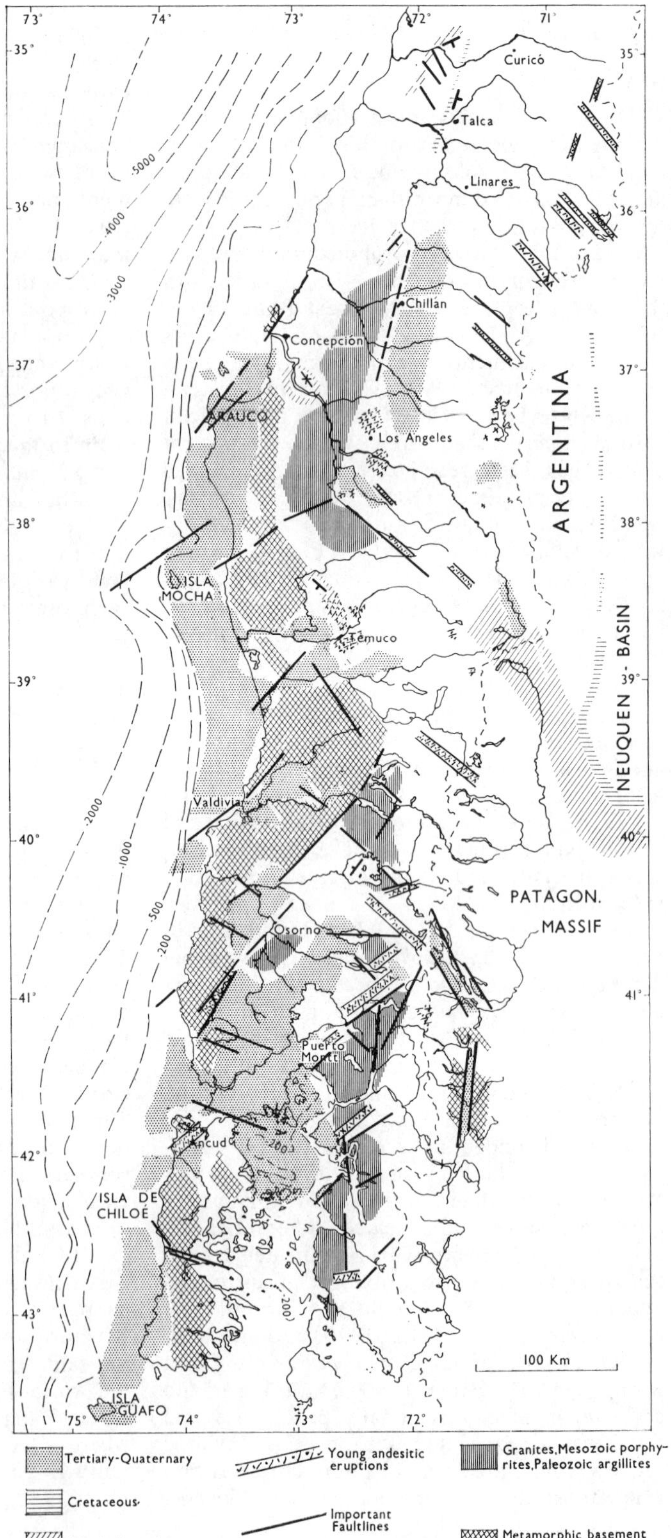

FIG. 1.—Structural synthesis cf central Chile, showing development of young, para-Andean marginal deeps in continental border area (late Tertiary to Quaternary). Basin and horst tectonics and alignment of andesitic eruption centers are within framework of reactivated structure pattern that is pre-Andean in origin. Offshore depth contours in meters.

Andes, mineralized fracture systems (Thomas, *in* Zeil, 1964, p. 131), major faulting, and the alignment of young volcanic centers have generally the same orientation oblique to coast and mountain range (Katsui and Katz, 1967; Miller, 1967; Katz, 1970). These alignments evidently reflect the main zones of weakness in a field of predominantly extensional stress, and it is concluded that a similar stress pattern must affect the entire area between coast and Andes.

The later history of this area as a whole shows a tectonic behavior of predominant extension, disruption by normal faulting, and a general break-up as is characteristic along the Pacific coast from Peru to Patagonia (Rüegg, 1960: Katz, 1962). Thus, geologically the continental margin shows no signs of being compressed, and there is no evidence of regional strike-slip faulting parallel with the coast; but the margin is an area of destruction and taphrogenic breakdown. Whereas extensive volcanism with large-scale extrusion of andesitic material dominates the main Cordillera, the tectonic movements between there and the coast, during the late Tertiary and Quaternary, are germanotype, with a tendency to total disintegration of the entire crustal segment. This entire area properly should be defined, in a tectonic sense, as the continental border governed by the same apparently characteristic stress conditions, tectonic style, and evolution. In central Chile this belt, which embraces the whole country from coast (or deep-sea trench) to the High Andes, is 200-300 km wide.

With regard to Andean orogeny, the Pacific borderland is backland, local sedimentary basins are para-Andean marginal deeps, and the Plio-Pleistocene volcanism in the Andes represents a subsequent magmatic phase. During successive stages Andean folding, involving compression and crustal shortening, everywhere migrated east away from the Pacific and toward the continental foreland (Gerth, 1955; Katz, 1964). Behind the eastward-migrating fold belts, the pre-Andean structure is rejuvenated and the old structural disposition of the continent is being revived; the post-Andean tectonic development is characteristically retrograde. In central and southern Chile the post-Andean tectonism has effectively erased most signs of Andean structure which at one time was superimposed (Fig. 1). Andean structure has been preserved only in the regional setting and arrangements of such first order elements as the Coast Range, Central Valley, and Andes.

Tectonic Pattern and Continental Margin

Apparently, the present state of retrograde development is related to a change in the stress field that occurred after the Andean orogeny, when the east-west compressive stress gave way to extensional stress. Although this is normal for later phases of an orogenic cycle, the question whether such an extension is consistent with, or related to, the particular tectonic environment of an active continental margin remains to be investigated.

If underthrust of the Pacific plate does occur, extension in the lithosphere must result where the oceanic plate bends downwards, *i.e.*, on the seaward side of the deep sea trench (Isacks *et al.*, 1968). Under the Chile Trench the crust is thinner than on either the continental or oceanic side (Worzel, 1965), a situation which is interpreted to indicate extension across the trench. Here subsidence is the result (Le Pichon, 1968), and there is no reason therefore to expect the trench sediments to be folded and compressed. A more recent model by Isacks and Molnar (1969) suggests that gravitational forces could exert a considerable downward pull on the part of the plate that remains on the surface, resulting in downdip extensional stresses in the sinking slab, parallel with the inclined seismic zone. If gravitational body forces are a major source of stress in the lithosphere, as Isacks and Molnar contend, the marginal area of the adjacent continental mass must likewise be affected by them. It can be assumed that downsucking of the sialic crust along the continental margin will exert a pull far inland; a measure of the minimum width across which extensional stress dominates the crust on the continental side is evidently given by the distance from the trench to the centers of andesitic volcanism. This width would depend on the rates and forces of continental drift and the opposing sea-floor spreading or downdip extension underneath the continent, as well as on the magnitude of the dip. The model in Figure 2 indicates the dimensions that exist in Chile, at about 40° south: an extensional stress would be generated in the crust on both sides of the trench across a total width of 400 km. of which about 300 km are on the continental side.

Savage and Hastie (1966) have shown that low-angle thrust faulting, as implied by this model, can in fact produce surface deformation of the kind here presented. In their comprehen-

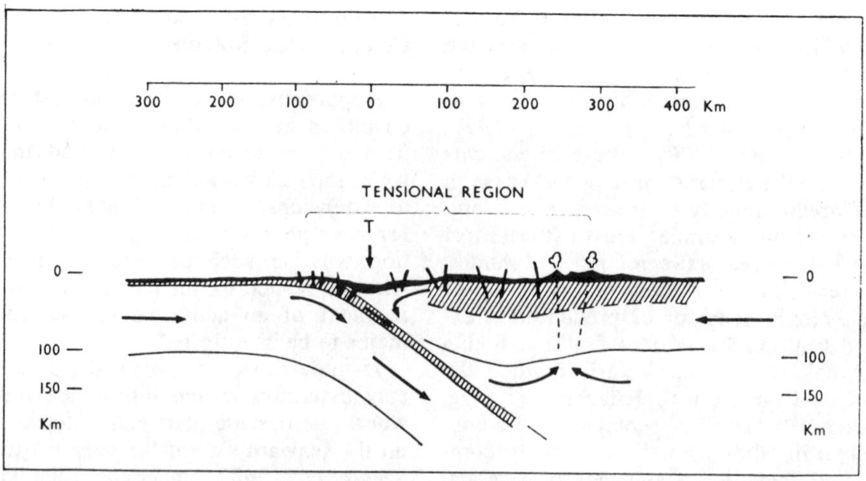

FIG. 2.—Diagrammatic cross section of continental margin in central Chile (modified after Isacks *et al.*, 1968). Crustal deformation on continental side is under extensional regime across belt 300 km wide, as measured from deep-sea trench *T* to centers of andesitic volcanism.

sive analysis of the 1960 Chilean earthquake, Plafker and Savage (1970) observed vertical displacements primarily caused by broad regional warping, whereas major faulting apparently was confined to the crust and upper mantle along the continental margin. In the part of the deformed region on land, the principal deviatory strains as calculated from triangulations before and after the earthquake clearly indicate east-west tension throughout the region (Plafker and Savage, 1970, Fig. 8, Table 2); greatest horizontal compression is directed parallel with the Andes and the coast. Thus, the relative seaward displacement of the overthrust block results in elastic extension in a zone adjacent to it; the probably most direct evidence of such an extension is seen in the earthquake-triggered, linear fissure eruptions near Puyehue volcano in southern Chile (Katsui and Katz, 1967). The model no doubt is far more complex in practice and would include, for example, much subsidiary slippage on dip-slip or strike-slip faults within the upper plate. Although this was not observed as a result of the 1960 earthquake, there is much evidence of similar faulting, particularly along northwest- and northeast-trending lineations, which affected older Quaternary and late Tertiary rocks and sedimentation (Illies, 1967; Katz, 1970). Thus, the present state of tectonic behavior appears in good agreement with the pattern observed from geologic evidence throughout the Neogene, and the same general model probably is valid and unchanged from about the Miocene to the present.

Conclusions

The tectonic style and evolution along the Pacific borderland in Chile indicate that extensional forces are dominant across a wide belt of sialic crust. This is explained in terms of subduction in the Chile Trench, causing tension and rupture in adjacent parts of the upper crust, particularly of the continental plate. The observed extensional deformation along the continental margin is thought to correspond closely to a stress condition that is related to the underthrust of the Pacific plate.

Thus, while the South American continent is being pushed westward, its margin is breaking up and disintegrating into blocks which are constantly moved up and down. These blocks eventually will disappear in response to the main forces of subduction along the deep sea trench. Regional surface compression in an east-west direction is virtually absent there; it only is possible farther inland. While the continental margin continuously is being destroyed by the movement toward each other of the oceanic and continental plates, the orogenic compression belt that originates farther inland will be displaced successively landward with respect to the continent, although keeping its more or less constant distance from the trench. The observed eastward migration of successive Andean folding stages is regarded, therefore, as another result of continental-margin tectonics; the distribution in time and space of the various Andean orogenic phases would be a direct measure of the forces along the active margin of the continent, and of the margin's respective positions.

Changes in the rate and direction of continental drift and sea-floor spreading as the principal driving forces would bring about changes in the lithosphere's stress field, and possibly directly influence the development and progress of orogenic cycles. Preliminary information from magnetic anomalies over the Chile Rise and off the coast of southern Chile suggests that spreading occurred during two different episodes (Herron and Hayes, 1969), possibly separated by a period of no spreading. Morgan *et al.* (1969) have mapped a change in the direction of active sea-floor spreading in this area, which probably happened 15–25 m.y. ago (early to middle Miocene). Its coincidence in time with an important change in stress conditions as revealed by geologic evidence on land seems more than fortuitous; it suggests that the taphrogenic breakdown which has been in progress in the actual continental border area since the beginning of the Neogene is indeed related to the processes of sea-floor spreading and, in particular, to its latest active phase.

References Cited

Brüggen, J., 1950, Fundamentos de la geología de Chile: Santiago, Chile, Inst. Geog. Militar, 374 p.

Ewing, M., R. Houtz, and J. Ewing, 1969, South Pacific sediment distribution: Jour. Geophys. Research, v. 74, no. 10, p. 2477–2493.

Gerth, H., 1955, Der geologische Bau der südamerikanischen Kordillere: Berlin, Gebrüder Borntraeger, 264 p.

Hayes, D. E., 1966, A geophysical investigation of the Peru-Chile Trench: Marine Geology, v. 4, no. 5, p. 309–351.

Herm, D., 1969, Marines Pliozän und Pleistozän in Nord- und Mittel-Chile unter besonderer Berücksichtigung der Entwicklung der Mollusken-Faunen: Zitteliana, no. 2, p. 3–158.

Herron, E. M., and D. E. Hayes, 1969, A geophysical study of the Chile Ridge: Earth and Planetary Sci. Letters, v. 6, no. 1, p. 77–83.

Illies, H., 1960, Geologie der Gegend von Valdivia/Chile: Neues Jahrb. Geologie u. Paläontologie Abh., v. 111, no. 1, p. 30–110.

——— 1967, Randpazifische Tektonik und Vulkanismus im südlichen Chile: Geol. Rundschau, v. 57, no. 1, p. 81–101.

Isacks, B., J. Oliver, and L. R. Sykes, 1968, Seismology and the new global tectonics: Jour. Geophys. Research, v. 73, no. 18, p. 5855–5899.

——— and P. Molnar, 1969, Mantle earthquake mechanism and the sinking of the lithosphere: Nature, v. 223, p. 1121–1124.

Katsui, Y., and H. R. Katz, 1967, Lateral fissure eruptions in the southern Andes of Chile: Hokkaido Univ. Fac. Sci. Jour., ser. 4, v. 13, no. 4, p. 433–448.

Katz, H. R., 1962, Fracture patterns and structural history in the sub-Andean belt of southernmost Chile: Jour. Geology, v. 70, no. 5, p. 595–603.

——— 1964, Some new concepts on geosynclinal development and mountain building at the southern end of South America: 22d Internat. Geol. Cong., India, Proc., pt. 4, p. 241–255.

——— 1970, Randpazifische Bruchtektonik am Beispiel Chiles und Neuseelands: Geol. Rundschau, v. 59, no. 3, p. 898–926.

Le Pichon, X., 1968, Sea-floor spreading and continental drift: Jour. Geophys. Research, v. 73, no. 12, p. 3661–3697.

Lomnitz, C., 1969, Sea floor spreading as a factor of tectonic evolution in southern Chile: Nature, v. 222, no. 5191, p. 366–369.

Miller, H., 1967, Zur jungen Tektonik in der Hochkordillere Nord-Chiles: Neues Jahrb. Geologie u. Paläontologie Monatsh., v. 9, p. 536–544.

Morgan, W. J., P. E. Vogt, and D. F. Fallis, 1969, Magnetic anomalies and sea floor spreading on the Chile Rise: Nature, v. 222, p. 137–142.

Paskoff, R., 1966, Terrasses littorales et tectonique récente entre l'embouchure du rio Limari et la baie Teniente, province de Coquimbo, Chili: Revista Geog. (Rio de Janeiro), no. 65, p. 57–67.

——— 1970, Recherches géomorphologiques dans le Chili semi-aride: Bordeaux, Biscaye Frères, 420 p.

Plafker, G., and J. C. Savage, 1970, Mechanism of the Chilean earthquakes of May 21 and 22, 1960: Geol. Soc. America Bull., v. 81, no. 4, p. 1001–1030.

Rüegg, W., 1956, Geologie zwischen Cañete-San Juan 13°00′–15°24′ Südperu: Geol. Rundschau, v. 45, no. 3, p. 775–858.

——— 1960, An intra-Pacific ridge, its continuation onto the Peruvian mainland, and its bearing on the hypothetical Pacific landmass: 21st Internat. Geol. Cong. Rept., Copenhagen, pt. 10, p. 29–38.

Savage, J. C., and L. M. Hastie, 1966, Surface deformation associated with dip-slip faulting: Jour. Geophys. Research, v. 71, no. 20, p. 4897–4904

Scholl, D. W., R. von Huene, and J. B. Ridlon, 1968, Spreading of the ocean floor: undeformed sediments in the Peru-Chile Trench: Science, v. 159, no. 3817, p. 868–871.

——— M. N. Christensen, R. von Huene, and M. S. Marlow, 1970, Peru-Chile Trench, sediments and sea-floor spreading: Geol. Soc. America Bull., v. 81, no. 5, p. 1339–1360.

von Huene, R., and D. W. Scholl, 1969, Tectonic implications of structures in stratified sequences at base of Pacific continental margins (abs.): Am. Assoc. Petroleum Geologists Bull., v. 53, no. 3, p. 747.

Worzel, J. L., 1965, Structure of continental margins and development of ocean trenches: Canada Geol. Survey Paper 66–15, p. 357–375.

Zeil, W., 1964, Geologie von Chile: Berlin, Gebrüder Borntraeger, 233 p.

Editor's Comments on Papers 24 and 25

24 **McCONNELL, MASSON SMITH, and BERRANGE**
Excerpts from *Geological and Geophysical Evidence for a Rift Valley in the Guiana Shield*

25 **KING**
Rift Valleys of Brazil

These papers are concerned chiefly with rift faulting in the shield areas east of the Andes. Paper 24 (by R. B. McConnell, U.K., D. Masson Smith, and J. P. Berrange, British Geological Survey) describes the North Savannas basin or rift valley in the Guiana shield, evidence for which combines geology and geophysics (aeromagnetic and gravity).

Brazil rift valleys discussed by L. C. King (of the University of Natal, South Africa) comprise: (a) the São Francisco rift valley lying west of the Serra Central and parallel to the Atlantic coast, here oriented NNE and followed by the Rio São Francisco; and (b) the much smaller Paraiba graben, also parallel to the coast, here with orientation NE-SW, and lying near the southern edge of the coastal Brazilian shield. Comparison is made with the geomorphology of the East African rifts.

The eastern continental margin of Brazil also contains the Reconcaro Basin, an "intracratonic semi-graben" and the Espirito Santo "asymmetric graben" (Asmus and Ponte, 1973).

On the southwest margin of the Brazilian shield in eastern Paraguay is the Chiquito "fracture zone" or graben, oriented NNW (Gansser, 1973, 104). In addition to these rift systems is the Ypacarai depression, a true graben 65 km by 8 km, oriented NNW, occupied in the north by the lake (Jenks, 1956, 111-113).

The Pampean Ranges of western Argentina are characterized by a "basin-range" structure marked by late Cenozoic sediments (Jenks, 1956, 155). The Argentine continental margin and the Malvinas plateau also contain rifted basins (Urien and Zambrano, 1973).

REFERENCES

Asmus, H. E., and F. C. Ponte, 1973, The Brazilian marginal basins, in *The Ocean Basins and Their Margins,* vol. 1, A. E. M. Nairn and F. G. Stehli, eds., Plenum, New York, pp. 87-133.

Burgh, H., 1967, The orogenesis in the Andean system of Colombia, *Tectonophysics* **4:**429-443.

Gansser, A., 1954, The Guiana Shield (S. America), geological observations, *Ecologae Geol. Helvetiae* **47:**77-112.

Gansser, A., 1973, Facts and theories on the Andes, *Geol. Soc. London Jour.* **129:**93-131.

Garner, H. F., 1959, Stratigraphic-sedimentary significance of contemporary climate and relief in four regions of the Andes mountains, *Geol. Soc. America Bull.* **70:**1327-1368.

Garner, H. F., 1974, *The Origin of Landscapes,* Oxford University Press, New York, 734p.

Gerth, H., 1955, *Der Geologische Bau der Sudamericanischen Kordillere,* Borntraeger, Berlin.

Gerth, H., 1956, Die Bedeutung der Alten Kerne für die geologische struktur der jungen Kordillere, *Geol. Rundschau* **45:**707-721.

Harrison, J. V., 1930, The Magdalena Valley, Colombia, South America, Rept. 25 *Int. Geol. Cong. South Africa, 1929, Part 2,* pp. 399-409.

Harrison, J. V., 1943, The geology of the Central Andes in part of the province of Junin, Peru, *Geol. Soc. London Quart. Jour.* **99:**1-36.

James, D. E., 1973, The evolution of the Andes, *Sci. American* **229:**60-69.

Jenks, W. F., 1956, Handbook of South America Geology, *Geol. Soc. America Mem.* **65:**378p.

Julivert, M., 1970, Cover and basement tectonics in the Cordillera Oriental of Colombia, South America, *Geol. Soc. America Bull.* **81:**3623-3646.

Levi, B., and L. Aguirre, 1981, Ensiatic spreading—subsidence in the Mesozoic and Paleogene Andes of central Chile, *Geol. Soc. London Jour.* **138:**75-81.

Myers, J. S., 1975, Vertical crustal movements of the Andes in Peru, *Nature* **254:**672-674.

Newell, N. D., 1949, Geology of the Lake Titicaca Region, Peru and Bolivia, *Geol. Soc. America Mem.* **36,** 111p.

Pitcher, W. S., and M. A. Bussell, 1977, Structural control of batholith emplacement in Peru: a review, *Geol. Soc. London Jour.* **133:**249-256.

St. Amand, P., and C. R. Allen, 1960, Strike slip faulting in northern Chile (abstract), *Geol. Soc. America Bull.* **71:**1965.

Urien, C. M., and J. J. Zambrano, 1973, Geology of the basins of the Argentine continental margin and Malvinas Plateau, in *The Ocean Basins and Their Margins,* vol. 1, A. E. M. Nairn and F. G. Stehli, eds., Plenum, New York, p. 135-169.

van Houten, F. B., and R. B. Travis, 1968, Cenozoic deposits Upper Magdalena valley, Colombia, *Am. Assoc. Petroleum Geologists* **52:**675-702.

24

Reprinted from pages 189, 194, 197, and 199 of *Geol. Mijnbouw* **48(2):**189-200 (1969)

GEOLOGICAL AND GEOPHYSICAL EVIDENCE FOR A RIFT VALLEY IN THE GUIANA SHIELD[1])

R.B. McCONNELL[2]) D. MASSON SMITH[3]) and J.P. BERRANGE[3])

ABSTRACT

Geological and geophysical evidence points to the existence in the Guiana Shield between latitudes 3° N and 4° N of a steep-walled graben structure which is filled with a thick packet of sediments of which the uppermost is the Takutu Formation showing rare outcrops beneath a blanket of laterite and Quaternary alluvium. The Takutu Formation has been dated on palynological evidence (v a n d e r H a m m e n and B u r g e r, 1966) as Jurassic to Cretaceous. A gravity reconnaissance was carried out over the western portion of the structure and revealed anomalies which indicate a depth to basement of several km. The strike of the geological structure appears to be ENE-WSW over 160 km, with a width of 50 km, and it is suggested that it is in the nature of a rift valley: it has been considered to constitute a major structural break which divides the folded Precambrian formations of Guyana into northern and southern geological provinces.

INTRODUCTION

The Guiana Shield forms the nucleus of the northeastern corner of the South American continent and consists chiefly of metasediments, metavolcanics and crystalline rocks of early Precambrian age, with a surface of low relief averaging 100-150 m in altitude and dominated by extensive forested pediplains rising towards their watersheds. Above this uniform surface a number of isolated hills, mountains and mountain ranges reach heights of up to 1,000 m, and, in the centre of the Shield the Pakaraima Mountains, formed by tabular Precambrian sandstones, display a succession of forested plateaus rising to isolated summits such as Ayanganna (1,900 m) and Roraima (2,770 m). On the western portion of the Shield, in the Rio Branco province of Brazil and the Rupununi District of Guyana, are large savanna areas covering a pediplain at the 100-150 m level (M c C o n n e l l, 1968). One branch of this plain forms the ca. 50 km wide flat-bottomed valley between latitudes 3° 10'N and 3° 50'N that extends to the east-northeast for about 160 km from the Takutu and Ireng Rivers. Although two thirds of this valley is forested (east of the Rupununi River) we propose to refer to it as the North Savannas Basin.

It has been shown (M c C o n n e l l, 1959; B a r - b o s a and A n d r a d e R a m o s, 1959) that this depression was the course of a proto-Berbice River which took the drainage of the upper Rio Branco basin to the Atlantic up to the end of the Tertiary, when the Rio Branco was captured by the Amazon.

[*Editor's Note:* Material has been omitted at this point.]

---

1) Published by permission of the Director, Institute of Geological Sciences

2) Streatwick, Streat near Hassocks, Sussex, U.K.

3) Institute of Geological Sciences, London.

STRUCTURE

The North Savannas Basin is bounded to the south by the north wall of the Kanuku Mountains, and where this scarp projects to the Takutu River the rocks are highly silicified and fractured over a distance of nearly a kilometre (Berrangé, 1967). As indicated on Fig. 1 the north face of the mountains is therefore assumed to be the eroded remnant of a fault scarp, and this can be traced east-northeast for 60 km as far as the Rupununi River. Further east the scarp is less well defined, but a combination of geological, photogeological and geophysical observations suggests that the boundary between the Apoteri Volcanic Formation and the Kanuku Group corresponds over part of its length to a fault striking E-W to ENE-WSW almost as far as the Rewa River. The western Kanuku Mountains form a deeply dissected highland mass the average elevation of which decreases gradually southward toward the South Savannas, the actual border between the two physical regions being a relatively straight fault scarp trending ESE-WNW. This is probably a normal fault with the downthrow on the South Savannas side. The western Kanukus are separated from the eastern Kanukus by the north-northeast flowing Rupununi River which possibly follows a zone of faulting. The western Kanukus are therefore considered to be a triangular-shaped horst bounded on all three sides by faults and sloping gently southwards. Although many strong faults have been recognized in the eastern Kanuku Mountains they merge gradually southwards into the forested 100 m pediplain.

The northern limit of the basin is not so well marked and the Quaternary cover embays the rocks of the Mazaruni Group which rise less steeply from the plain. Geophysical investigations indicated the existence, underneath the alluvium near the northern mountain front, of faults striking ENE-WSW. Subsequent photogeological studies and field work (Berrangé, 1967) have locally confirmed the existence of these faults which in some cases apparently join up with a very strong, closely spaced, set of faults striking NE-SW through the Mazaruni metavolcanics and granophyric granites.

The Takutu Formation has not been mapped on the ground further east than longitude 58°30′W, but photogeological observations suggest that basic lavas form mesas at least as far east as the Berbice River, and that faulting continues up to the Surinam border. However, the graben shallows markedly towards the east in this area and its place seems to be taken by a belt of biotite granites (Barron, 1962b) which continues at least as far as the Corentyne River. These granites are believed to belong to the Younger Granite Group and to separate rocks of the Mazaruni Group belonging to the northern geological province (McConnell et al., 1964) from the uncorrelated gneisses and gneissose granites forming the southern geological province of Guyana: the photogeological work now in progress may provide an answer to this problem.

[*Editor's Note:* Material has been omitted at this point.]

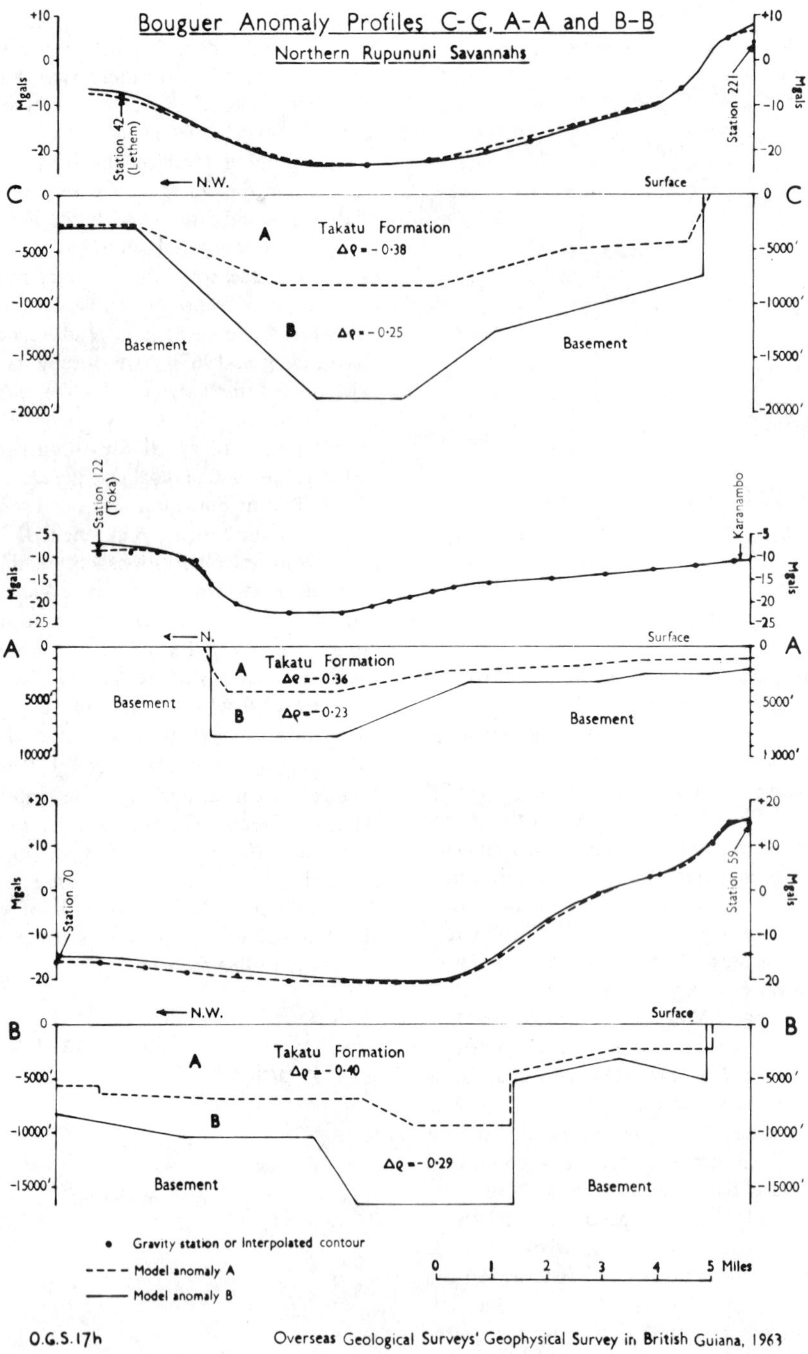

Fig. 4
Bouguer anomaly profiles, North Savannas Basin, Rupununi District, Guyana.

[*Editor's Note:* Material has been omitted at this point.]

RIFT VALLEY STRUCTURE

The term "rift valley" was first applied by J.W. Gregory (1896) to the remarkable faulted valley structures which extend some 3,500 km from Ethiopia to Mozambique down the eastern side of the African continent. These rift valleys are in part steep-sided graben structures, and are characterized by (1) great length of strike, (2) an average width of 30-60 km, and (3) localized volcanism with alkali-rich magmas associated with nearly half of the total length of the overlapping rift valleys.

It is therefore concluded that the North Savannas Basin, between the Takutu and Ireng rivers in the west and longitude 58°30′W in the east, is a true rift valley for the following reasons:

(a) It is an elongate graben, about 50 km wide and 160 km long, bounded by steeply dipping faults.
(b) The depression is the site of low magnetic relief and negative gravity anomalies, which are interpreted as being due to a sedimentary infilling attaining several km in thickness.
(c) Andesitic and basaltic lavas of probable post-Proterozoic age occur along the southern and eastern margin of the basin.
(d) In view of the characteristic association of alkali-rich volcanism with rift valleys it is of interest that the riebeckite granite of Makarapan Mountain, provisionally dated as ca. 2,500 m.y. B.P., is the only alkali-rich rock so far reported in Guyana.
(e) The feature appears to form part of an important fracture system dividing two geological provinces of the early Precambrian. If the Bakhuis Mountains horst can be correlated with the Kanuku Moutains horst, then the fracture system extends for at least 450 km.

[*Editor's Note:* Only the references cited in the preceding excerpts are reproduced here.]

REFERENCES

Barbosa, O. & Andrade Ramos, J. R. de (1959)—Territorio do Rio Branco (Aspectos principais da geomorphologia, da geologia e das possibilidades minerais de sua zona setentrional); Div. Geol. Miner. Brasil, Bol. 196, 46p.

Barron, C. N. (1962b)—Kurupukari SE Quarter Degree Sheet 0548 SE Geol. Surv. Brit. Guiana, Geological Atlas, 1 : 200,000.

Berrange, J. P. (1967)—Progress report on photogeological mapping of Guyana south of 4° latitude— Operation El Dorado. Inst. Geol. Sciences, London, Photogeological Div., Rep. No. 24, Unpublished.

Gregory, J. W. (1896)—The Great Rift Valley. London, John Murray, 405 pp.

Hammen, T. van der & Burger, D. (1966)—Pollen flora and age of the Takutu Formation (Guyana). Leidse Geol. Meded., 38, p. 173-180.

McConnell, R. B. (1959)—Fossils in the North Savannas and their significance in the search for oil in British Guiana. Timehri (Georgetown), 38, p. 65-84.

McConnell, R. B. (1968)—Planation surfaces of Guyana. Geogr. Journ., 134, p. 506-520.

McConnell, R. B., Williams, E., Cannon, R. T. & Snelling, N. J. (1964)—A new interpretation of the geology of British Guiana. Nature, 204, p. 115-118.

25

Reprinted from *Geol. Soc. South Africa Trans.* **59:**199–209 (1956)

RIFT VALLEYS OF BRAZIL

by Lester King

[PLATES XXVIII–XXXII]

[*Editor's Note:* The plates have been omitted.]

ABSTRACT

Rift valleys of Pleistocene age are shown to exist in Brazil along major crustal arches affecting early and late Tertiary cyclic erosion surfaces. They show great similarity to the African rift valley system.

CONTENTS

INTRODUCTION

The rift valley system of Africa extending from the Zambezi to the river Jordan is familiar, by description at least, to all students of the earth-sciences. By many it is regarded as unique ; but except for its length, which is indeed remarkable, covering one-sixth of the earth's circumference, the African rift-valley system is not without its compeers in other parts of the world (*e.g.* the San Joaquin-Sacramento valley of California, and the Baikal trough of Siberia).

A few years ago, Brazilian investigators (R. O. de Freitas, 1951) began to apply the title of rift valley to the São Francisco valley in that country where the river flows in a broad flat-floored depression parallel to the coast for several hundreds of miles before breaking through to the sea. F. Ruellan (1947) has indeed remarked on the possibility of the breaking of a crustal arch in the upper course of the valley with subsidence between the Serra Espinhaço and Minas Gerais on the east and Goiaz on the west. Others have regarded the Paraiba valley at the foot of the Serra da Mantiquera in eastern São Pãulo as a graben, though this opinion has not gone uncontested.

On a recent geomorphological tour in Brazil, I was able to make further field observations and to note the truly remarkable similarities in situation, size, age, and relation to other continental structures that exist between the Brazilian and African rift-valley systems.

SALIENT FEATURES OF AFRICAN RIFT VALLEYS

Students of African rift valleys, however they have begun their observations, have always ended by studying the adjoining plateaus equally with the rift structures (e.g. Willis, 1936), for in many ways these plateaus are more significant than the rifts, and they afford decisive data without which the history and tectonics of the rifts could not be satisfactorily interpreted. " It is perhaps fair to say that, though the scarps are clear and striking, they are not more so than fault-scarps in other parts of the world. The astonishing fact about rift valley topography is its flatness. The crest-lines of scarps run with almost perfect horizontality mile after mile. The floor, also, is astonishingly even. In places minor faulting conferred a degree of irregularity upon the original floor ; but sedimentation in the once more extensive lakes (Naivasha, Elmenteita and others) has eliminated much of this. Only volcanic accumulations, recognizable as such from afar, interrupt the almost universal, horizontal straight line along great lengths of the eastern rift. It is these circumstances which confer on the scarps of the East African rift valleys their striking appearance. Not the escarpments, but their setting is remarkable. Within the wide erosional plains of Africa the last thing one expects to see is one of these gigantic trenches " (King, 1951).

The chief landform across which the rift valleys gash their way is a mid-Tertiary denudational plain of exceptionally smooth planation, which provides a reference datum. In the African interior this plain stands usually between 3,000 and 4,000 feet above sea-level, but for perhaps a score of miles on either side of the rift valley the plateau warps steadily upwards until at the edge of the rift it attains a maximum commonly above 6,000 feet and occasionally reaches almost to 10,000 feet. This broad arching constitutes what is generally called the " rise to the rift " and is a tectonic feature far larger than the rift itself. Occasionally, as at Naivasha, the rise is so great that even the floor of the rift (here at 6,000 feet) is carried well above the original level of the plateau. Thus it is not the rifts which are important but the rise of a broad arch of country to either side of the rift which itself represents, to use J. W. Gregory's phrase, " the subsidence of the keystone of the arch." Upon these basic principles all modern observers of the rifts are agreed. The accompanying phenomena of predominantly tensional, and subordinately compressional and transcurrent, faulting arise upon the general plan appropriately to local conditions and controls such as the existence of ancient strikes and structures in the African basement.

F. Dixey (1946 ; 1955) and McConnell (1951) have, indeed, demonstrated clearly that the rift valley systems are by no means novel, and that the modern troughs are situated closely along lines of previous fracture belonging to Cretaceous, post-Karroo, and even pre-Cambrian tectonic episodes. They are sited, indeed, along zones that have shown in the past a repeated tendency to rise and many of the details of rift-valley faulting follow closely the lineation of the pre-Cambrian basement gneiss. The modern African rift valleys date principally from the mid-Pleistocene, though of course the movements began significantly before that date, and some still continue.

THE BRAZILIAN RIFT VALLEYS

The Brazilian rift valleys also dislocate a mid-Tertiary cyclic landsurface of extraordinarily smooth planation, which, though often dissected by later erosion, is still patently the original plateau surface over the greater part of Brazil from Rio Grande do Norte to Rio Grande do Sul. This magnificent datum is transected and cast down at fault-scarps in exactly the same manner as its African counterpart (Plate XXVIII) rendering the same horizontal skylines and the same broad, sometimes alluviated, longitudinal depressions. As the general width and depth of the São Francisco rift valley (10–20 miles, and several hundreds of feet respectively) are similar to the dimensions of the African rifts, the resemblance between them is often extraordinarily close.

The Brazilian rift valleys, though belonging to a single tectonic framework, may be studied conveniently in two separate systems (*a*) the São Francisco system which is directed somewhat east of north, parallel to the coast of Baia and Espirito Santo, and (*b*) the Paraiba graben which is directed from north-east to south-west parallel to the coast of the Serra do Mar between Rio de Janeiro and Santos (Fig. 1).

(*a*) *The São Francisco Rift Valley.*—The São Francisco rift begins about Itabira, to the east of Belo Horizonte where two systems of fractures intersect in a broad angle reminiscent of the crossing of eastern and western African rifts in Tanganyika behind Mbeya. The eastern fractures are represented by scarps directed to the north-east. The major scarp continues almost to Govenador Valadares, and overlooks the broad depression occupied by the Rio Doçe. As the downstream block has been tilted northwestwards, the fault angle depression has been largely alluviated in the Parque Florestal. So far as I could ascertain, this depression does not continue beyond Govenador Valadares, at which point the Rio Doçe turns at a right angle and flows directly to the sea. In other words, the northeastward course of the Rio Doçe, anomalous among rivers of the eastern seaboard, is controlled by Quaternary faulting.

The main line of rifting is directed, however, northwestward from Itabira, along the line of the Rio Cipo towards the Rio São Francisco near Pirapora. To begin with the scarp is modest, a few score feet in height, but it steadily increases in height to the northwest. Where the road from Belo Horizonte to Serro leaves the rolling lowland of the Rio Cipo drainage to ascend the watershed of the Serra Geral, the fault-scarp is to be observed (Plate XXIX) as a wall-like scarp, three hundred feet or more in height, making the foot of the range and extending to the northwest almost as a ruled line. It here transects the basement structures. Grown magnificently in elevation, the scarp continues thence as the impressive face of the Serra do Cabral, one of the major topographic features of this part of Minas Gerais. Behind Varzea das Palmas the dislocation of the horizontally planed landsurfaces is very clear, and the fault scarps make a striking contrast with these lines. Ultimately the fault scarp reaches the Rio São Francisco about 12 miles downstream of Pirapora where it helps to define the valley. My travels did not continue beyond this point so that the final extent of this great fracture was not determined, but it is proved to exist for a length of nearly 200 miles. Along the whole length of the scarp so far described appears but a single gap—to the northeast of Corinto. Here, beginning at the " nose " of the Serra do Cabral, is a gateway leading to a true two-sided rift valley that broadens northward (see Fig. 1) to Montes Claros and Francisco Sa where its eastern margin, displacing the mid-Tertiary planation, is depicted in Plate XXVIII. This is the rift valley of the São Francisco, in the strict sense, that continues northward for nearly 500 miles until the river breaks a passage through a youthful valley, with gorges, to the Atlantic Ocean. The eastern side of the rift is magnificent, with scarps many hundreds of feet in height continuously for almost the whole of the way. The outline is not smooth but broken by bastions and re-entrants according to the pattern of faulting (Fig. 1). Descent from the adjoining plateau is always abrupt. Scenically, this side of the rift is every bit as impressive as the African rift-scarps.

Between 12° and 13° south latitude, a lateral rift branches off and extends southward to Paramirim. This is a true two-sided rift-valley with superb lateral scarps a thousand feet and more in height (Plate XXX) which occasion considerable difficulty to road-engineers. The grades and heavily-cobbled surfaces (as on the descent from Ibitiara to Santa Maria) are deterrent to traffic other than an occasional lorry or "Jeepi." Owing to its depth and inland situation, the whole of the São Francisco rift valley experiences a very dry climate. The stamp of aridity, as expressed in the minor landforms and the sparse and xerophytic vegetation, engenders again a whole series of correspondences with the detailed aspect of the East African rifts, and of no portion is this more true than of the lateral rift leading to Paramirim.

On the western side of the São Francisco rift little information is available beyond the fact that scarps do exist, and the requirements of my main geomorphologic study in Brazil did not allow time for investigation in that area. A most rewarding study there awaits some future investigator.

Well to the eastward of the São Francisco rift in northern Baia is a long, and apparently faulted, scarp extending for over 200 miles southward from Senhor do Bonfim. Fronting to the east, and with a difference of elevation of 200–300 feet, or more in places, it assists in the rise of the mid-Tertiary plateau towards the rift valley, to which it is, in all probability, tectonically related (Fig. 1).

The scarp to the east of Tucano also appears very fresh and may well be a rejuvenation upon an older, Cretaceous graben that formerly existed in this locality. Sundry short fault-scarps of recent aspect have also been described to the north of the lower course of the São Francisco.

The profound topographic changes induced by rifting have provoked heavy alluviation along the course of the Rio São Francisco, and many of its larger tributaries, between Pirapora and Joazeiro. The total thickness of sediments exceeds 200 feet. Fossils recovered from the deposits at Bom Jesus de Lapa are said to be of Plio-Pleistocene date. This is in accordance with the geomorphology, for the fracturing post-dates the scoring of shallow valleys belonging to a late Cainozoic denudation cycle across the mid-Tertiary plateau. The rift valley is thus latest Pliocene or Pleistocene in age.

(*b*) *The Paraiba Graben and related fractures.*—Being more accessible (indeed it is traversed from end to end by the main road linking the two largest cities of Brazil) the Paraiba graben has been much more fully studied than the São Francisco rift valley. It extends in a north-easterly direction and has a width of fifteen miles or less. Flanked on the north-west by the spectacular

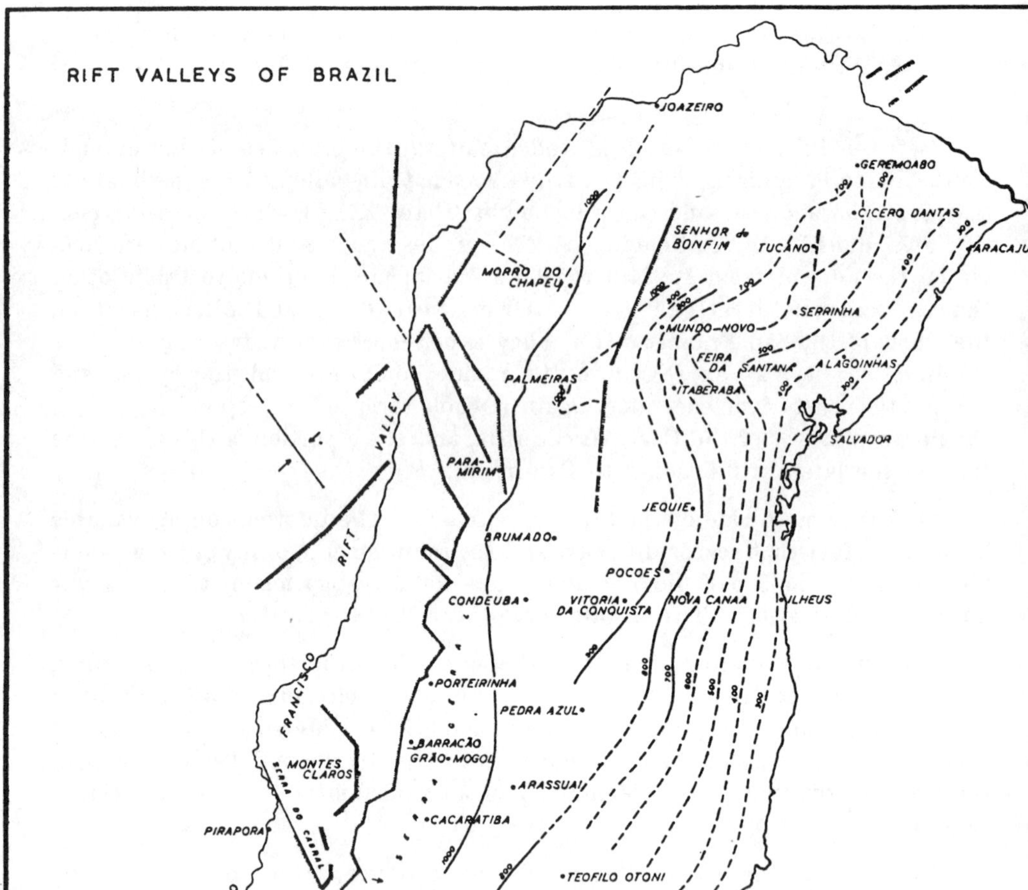

Fig. 1.—The rift valleys of Brazil and their tectonic setting. Contours (in metres) are upon the mid-Cainozoic landsurface. Elevations calculated from other cyclic surfaces are shown in broken line.

scarp of the Serra da Mantiquera which culminates in Agulhas Negras at 9,255 feet above sea-level, the floor of the valley is smoothly alluviated at about 1,500 feet of elevation. The general absence of precise evidence of fracturing, and the lack of gravels at the foot of the scarp led Moraes Rego and A. R. Lamego to postulate that the Mantiquera scarp was wholly erosional ; but Washburn has reported tectonic disturbance of the (late Cainozoic ?) infill at the south-western end of the graben, and the attitude of cyclic erosional landsurfaces mapped in reconnaissance (King, 1956) proves beyond doubt that the lower levels of the great mountain front originated as a Quaternary fault-scarp (Plate XXXI). The greater heights, however, do belong to an earlier watershed of rugged hills that existed prior to the fracturing.

The south-eastern flank of the graben is formed by a dissected plateau similar in all respects to the dissected upland which forms the greater part of the Parana hydrographic basin. The ridge crests are everywhere accordant at the level of the mid-Cainozoic denudational plain ; but this was dissected by valleys of the late-Cainozoic cycle before the subsidence of the Paraiba graben at or after the close of the Cainozoic era. The age of the Paraiba graben is thus the same as that of the São Francisco rift valley.

The scarps separating the flanking plateau from the graben are clear and straight, though a distinct block projects into the valley at Guaratingueta, causing also a rock barrier across the river bed.

Within the Paraiba lowland, especially towards the southwestern end, there accumulated a volume of arenaceous and argillaceous sediments of late, but not accurately defined date. These are the beds described by Washburn as tectonically disturbed towards the Mantiquera front. They have been partly removed as the river later lowered the rock sills at the northeastern end, where it leaves the graben, and the present valley floor is an alluvial plain at lower level (Plate XXXII).

In addition to the fault-scarps responsible for the Paraiba graben we must take cognizance of a few other such scarps in Rio de Janeiro and southern Minas Gerais. Near Viuva da Graça are several fault splinters rising from the alluvium which extends from the head of Guanabara Bay, and these are backed by a pronounced scarp up which the highway climbs by Monumento Rodavaria. These scarps, facing to the southeast, are apparently the only fresh fault-scarps that approach the lowland about Guanabara. The huge scarps which elsewhere rise from the flats are, by common consent among later geomorphologists, purely erosional. Another scarp to the south of Juiz de Fora extends the line of the Mantiquera front far to the northeast (Fig. 1) ; indeed I was unable to determine its full extent in that direction. But although the two scarps are collinear they are not continuous. Between them, as on the map, a broad swathe of country crosses obliquely from the interior to the coast without dislocation of the mid-Cainozoic landsurface. This undisturbed swathe, together with the fault splinters at Monumento Rodavaria, indicates that the forces responsible for the fracturing operated in part as a couple. The most northerly faults of the Paraiba system determined in the recent survey are two parallel faults, with the usual north-eastward orientation, just north of Santos Dumont. The full lateral extent of these was not followed, but both cast down the mid-Tertiary summit plane to the southward.

TECTONIC FRAMEWORK OF THE BRAZILIAN RIFT VALLEYS

Both the São Francisco Rift and the Paraiba graben dislocate the mid-Cainozoic plateau and the valleys of the late Cainozoic erosion cycle which dissect the plateau almost to the headwaters of the river systems. The rifts are thus both of the same age, which can scarcely be older than Quaternary. They must therefore have been generated not only simultaneously but by the same set of tectonic forces. Yet the two rift systems are neither parallel nor connected. Each is oriented independently parallel to the nearest margins of the continent : the São Francisco rift parallel to the east coast between Vitoria and Aracaju, and the Paraiba graben parallel to the Serra do Mar fronting the sea between Rio de Janeiro and Santos. We shall study each in its tectonic environment.

Geomorphic reconnaissance (King, 1956) has revealed that the mid-Cainozoic denudational plateau, carved originally near sea level, has been strongly arched in late Cainozoic and Quaternary time (Fig. 1). Standing at about 250 metres along the present coast (beyond which it flexes steeply downward into the continental shelf) it rises to approximately 1,000 metres along the line of the Serra Geral, which constituted the mid-Cainozoic watershed. But the rise of the mid-Cainozoic **chapada* does not now cease at the Serra Geral ; west of the residual ranges of quartzite marking the former waterparting, the *chapada* continues to ascend westward until the ground suddenly falls away at the scarp of the rift valley. These relations can be well seen in east-west section through Grão Mogol (Plate XXVIII). The head of the direct drainage to the coast now stands at the crest of the fault scarp and the rivers (*e.g.* Rio Ticororo) flow eastward in shallow valleys incised into the *chapada* until they pass through the Itacolomi quartzites of the Serra Geral in youthful gorges. The drainage of the region between the Serra Geral and the rift has thus been reversed in direction following the rise of the arch, of which the rift forms the broken crest.

The Serra Geral is itself the site of repeated earlier uplifts dating back at least to the early Palaeozoic era, so that—as with the African rifts—the São

**Chapada*, a tableland, is a term applied to both structural and denudational plateaus in Brazil.

Francisco valley not only follows the crest of an arch formed during late Cainozoic and Quaternary time, but also follows closely the alignment of former episodes of differential uplift. It is moreover clearly to be understood in relation to the great coastal monocline (Fig. 1) by which the denudational landforms of eastern Brazil plunge beneath the sea.

Topographic relations at the Paraiba graben are essentially similar, though the inland flank of the rift differs in that it was mightily uplifted along the Mantiquera scarp (Plate XXXI). Between the Paraiba rift and the great cliffed coast of the Serra do Mar is a dissected upland of the mid-Cainozoic plateau standing at about 900 metres above sea-level. It is the eroded edge of this plateau that forms the magnificent coastline of the Serra do Mar. Several authorities have cited this margin as a fault coast, but Rich (1953) has more recently demonstrated that the scarps facing the sea are wholly erosional and that the basic structure is a steep coastal monocline. This interpretation I have verified, finding even the magnificent scarps behind Guanabara Bay to be wholly erosional in origin. The deeply penetrating erosion of that Bay, and of Baia de Ilha Grande, however, may well be initiated by tensional openings along the axis of the monocline.

Like the São Francisco rift valley, the Paraiba graben is a tectonic appendage of the mighty flexing which has afflicted the coastal regions of Brazil at about

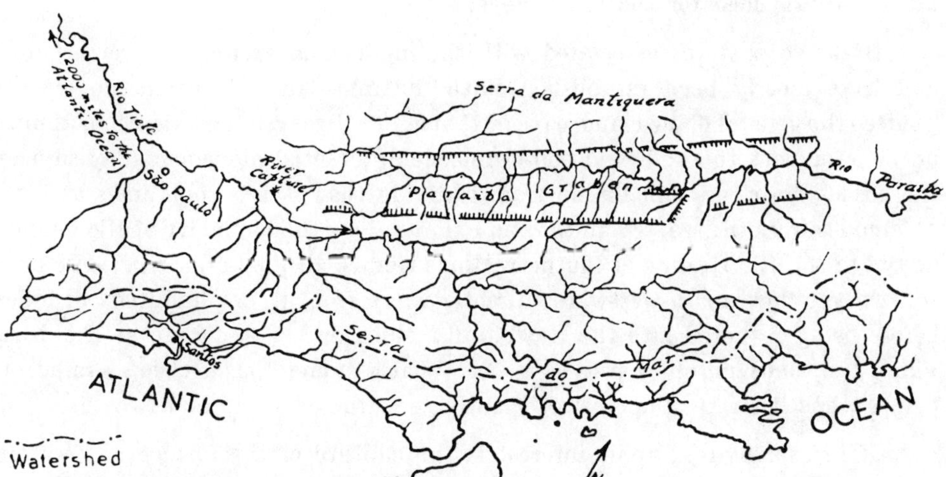

Fig. 2.—Drainage patterns about the Serra do Mar, one of the world's most asymmetrical divides. Capture of the former headwaters of the Rio Tiete by the Rio Paraiba.

the close of the Cainozoic epoch. Like the São Francisco rift, too, it follows closely the line of repeated earlier strong upwarpings and coastal flexings. Indeed the strongly gneissose Archean basement which rises to great heights (2,200 metres) in this region suggests that, like parts of the rift valley system of Africa, it may have functioned repeatedly as an axis of major elevation since the pre-Cambrian era.

The late Cainozoic axis of flexing appears to have passed closely east of Guanabara Bay in a north-east direction. Eastward of this line, to Cabo Frio, are extensive coastal lagoons and salinas shut in behind long trailing sand spits, indicating a subsided shoreline. Westward of the line, however, tower the great cliffs of a strongly uplifted shoreline (allowance being made for the recent eustatic rise of sea-level which affects all the shoreline of Brazil and drowned Guanabara Bay) heavily retrograded by marine erosion and with a deep offshore zone.

How far does this late Cainozoic axis continue to the south-west ? It may well re-enter Brazil at the southern end of Santa Catarina where the Rhaetic basaltic lavas warp down from the plateau to the sea, and where the Serra do Mar fades away to be replaced on the south side by coastal lagoons (increasing in size southward along the littoral of Rio Grande do Sul) and the alluviated lowland of the Rio Jacui. Both the contrasted types of coastline, lagoonal south of the axis and steeply cliffed north of it, show by their youthfulness the relatively recent date of the monoclinal flexing, in conformity with the dating already established for the rift valleys.

Drainage patterns associated with the monoclinal arching are remarkable, and have mostly been established since the mid-Cainozoic earth movements uplifted the general planed landscape of that age. Because the axis of maximum upwarp for both the mid- and end-Cainozoic differential movements lay so near the coast, the rivers run inland from a divide that stands sometimes as little as 10 miles from the sea. From the Rio Pelotas along the boundary of Rio Grande do Sul to the Rio Grande of southern Minas Gerais, all the major rivers rise near the sea yet flow north-westward, directly away from it, for hundreds of miles before being gathered into the Rio Parana which enters the ocean at Rio de la Plata, two thousand miles away (Fig. 2). What a momentous decision a raindrop makes when it strikes the crest of the divide of the Serra do Mar !

Of further hydrographic interest is the capture of the former headwaters of the Rio Tiete effected by the Rio Paraiba (Fig. 2) causing them, after a westward passage of 80 miles to return eastward along the Paraiba graben and ultimately to the sea near Campos. The course of the Paraiba itself is clearly locked in behind the arch of the coastal monocline.

The São Francisco and Paraiba rift valleys are each related structurally to the differential tectonic movements that affected the adjacent coasts. What, now, are the mutual relations of the two fracture systems that we already know to be of the same age ? According to the attitudes of cyclic erosional surfaces in the eastern part of Estado do Rio de Janeiro, it is not difficult to adduce that the two coastal monoclines join into one by simple curvature (Fig. 1). Inside the bend is an elongated dome of " super-uplift," where the mid-Cainozoic cyclic landscape rises to heights in excess of 1,400 metres, and the highest country, of Archean gneiss, exceeds 2,200 metres in the mountains west of Teresopolis (Plate XXXII). The great erosional scarp by which the road ascends from Rio de Janeiro to Petropolis is another secondary result of the excessive uplift of this area.

With such abrupt curvature of a major continental flexure due to differential earth-movement, features due to transcurrent faulting may be expected within the angle of deflection, but none could be identified in the short time available to me here for observation. Farther north, the Pico de Bandeira also stands in a position where torsional stresses could have operated to raise it to its unusual elevation (9,396 feet ; the highest point in Brazil). Such forces with slight transcurrent displacement are now understood to be responsible for the uplift of Ruwenzori in the curving Albertine rift of Central Africa, there a portion of the African plateau has been cast up from its local elevation of 5,000 feet to 14,000 feet.

So in situation, size, scarp form, relation to cyclic erosion surfaces, age, relation to recent arching, close alignment with ancient axes of elevation, drainage anomalies, and valley-floor sediments, the valleys of the Rio São Francisco and western Rio Paraiba are true rift valleys, comparable in all these respects with the classic rift valleys of Africa, and lacking only associated Pleistocene volcanic phenomena to render the comparison complete.

ACKNOWLEDGMENT

These observations were made during a study tour in Brazil conducted under the auspices of the *Conselho Nacional des Pesquisas* and the *Instituto Tecnologia* of Belo Horizonte, to which bodies I am indebted for generous research facilities in a country of magnificent opportunity for geomorphic studies.

REFERENCES

RUELLAN, F. Algunas Aspectos do Relevo no Planalto Central do Brasil. *Anais Ass. Geog. Brasil*, Vol. 2, pp. 17–28 (1947).

DE FREITAS, R. O. Ensaio Sôbre a Tectonica Moderna do Brasil. *Bull. Fac. Ciencias Univ. S. Paulo*, No. 130, 120 pp. (1951).

DIXEY, F. Erosion and Tectonics in the East African Rift System. *Quart. Journ. Geol. Soc., Lond.*, Vol. 102, pp. 339–388 (1946).

DIXEY, F. Some Aspects of the Geomorphology of Central and Southern Africa. Du Toit Memorial Lecture (1955). *Proc. Geol. Soc. S. Afr.*, Vol. 55 (in the press).

WILLIS, B. East African Plateaus and Rift Valleys. *Carnegie Inst.* Publ. 470, (1936).

KING, L. C. *South African Scenery.* Edinburgh (1951).

KING, L. C. The Geomorphology of Eastern Brazil. *Revista de Geografia*, Bras (in the press) (1956).

RICH, J. L. Problems in Brazilian Geology and Geomorphology suggested by Reconnaissance in Summer of 1951. *Fac. Ciencias. Univ. S. Paulo*, No. 146, 80 pp (1953).

MCCONNELL, R. B. Rift and Shield Structure in East Africa. Rep. XVIII. *Intern. Geol. Cong.*, Part XIV, pp. 199–208 (1951).

Part VII

AUSTRALIA

Editor's Comments on Papers 26, 27, and 28

26 FAIRBRIDGE and FINKL
Excerpts from *Geomorphic Analysis of the Rifted Cratonic Margins of Western Australia*

27 TWIDALE
Fault Blocks and Rifts in South Australia

28 WELLS et al.
Geology of the Amadeus Basin, Central Australia

Australia is known as the quiet continent because of its relative tectonic stability and the absence of appreciable seismic activity (Williams and Corliss, 1982). For orientation purposes refer to Tectonic Map of Australia (Geol. Soc. Australia, (1971), Khain (1981, in Russian), Plumb (1979), and Rutland (1981). Like both North and South America, Australia has wide belts of folding and orogeny. These meridianally oriented belts are on the Pacific side of all three continents, while the opposite sides have typical trailing-edge or Atlantic-type margins. On the west and northwest of Western Australia are rifted cratonic margins where the Indian plate and a number of platelets separated finally during the early Cretaceous (Shields, 1977). The separation had been preceeded by rifting in the Ordovician (Fairbridge, 1982). Along this western margin are structures typical of continental separation—basins, aulacogens, and rifts (Veevers and McElhinny, 1976; Powell, 1976; Branson, 1978; Khain, 1981). Paper 26 is about these features and in particular the "Westralian geosyncline" (Teichert, 1939, renamed the Westralian Aulacogen by R. W. Fairbridge and C. W. Finkl. Fairbridge is professor emeritus at Columbia University, New York City, and Finkl is at Florida Atlantic University in Boca Raton). The Westalian Aulacogen gradually deepened, and, in the course of time, except for the breakaway Naturaliste block, its eastern flank, the Darling fault, became the cratonic margin. This pattern was repeated northward (Finkl and Fairbridge, 1979; Veevers, 1974) from the Perth Basin to the Coolcallaya onshore basin and the Carnarvon basin.

Around the Pilbara Block are smaller basins, also no doubt extensional features, and the Canning basin (Veevers and Wells, 1961) has the Fitzroy trough or graben along its northeastern flank (Guppy et al. 1958). The Bonaparte Gulf with the Petrel graben is north of the Kimberley basin (Doutch and Nicholas, 1978).

The southern margin of Australia, west of Victoria, is of Atlantic type but without the separation characteristics of aulacogens and basins (Talwani et al., 1978; Veevers, 1976). It is backed by two continental basins and has a wide shelf. Antarctica separated in the late Paleocene. Toward the eastern end is the zone block faulting and penetrating rifts, the subject of Paper 27, the South Australia rift zone by Professor C. R. Twidale who was with the University of Adelaide, South Australia. The faulting appears to be rejuvenated from the Paleozoic (Stewart and Mount, 1972; Crawford, 1963; Fenner, 1930; Rutland et al., 1981). This zone lies between the Gawlor block and the Adelaide geosyncline (Falvey and Mutter, 1981; Williams and Corliss, 1982 p. 548). Reference to the latter as an aulacogen appears to be justified.

The older interior basins of Australia (Doutch and Nicholas, 1978), which were all probably originally extensional structures, include the Amadeus basin, described in Paper 28 (by A. T. Wells and colleagues; he is at the Bureau of Mineral Resources in Canberra). Both the Amadeus and Ngalia, basins (Wells and Forman, 1979) have been referred to as aulacogens, but the Officer basin, also sometimes included, is certainly not one. If the Amadeus and Ngalia troughs are to be considered aulacogens, they belong to the "inner" or "blind" class of Bogdanov (see Paper 4). It is the history of their sedimentation and the nature of the faulting of the margins (overthrust inward from north and from south) that raises doubts (Rutland, written communication, 1984). However, as the Amadeus basin is "the most interesting and unusual feature of the entire Australian platform" (Khain, 1981), this summary description is included as Paper 28.

The remaining domain where extensional structures are found is the Tasman geosynclinal province or "Tasmanides" (Scheibner, 1978). Grabens and tectonically derived troughs are found in (1) Tasmania (Williams, 1978), where there are five grabens of differing dimensions, all oriented north or northeast: the Dundas on the west, which contains ophiolites; the MacQuarrie, the Cressy, the Tamar; and the Derwent grabens (Gill, 1962); (2) Victoria, where are the Melbourne, Lachlan, Ballarat, and Cowra troughs; the Lachlan extends into New England, where an extensional phase is indicated; and (3) Queensland —the Upper Brisbane valley or Esk rift. This is clearly defined, 65 km long and 10 km wide (Hill, 1960; Murphy, 1976).

REFERENCES

Branson, J. C., 1978, Evolution of Sedimentary basins from Mesozoic times in Australia's continental slope and shelf, *Tectonophysics* **48:**389–412.

Crawford, A. R., 1963, The geology of Yorke Peninsula, *Geol. Survey S.Aust. Bull.* **39**.

Doutch, H. F., and E. Nicholas, 1978, The Phanerozoic sedimentary basins of Australia and their tectonic implications, *Tectonophysics* **48:**365–388.

Fairbridge, R. W., 1982, The fracturing of Gondwanaland, in *The Ocean Floor,* R. A. Scrutten and M. Talwani, eds., Wiley, New York, pp. 228–235.

Falvey, D. A., and J. C. Mutter, 1981, Regional plate tectonics and the evolution of Australia's passive continental margin, *Bureau of Mineral Resources Journal of Australian Geology and Geophysics* **6:**1-30.

Fenner, C., 1930, The major structural physiographic features of South Australia, *Roy. Soc. South Australia Trans.* **54:**1-36.

Finkl, C., and R. W. Fairbridge, 1979, Paleogeographic evolution of a rifted cratonic margin: S. W. Australia, *Palaeogeography, Palaeoclimatology, Palaeoecology* **26:**221-252.

Geological Society of Australia, 1971, *1971 Tectonic map of Australia and New Guinea,* 1:5,000,000, Sydney.

Gill, E. D., ed., 1962, Cainozoic, Geology of Tasmania, *Geol. Soc. Australia Jour.* **9:**233-254.

Guppy, D. J., A. W. Lindner, J. H. Rattigan, and J. N. Casey, 1958, *The Geology of the Fitzroy Basin,* Australia Bur. Mineral Resources Bull. 36, 116p.

Hill, D., 1960, The upper Brisbane fault trough, *Geol. Soc. Australia Jour.* **7:**269-274.

Khain, V. E., 1981, *Regional Geotectonics,* Vol. 3: *Extra-Alpine Eurasia, Australia,* Nedra, Moscow, 340p. [in Russian; translation unpublished].

Murphy, P. R., 1976, *Geology of the Gympie sheet area: Esk Trough,* Geol. Surv. Queensland, Pub. 96, p. 87.

Nairn, A. E. M., and F. G. Stehli, eds., 1982, *The ocean basins and margins, vol. 6: The Indian Ocean,* Plenum Press, New York, 776p.

Plumb, K. A., 1979, The tectonic evolution of Australia, *Earth Sci. Rev.* **14:**205-249.

Powell, D. E., 1976, The geological evolution of the continental margin of northwest Australia, *APEA Jour.* **16:**13-23.

Rutland, R. W., 1981, Structural framework of the Australian Precambrian, in *Precambrian of the Southern Hemisphere,* D. R. Hunter, ed. Elsevier, Amsterdam, pp. 1-32.

Rutland, R. W. R., A. J. Parker, G. M. Pitt, W. V. Preiss, and B. Murrell, 1981, The Precambrian of South Australia, in *Precambrian of the Southern Hemisphere,* D. R. Hunter, ed., Elsevier, Amsterdam, pp. 309-360.

Scheibner, E., 1978, Tasman fold belt system or orogenic system—introduction, *Tectonophysics* **48:**153-157.

Shields, O., 1977, A Gondwanaland reconstruction for the Indian Ocean, *Jour. Geol.* **85:**236-242.

Stewart, I. C. F., and T. J. Mount, 1972, Earthquake mechanisms in South Australia in relation to plate tectonics, *Geol. Soc. Australia Jour.* **19:** 41-52.

Talwani, M., J. Mutter, R. Houtz, and M. Konig, 1978, The margin south of

Australia, a continental margin paleorift, in *Tectonics and Geophysics of Continental Rifts,* I. B. Ramberg and E.-R. Neumann, eds., D. Reidel, Dordrecht, pp. 203-220.

Teichert, C., 1939, Research work in the Permian of Western Australia, *Australian Jour. Sci.* **2:**5-7.

Veevers, J. J., 1974, Western continental margin of Australia, in *Geology of Continental Margins,* C. A. Burk and C. L. Drake, eds., Springer-Verlag, Berlin, pp. 606-616.

Veevers, J. J., 1976, Early Phanerozoic events on or alongside the Australian-Antarctic platform, *Geol. Soc. Australia Jour.* **23:**183-206.

Veevers, J. J., 1982, Western and Northwestern margin of Australia, in *The Ocean Basins and Margins* vol. 6, A. E. M. Nairn and F. G. Stehli, eds., Plenum, New York, pp. 513-544.

Veevers, J. J., and A. T. Wells, 1961, The geology of the Canning Basin, Western Australia, *Bur. Mineral Resources Bull.* **60:** 323p.

Veevers, J. J., and M. W. McElhinny, 1976, The separation of Australia from other continents, *Earth Sci. Rev.* **12:**139-159.

Wells, A. T., and D. J. Forman, 1979, Ngalia Basin study, *Bur. Mineral Resources Yearbook* (Canberra) **78:**16-21.

Williams, E., 1978, Tasman fold belt system in Tasmania, *Tectonophysics* **48:**159-205.

Williams, F. D., and B. H. Corliss, 1982, The South Australian margin, in *The Ocean Basins and Margins,* vol. 6, A. E. M. Nairn and F. G. Stehl, eds., Plenum, New York, pp. 545-585.

26

Reprinted from pages 369-372, 373, and 385-386 of *Zeitschr. Geomorphologie N. F.* **22**:369-389 (1978)

Geomorphic analysis of the rifted cratonic margins of Western Australia

by Rhodes W. Fairbridge, New York,

and Charles W. Finkl, jnr., Fort Lauderdale

with 4 figures

Zusammenfassung. Durch die geomorphologische Analyse im südwestlichen Teil des Australischen Kratons wurde ein permisches Flußsystem erschlossen, das charakteristische glaziale Erratika enthält, infolge Reliefumkehr liegt es heute auf den Wasserscheiden des Oberlandes. Wenig tiefer in der Meereshöhe liegen die weitverbreiteten mesozoischen Ablagerungen verzweigter Flüsse. Überlappend oder eingeschnitten darin sind dendritische, mäandrierende Läufe von spätkretazischen bis eozänen Flüssen, die an 280 m Knickpunkten enden, was dem mitteleozänen eustatischen Meeresspiegelstand entspricht.

Der westliche Rand des Kratons entwickelte sich durch Abschiebung eines taphrogenen Troges (des Westralian Aulakogen), der über 10 km einer sedimentären Füllung in einer fast kompletten Abfolge vom Ordovizium bis zum Quartär enthält. Im Rand des Kratons, während des Aufbrechens aufgebogen, entstanden tiefe Klammen, ursprünglich bis zu 500 m Tiefe, die Fjorde während der permischen Vereisung gewesen sein mögen. Durch das Wiederbeleben des Aufbrechens und die Ausweitung des Ozeanbodens im Mesozoikum wurden sie zu Rias umgewandelt, und diese tiefen Einschnitte wurden in der oberen Kreide und im Eozän weitgehend mit Delta- und Ästuar-Sedimenten aufgefüllt, aber zum größten Teil wieder freigelegt.

Der südliche Rand des Kratons wurde aufgebogen und nach Süden gekippt, als im Eozän die australische Platte von der Antarktis getrennt wurde. Die neuen, nach Süden gerichteten Flüsse bewirkten eine weitverbreitete Anzapfung der vorher nach Nordwesten ziehenden Entwässerung. Eine ästuarzine Fazies reicht auf Grund des hohen eozänen Meeresspiegels bis zu 100 km landeinwärts. Ein niedriger Meeresspiegelstand im Oligozän korreliert mit einer klimabedingten Austrocknung und der weitverbreiteten Verhärtung der tiefverwitterten Oberflächen, so daß der harte Laterit entstand, der heute einen Schutzschild über weiten Teilen des Kratons bildet.

Summary. Geomorphic analysis of an area in the southwestern part of the Australian craton discloses a Permian river system containing characteristic glacial erratics; it occupies an inverted relief position on the upland divides. Slightly lower in elevation are widespread braided stream deposits of Mesozoic age. Overlapping or incised in them are dendritic meandering patterns of Late Cretaceous-Eocene streams that terminate at 280 m knickpoints, the level of mid-Eocene eustasy.

The western edge of the craton developed by the down-faulting of a taphrogenic trough (the "Westralian Aulacogen") that contains over 10 km of sedimentary fill, an almost complete section from Ordovician to Quaternary. The cratonic margin, uplifted during rifting, developed deep gorges, originally up to 500 m deep, that may have been fjords in Permian glacial time. Converted to rias by revived Mesozoic rifting and sea floor spreading, these gorges were largely infilled by deltaic and estuarine deposits in the Late Cretaceous and Eocene, but have subsequently been largely re-excavated.

The southern margin of the craton was upwarped and tilted southward when the Australian plate was severed from Antarctica in Eocene time. The new, southerly directed streams show extensive piracy of the formerly northwest trending drainage. Estuarine facies extend up to 100 km inland because of the high Eocene sea level. Low eustatic sea level in the Oligocene correlates with a climatic desiccation and widespread induration of deep weatherring surfaces to form a hard laterite that today forms a carapace over most of the craton.

Résumé. L'analyse géomorphologique d'une région de la partie sudouest du craton d'Australie a dévoilé un système de rivières permiennes contenant des erratiques glaciaires; les terrains de ces rivières occupent une position en inversion de relief sur les parties élevées des lignes de séparation des eaux. A une altitude légèrement inférieure se trouvent des dépôts de rivières anastomosées d'âge mésozoïque. Les recouvrant ou incisé dans ces formations, on reconnaît un réseau dendritique de rivières méandrantes existant à la fin du Crétacé ou à l'Eocène, rivières qui se terminent par des ruptures de pente vers 200 m, altitude du niveau marin au milieu de l'Eocène.

La bordure occidentale du craton est développée par la cassure d'une fosse taphrogénique (l'aulacogène westralien) qui contient plus de 10 km de remplissage sédimentaire en une série presque complète de l'Ordovicien ou Quaternaire. La marge du craton, soulevée pendant le déplacement du continent, a permis le façonnement de gorges profondes, au départ de plus de 500 m de profondeur, qui peuvent avoir été des fjords lors de l'époque glaciaire permienne. Converties en rias par le déplacement du continent au Mésozoïque et l'extension du fond de la mer, ces gorges furent largement remplies par des dépôts deltaïques et d'estuaire à la fin du Crétacé et de l'Eocène; elles ont été, par la suite, légèrement recreusées.

La bordure méridionale du craton a été soulevée et inclinée vers le sud quand la plaque australienne a été séparée de l'Antarctique à l'époque Eocène. Les nouvelles rivières, dirigées vers le sud, provoquèrent de nombreuses captures au détriment du drainage dirigé précédemment vers le nord-ouest. Le facies d'estuaire se retrouve jusqu'au-delà de 100 km à l'intérieur des terres suite au niveau marin élevé de l'Eocène. Le niveau marin peu élevé de l'Oligocène doit être mis en relation avec un assèchement climatique et une induration largement répandue des surfaces profondément altérées jusqu'à former une couverture latéritique dure qui recouvre encore de nos jours la majeure partie du craton.

1 *Introduction*

The dynamic geomorphic evolution of "Atlantic type" continental margins, i. e. those affected by rifting and continental drift (FAIRBRIDGE 1968, p. 34, 174), has been considered already by JESSEN (1943), OBST & KAISER (1949), and by DU TOIT (1940). On the basis of geomorphic evidence rifted continents are recognized as developing an uplifted rim, the "Randschwelle" of JESSEN. The discoveries associated with plate tectonics and sea floor spreading have introduced the concept of the "trailing" plate margin (in contrast to a subducting one), with its associated heat flow and consequent vertical crustal motions (RONA 1974; KINSMAN 1975). The feature is particularly well developed in southern Brazil

(Serro de Mar), South Africa ("Great Escarpment" and Drakensberg), and Western Australia (Darling Range and Jarrahwood Axis).

While the first two examples have been adequately described, the geomorphic features of southwestern Australia are less well recognized. A new understanding of the situation has been made possible by close field mapping of the alluvial formations in a selected area between Donnybrook and the south coast, and subsequently coordinated with results of stratigraphic studies by members of the Geological Survey of Western Australia, the faculty of the University of Western Australia, and exploratory boring by WAPET (West Australian Petroleum Pty. Ltd.). Warm appreciation is due to former colleagues and erstwhile students in respective organizations.

2 *Structural setting*

The southwestern part of the Australian craton consists largely of an Archean nucleus, the Yilgarn Block, overlapped or bordered in places by slightly metamorphosed Proterozoic sediments. The western margin is marked by a 1500 km N-S lineament of great antiquity, the Darling Fault, along which in places the Proterozoic formations are deformed up to vertical attitudes and what were formerly carbonate stromatolites are totally replaced by silica (FAIRBRIDGE 1951). The Darling Fault was probably activated in Ordovician times (or earlier), again during the Permian, mid-Mesozoic, in the Early Cretaceous (Neocomian), and in younger times. Some early motions may have been partly strike-slip but younger activity has been strictly normal faulting (PLAYFORD et al. 1975). Small wedges of Precambrian cratonic rocks outcrop here and there in scattered fault blocks also west of the Darling line (Leeuwin Block, Northampton Block, etc.).

The southern coast is not truncated in the same manner as the west coast, but bordered by an E-W trending 50–100 km wide belt of Middle Proterozoic granitic rocks, volcanics and metasediments (the last only in the Stirling Range) that extend as far east as Bremer Bay. Farther east drowned inselbergs of the same rocks, forming the Archipelago of the Recherche, extend almost to the edge of the continental shelf.

The plate tectonic history of the western borders of the Australian craton evolved in three principal steps. These were:

a) An aulacogen, the "Westralian Geosyncline" of TEICHERT (1938, 1958), that opened as a wedge coming in from the north to south, probably in Ordovician times, forming a narrow gulf partly separating the western border of the craton from the rest of Gondwanaland. The southern part of this gulf now constitutes the Perth Basin. The oldest formation here is the *Skolithos*-bearing Tumblagooda Sandstone, the age of which is variously estimated as Cambrian to Silurian. (Paleomagnetic interpretation indicates a Cambrian age: SCHMIDT & MORRIS 1977). In the northwest an almost complete marine section can be traced from Ordovician to Cenozoic in considerable thickness (over 10,000 m). In the southwest, the thickness exceeds 10,500 m (PLAYFORD et al. 1975, p. 258). Much of the sedimentary fill consists of subgraywacke suites, which suggests an orogenic provenance, possibly derived from a now-remote Tibetan source area, but difficult to explain as shield rocks, either in terms of petrology or volume.

b) Sea-floor spreading began within this trough in Late Jurassic time, separating the Australian plate from India-Tibet (FALVEY 1972; VEEVERS & MCELHINNY 1976; JOHNSON et al. 1976; VEEVERS & COTTERILL 1978), while the Antarctica/Australia connection remained intact. Magnetically "striped" segments of Late Jurassic sea floor have been mapped to the northwest and southwest of Australia (LARSON 1975). Basalt flows, the Bunbury Basalt, dated 136 ± 3 m. yr. B. P. (uppermost Jurassic) are found in the southwest (PLAYFORD et al. 1975, p. 252), and there were important motions of the Darling Fault that appear to be mid-Jurassic-Neocomian. Since that time there has been discontinuous subsidence along the fault, particularly in the middle part of the Perth Basin.

c) The continental shelf margin of the southern coast appears to have been activated by strike-slip faulting in Ordovician times (CRAWFORD & CAMPBELL 1973). This strike-slip motion was revived during Late Jurassic / Early Cretaceous times, as would be required by the geometry involved in the break-up of Gondwanaland (VEEVERS 1971; FALVEY 1974; DEIGHTON et al. 1976; NORTON & MOLNAR 1977). A shallow marine connection then reached from the newly opened southern Indian Ocean to the Eucla Basin in southern Australia and thence by epicontinental seas to the Great Artesian Basin and the western Pacific. A right-lateral strike-slip displacement must have continued from latest Jurassic until Eocene time when the rift began to open in a N-S sense and Australia was ruptured from Antarctica. This rupture had a profound effect on the paleoclimatic and geomorphic history of Australia, because, continuously from Permian time onwards the South Pole has remained within the general framework of Antarctica and thus southwest Australia lay entirely within paleolatitudes 50—55° S. Subsequent to the rupture the Australian continent has migrated progressively northward, thus into subtropical to tropical latitudes (EMBLETON 1973; KEMP 1978).

[*Editor's Note:* Material has been omitted at this point.]

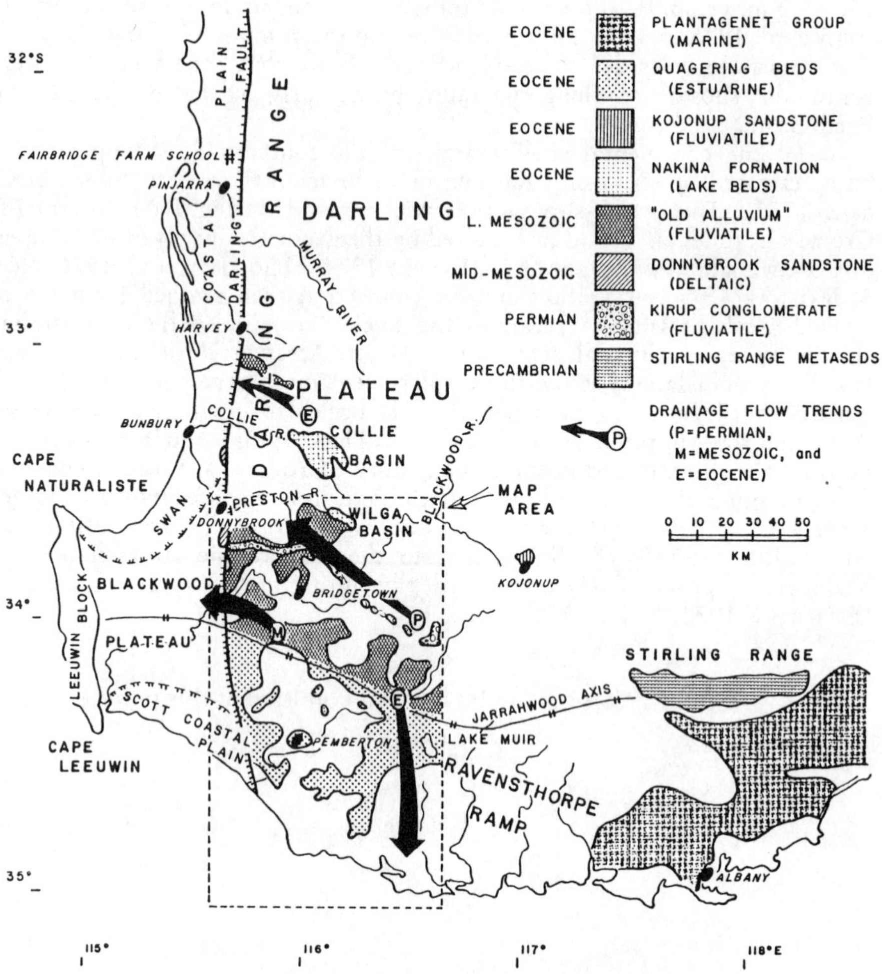

Fig. 1. Locality sketch-map of southwestern Australia showing the area mapped (see Fig. 2) and the principal drainage trends: P = Permian glaciation and fluvioglacial trends (erratics probably from the Stirling Range); M = Mesozoic trends (accelerated during the Late Jurassic rupture and sea-floor spreading); E = Eocene trends (to the west truncated by Oligocene eustatic drop in sea level and during younger oscillations; to the south, modified by epeirogenic warping — Jarrahwood Axis — associated with rupture and sea-floor spreading in the Southern Ocean). — Correction note: the Kojonup Beds should be SW of the township.

6 *Conclusions*

(1) The Australian craton has remained remarkably stable for over 10^9 yr as evidenced by flat-lying Proterozoic rocks unconformable on a peneplaned Archean basement. Its southwestern part, the Darling Plateau, has a very constant surface elevation of 300–500 m above MSL.

(2) The western margin was truncated along the Darling Fault in Early Paleozoic time and an aulacogen was created forming a wedge between Australia and (?) India-Tibet. Mainly marine seas occupied this taphrogenic trough (TEICHERT's "Westralian Geosyncline"), accumulating over 10 km of sediments that are almost continuous from Ordovician to Quaternary.

(3) According to plate tectonic theory, the opening of the Westralian Aulacogen would be accompanied by secondary uplift of the cratonic margins, where fluvial erosion would create incised river gorges.

(4) The Permo-Carboniferous glaciation led to further relief modification, and some scour depressions within the craton (selective lithologic control) reached 1500 m below the mean level of the plateau. These troughs became the site of subsequent Permian coal swamps and fluvial accumulation.

(5) A Permian river channel ("Kirup Drainage System") has been traced from SE to NW, the trend of the glacial grooving, and today, with inverted relief, occupies upland divides. The alluvial fill (Kirup Conglomerate) includes large faceted clasts of quartzite apparently derived from ranges 80 km to the southeast.

(6) A late-Jurassic sea-floor spreading is now identified in the eastern Indian Ocean, converting the Westralian Aulacogen to a "trailing edge" continental margin, at the same time reviving activity along the Darling Fault and the fluvial incision of the marginal gorges. Basalt flows that partly fill the lower fluvial channels are dated 136 ± 3 m. yr. (latest Jurassic).

(7) On the craton a westward flowing braided stream regime ("Old Alluvium") occurs at levels of 10–20 m below the Kirup Conglomerate. This "Westralian Drainage System" feeds into the deep gorges and appears to interfinger with estuarine and marine facies (Donnybrook Sandstone and related formations) of Early Cretaceous age.

(8) World-wide rises of sea level in the Late Cretaceous-Eocene interval led to thalassostatic reduction of effective relief and a dendritic meandering fluvial regime was instituted on the craton ("Darling Drainage System"). The fine-grained fill (Boronia Swamp Alluvium) is today sharply truncated at the heads of deep gorges by knickpoints around 280 m, some with dry valleys while other still active streams provide spectacular waterfalls. A eustatic high sea level at 280–300 m of mid-Eocene age is widely developed in Australia.

(9) The Eocene separation of Australia from Antarctica led to a marginal upwarp developing along the south coast. Henceforth a fluvial pattern developed, the "Ravensthorpe Drainage System", involving much stream piracy, with replacement of northwesterly trends by southern directions. Deep gorges are absent in this partly drowned south-sloping margin of the craton, and the

Eocene marine sequence interfingers with a fluvial and estuarine sedimentary facies (Quagering Beds).

(10) A world-wide Late Eocene and Oligocene fall of sea level, accompanied by glaciation in Antarctica, led to widespread desiccation, loss of vegetative cover and decrease in surface runoff. The deeply weathered regolith became now massively indurated forming extensive duricrusts (laterite, silcrete, etc.). The lateritic cap has played an important role in preserving the unconsolidated alluvial formations on upland surfaces.

(11) In Late Cenozoic time, there appears to have been continued subsidence along the Ravensthorpe Ramp with evident drowning of the southern coast. The laterite can be traced passing below the coastal eolianites and dunes.

(12) This survey highlights several important geomorphic principles: (a) the extreme stability of the cratonic interiors and their planation surfaces; (b) the slow rate of denudation there, c. 10–20 cm/m. yr.; (c) the intimate relationship of cratonic margins with plate tectonic events, which temporarily upset the stability; (d) the dependence of fluvial regimes upon eustatic history; (e) the inheritance of major landforms (e. g. Permo-Carboniferous glacial features, ? fjords); (f) the remarkable preservation of ancient relief features (e. g. the Permian river gravels and their thalwegs) due to massive pedogenic silicification; and (g) the increasing relative relief of ancient inselberg landscapes.

[*Editor's Note:* Only the references cited in the preceding excerpts are reproduced here. In the original article, Teichert (1938, 1958) was omitted.]

REFERENCES

Crawford, A. R. & K. S. W. Campbell (1973): Large-scale horizontal displacement within Australo-Antarctica in the Ordovician.—Nature, Phys. Sci. **241:**11-14.

Deighton, I., D. A. Falvey & D. J. Taylor (1976): Depositional environments and geotectonic framework: southern Australian continental margin.—Aust. Petrol. Explor. Assoc. J. **16:**25-36.

Du Toit, A. L. (1940): An hypothesis of submarine canyons. Geol. Mag. **77:**395-404.

Embleton, B. J. J. (1973): The paleolatitude of Australia through Phanerozoic time.—J. Geol. Soc. Aust. **19, 4:**475-482.

Fairbridge, R. W. (1951): Some recent advances in the geology of Western Australia.—Geol. Rdsch. **59, 1:**282-292.

Fairbridge, R. W. & V. N. Serventy (1968): Atlantic and pacific coasts; and continental flexure.—In: Encyclopedia of Geomorphology (R. W. Fairbridge, ed.). Pp. 34-35 and 174-177 Reinhold, New York.

Falvey, D. A. (1972): Sea-floor spreading in the Wharton Basin (northeast Indian Ocean) and the break-up of eastern Gondwana.—Aust. Petrol. Explor. Assoc. J. **12, 2:**86-88.

Falvey, D. A. (1974): The development of continental margins in plate tectonic theory.—Aust. Petrol. Explor. Assoc. J. **14: 1:**95-101.

Jessen, O. (1943): Die Randschwellen der Kontinente.—Petermanns Geogr. Mitt., Erg. **241:**

Johnson, B. D., C. McA. Powell & J. J. Veevers (1976): Spreading history of the eastern Indian Ocean und greater India's northward flight from Antarctica and Australia.—Geol. Soc. Amer. Bull. **87, 11:**1560-1566.

Kemp, E. M. (1978): Tertiary climatic evolution and vegetation history in the southeast Indian Ocean region.—Palaeogeogr., Palaeoclimatol., Palaeoecol. **24, 3:**169-208.

Kinsman, D. J. J. (1975): Rift valley basins and sedimentary history of trailing continental margins. In: Petroleum and Global Tectonics (A. G. Fischer, ed.). Pp. 83-126. Princeton University Press, Princeton.

Larson, R. L. (1975): Late Jurassic sea-floor spreading in the eastern Indian Ocean.—Geology **3:**69-71.

Norton, I. & P. Molnar (1977): Implications of a revised fit between Australia and Antarctica for the evolution of the eastern Indian Ocean.—Nature, Lond. **267:**338-340.

Obst, E. & K. Kaiser (1949): Die große Randstufe auf der Ostseite Südafrikas und ihr Vorland.—Geograph. Gesellschaft, Hannover, 292 p.

Playford, P. E., R. N. Cope, A. E. Cockbain, G. H. Low & D. C. Lowry (1975): Perth Basin; Collie and Wilga Basins; Superficial Phanerozoic deposits on the West Australian Shield. West. Aust. Geol. Surv. Mem. **2:**227-259; 420-431; 432-433.

Rona, P. A. (1974): Subsidence of Atlantic continental margins.—Tectonophysics **22:**283-299.

Schmidt, P. W. & W. A. Morris (1977): An alternative view of the Gondwana Paleozoic apparent polar wander path.—Can. J. Earth Sci. **14, 1:**2674-2678.

Veevers, J. J. (1971): Phanerozoic history of Western Australia related to continental drift.—J. Geol. Soc. Aust. **18:**87-96.

Veevers, J. J. & M. W. McElhinny (1976): The separation of Australia from other continents.—Earth Sci. Rev. **12:**139-159.

Veevers, J. J. & D. Cotterill (1978): Western margin of Australia; Evolution of a rifted arch system.—Geol. Soc. Amer. Bull. **89:**337-355.

27

Reprinted from pages 181-186 of *The World Rift System,* Report of the Upper Mantle Committee Symposium on the World Rift System, T. N. Irvine, ed., Geol. Surv. Canada Paper 66-14, 1965, 471p.

FAULT BLOCKS AND RIFTS IN SOUTH AUSTRALIA[1]

C.R. Twidale
University of Adelaide. At present, Rensselaer Polytechnic Institute, Troy, N.Y.

Abstract

The pattern of faulting in the South Australian rift area is described and is related to the continental pattern of lineaments outlined by E.S. Hills. Although the framework is an ancient one, the faults have been active in the late Cenozoic, and indeed, their development continues.

TEXT

Although the Australian rifts are most obviously manifested in southern South Australia, a pattern of faulted blocks of which the rifts are an inherent part extends over the entire continent. This continental tectonic framework is expressed in long, continuous, rectilinear or gently arcuate structures called lineaments which are the traces of steeply dipping faults and joints, as well as warp axes, and which are fundamentally of great antiquity although subject to recurrent movements. They determine the major features of Australian topography, display limited trend-ranges, and according to some authorities, they are part of a global pattern.

In South Australia, the rift system is most marked about the Mt. Lofty-Flinders Ranges, and at the eastern margin of the West Australian Shield (Fig. 1). Although the Ranges are represented on topographic maps as having curved outlines, in reality, they consist of a series of juxtaposed blocks principally delineated by NW-SE and SW-NE trending faults and warps, although in some areas N-S and E-W trends are apparent. To the west of the upland backbone, which has an offshoot in the south in the horst block of Yorke Peninsula, a number of rift valleys are arranged like an inverted and distorted Y. In the south, the two arms of the Y, the Spencer and St. Vincent Gulfs, are inundated by the ocean. The western side of the Spencer Gulf (the eastern margin of Eyre Peninsula) is delineated by the Lincoln fault zone, which trends SW-NE but within which are important faults trending N-S. This zone continues to the NE beyond the head of the Gulf and can be

---

[1]Contribution No. 66-1, Rensselaer Polytechnic Institute, Troy, New York.

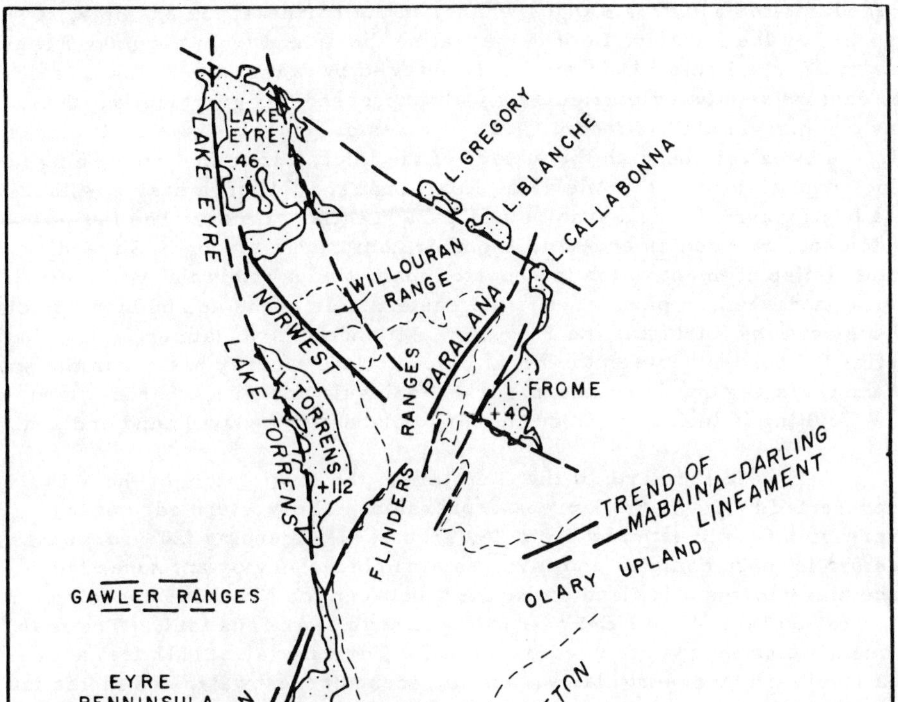

Figure 1. Sketch map of the principal faults in the vicinity of Adelaide.

traced into the Flinders Ranges where, in the northeast for example, it appears as the Paralana fault demarcating the upland from the Lake Frome basin. Yorke Peninsula is similarly outlined by faults, the Kulpara fault on its eastern side being particularly well expressed topographically. On the eastern margin of St. Vincent Gulf, a series of gently arcuate fault blocks dip southwesterly beneath the waters of the inlet, and in the intervening fault block depressions — the Adelaide, Noarlunga and Willunga basins — there has been spasmodic deposition since early Tertiary times. The faults, which are arranged en echelon, together comprise a strong N-S trending zone of displacement which is repeated many times within the Mt. Lofty block, and which is paralleled in the eastern side of the upland by a complex of dislocations (including the Florieton, Milandella and Palmer faults) that define the border between the horst block and the Murray basin. In the south of the Gulfs region, in southern Yorke Peninsula and on Kangaroo Island E-W faulting is important in delineating upland and lowland, land and sea.

To the north of the Gulfs area, the single stem of the Y is represented by the Lake Torrens depression. The western edge of the depression is delineated by a single, arcuate N-S trending fault zone; the eastern is more complex and involves a number of important zones but essentially forms a lowland embayment between the NNW-SSE trending Norwest fault, and the NE-SW trending Lincoln-Paralana fault. The fault trough, which is now occupied by the Lake Torrens salina (112 feet above sea level) and by desert plains, was the scene of freshwater sedimentation in the middle Tertiary. Still farther to the north, in a large westerly extension of the Great Artesian Basin, is the lowest part of the Australian Continent — the bed of Lake Eyre, a salina about 3,600 square miles in area and 46 feet below sea-level. Not only is the whole Lake Eyre depression a negative tectonic area, but the Lake Eyre salina itself appears to be a downfaulted block. The western margin of the Lake is, disregarding minor crenulations due to stream erosion, a remarkably rectilinear feature. On the bed of the lake, mound springs display alignment along N-S lines; earth tremors are restricted to the neighbourhood of the western scarp; and Wopfner has suggested that the gypsite which caps the western escarpment of the lake basin is the same formation which occurs at shallow depth beneath the bed of the salina. Moreover, this postulated major fault lines up with fractures to the south. Thus subsidence, rather than deflation, appears to be primarily responsible for the Lake Eyre salina.

Other major salinas, including Lake Frome (about 40 feet above sea-level), either lie along, or are to a significant degree defined by, faults of the lineament pattern (Fig. 1).

Thus the central N-S zone of South Australia, essentially at the margin of the Shield, comprises a series of positive and negative faulted or warped blocks, some of them long and narrow and of the horst and graben type, and others having a more complex pattern.

In eastern South Australia, in the Olary upland and Murray basin, strong lineament trends (many of them the terminations of structures strongly developed in adjacent States) determine the major geomorphological features. The Anabama-Darling lineament, for example, continues in the ENE-WSW trend of the Olary and its regional structures, and the course of the Murray owes much to lineaments in the underlying bedrock. The Padthaway ridge (or horst), separating the Murray basin with its late Mesozoic and Tertiary marine sediments from the Southern Ocean, trends NW-SE, is only shallowly covered by unconsolidated deposits (so that granitic rocks, for example, are exposed) and is in part at least defined by faults.

To the west of the Mt. Lofty-Flinders upland, the whole of Eyre Peninsula is part of the rectangular Gawler block, the borders of which trend NW-SE and NE-SW. The Eucla basin, containing Cretaceous and Tertiary marine sediments, is bounded on the NE by the lineament defining the SW edge of the Gawler block, and on the NW, it is probably bounded by another strong structural trend which turns NNE-SSW and which is paralleled by sections of the West Australian south coast. In the far NW, the Everard, Musgrave, Mann and other ranges form part of the E-W structure that is so distinctive a feature in central Australia. But there is an obvious and marked concentration of major faults within and bordering the upland backbone of South Australia, and immediately to the west, at the margin of the Shield. This is the South Australian rift zone.

In most areas the nature of the faulting is not clear. The Norwest and Paralana faults are thrust faults, and although those in the Adelaide area are generally regarded as normal, there is some evidence that, locally at least, they are of reverse type. A theory of the mechanics of faulting on the South Australian rift area has been elaborated by O'Driscoll (1962), but much geophysical work remains to be done in the area.

The framework of lineaments is ancient, but there is ample evidence and suggestion of recent displacement of the fault blocks. For example:

(a) Earthquakes and earth tremors commonly occur in the vicinity of known faults.
(b) Many fault escarpments are so little dissected and have receded so little that the faulting must have occurred in geologically recent times.
(c) This argument can in some areas be substantiated stratigraphically. From evidence exposed on the western side of the Mount Lofty Ranges near Strathalbyn, Horwitz (1960) has argued that the faulting is

post-Miocene. In the Adelaide area, Glaessner (1953) has been able to demonstrate several phases of dislocation, including one of Pliocene or later age. Near Whyalla, marine Pliocene sediments were deposited at the base of the Lincoln fault zone escarpment. The possible displacement of Pleistocene gypsite at the western margin of Lake Eyre has already been mentioned.

Other possible demonstrations of recent and continued dislocation, for example in the Adelaide and the Gulfs areas, are uncertain because of difficulties in distinguishing between tectonic and eustatic effects, but in all probability, the rift system in South Australia is continuing to develop.

SELECTED REFERENCES

Geol. Atlas of South Australia, S.A. Geol. Survey, Adelaide.

Glaessner, M.F. 1953. Some problems of Tertiary geology in southern Australia. Jour. and Proc. Royal Soc., N.S.W. 87, 31-45.

Hills, E.S. 1956. A contribution to the morphotectonics of Australia. Jour. Geol. Soc. Austr. 3, 1-15.

Hills, E.S. 1961. Morphotectonics and the geomorphological sciences with special reference to Australia. Quart. Jour. Geol. Soc. London 117, 77-89.

Horwitz, R.C. 1960. Géologie de la région de Mt. Compass (feuille Milang), Australie Méridionale. Eclog. Geol. Helv. 53, 211-263.

Johns, R.K. 1963. Investigation of Lake Eyre. Pt. I, Report of Investigations No. 24, Dept. Mines, S. Austr.

O'Driscoll, E.S. 1962. Fold interference patterns in model experiments. Nature 193 (4811), 115-117.

Sprigg, R.C. 1962. On the structural evolution of the Great Artesian Basin. 37-56 in Austr. Petroleum Expl. Assoc. Conf. on geology of southern and eastern Australia, Melbourne, 1961.

Sprigg, R.C. 1962. Oil and gas possibilities of the St. Vincent Gulf graben, 71-88 in Austr. Petroleum Expl. Assoc. Conf. on geology of southern and eastern Australia, Melbourne, 1961.

Webb, B.P. 1958. Summary of tectonics and sedimentation, 136-143 in Geology of South Australia (Ed. M.F. Glaessner and L.W. Parkin, Melbourne.

Wopfner, H. and Twidale, C.R. 1966. Geomorphological history of the Lake Eyre Basin, in Landform studies from Australia and New Guinea (Ed. J.N. Jennings and J.A. Mabbutt) A.N.U. Press, Canberra.

28

Reprinted by permission of the Director, Bureau of Mineral Resources, Geology, and Geophysics, Canberra, Australia from Australia *Bur. Mineral Resources Geol. Geophys. Bull.* **100:**1–4 (1970)

GEOLOGY OF THE AMADEUS BASIN, CENTRAL AUSTRALIA

A. T. Wells, D. J. Forman, L. C. Ranford, and P. J. Cook

SUMMARY

The Amadeus Basin is a 500-mile-long intracratonic depression, with an area of about 60,000 square miles, which lies across the Northern Territory south of latitude 23°30′S and extends into Western Australia. It contains Proterozoic, Ordovician, possibly Silurian, and Devono-Carboniferous and minor Permian and Tertiary freshwater deposits. About 30,000 feet of sedimentary rocks are preserved.

The older Precambrian basement comprises the Arunta Complex along the northern margin of the basin, the Musgrave-Mann complex and Olia Gneiss in the south, and other unnamed Precambrian rocks. The Arunta Complex of igneous and metamorphic rocks was deformed during the Arunta Orogeny before deposition of the sedimentary rocks of the Amadeus Basin. In the southwest, thick sequences of younger Precambrian basement sediments and volcanic rocks lie between the older Precambrian rocks and the oldest sediments in the basin.

The first sediments deposited in the basin were about 2000 feet of sand (Heavitree and Dean Quartzites) laid down on a stable epicontinental shelf. After mild epeirogenic movement, evaporites were laid down in a shallow and restricted sea, and were succeeded by penesaline sediments, and finally by marine stromatolitic carbonate rocks and shale. The sediments of this evaporite cycle have a maximum thickness of 3000 feet, and have been named the Bitter Springs Formation and equivalent Pinyinna Beds. The only known volcanics interbedded with the sediments occur in the northeast in the upper part of the Bitter Springs Formation, and some poorly exposed mafic rocks in the southwest.

The tectonic events which followed represent the first phase in the development of an intracratonic depression roughly parallel to the present trend of the basin. There were two important periods of diastrophism during this tectonic phase, each recorded by unconformities in the Proterozoic sediments. The first, the Areyonga Movement, occurred after the deposition of the Bitter Springs Formation, and the second, the Souths Range Movement, after the deposition of the overlying Proterozoic rocks. Areas to the south were uplifted during each movement and became the main provenance of the Proterozoic sediments which poured into the subsiding area to the north, in which about 15,000 feet of clastic sediments accumulated. The depression was separated from a northern shelf area by an east-west linear zone, or hinge-line, along which the rate of subsidence changed. A thick sequence of Proterozoic sediments along the zone and to the south was uplifted and subsequently eroded in the late Proterozoic or early Cambrian Petermann Ranges Orogeny. The average thickness of the equivalent sediments on the shelf is about 3000 feet.

Two periods of Proterozoic glaciation are recognized. The older and more widespread resulted in deposition of the glacial sediments of the Areyonga Formation on the shelf area. The equivalent Inindia Beds of the depression in the south contain thin tillitic beds near the top, and correlation of these beds with the glacial deposits in the Areyonga Formation suggests that after the Areyonga Movement Proterozoic sedimentation probably began in the depression.

The younger glacial period was restricted to the northeastern part of the shelf, and during the deposition of the Pertatataka Formation a thickened lobe of sediments, partly glacial in origin, formed. The equivalent Winnall Beds of the depression have no identifiable tillitic beds.

The Petermann Ranges Orogeny uplifted and overfolded a large area in the southwest in the late Proterozoic or early Cambrian. The Pinyinna Beds and Dean Quartzite were complexly infolded with the basement in the Petermann Ranges Nappe, but the younger Proterozoic sediments slid northwards over the incompetent beds in the Bitter Springs Formation and were tightly folded along the margin of the nappe. The intensity of folding decreased to the north and its effect was almost negligible north of the Proterozoic hinge-line. The main effects of the Petermann Ranges Orogeny appear to be localized along the original axis of maximum sedimentation in the depression.

Isostatic uplift followed the Petermann Ranges Orogeny in the southwestern margin of the basin and provided the source of the Cambrian sediments. Thick conglomerate and arkose (Mount Currie Conglomerate and arkose at Ayers Rock) were deposited in the southwest, adjacent to the uplifted area. Farther north, deltaic sands (Cleland Sandstone) were deposited; to the northeast, the sands interfinger with sand, shale, and some marine sediments. The predominantly clastic sediments are thinner in a north-south zone near the centre of the basin, which was probably being uplifted during the sedimentation. To the east, about 7000 feet of predominantly carbonate rocks and lutites were deposited in the northeastern part of the basin. The Arumbera Sandstone, the basal formation of the Pertaoorrta Group in the northeast, was succeeded by evaporite deposits (Chandler Limestone) which were followed by penesaline and marine stromatolitic carbonate rocks (Giles Creek Dolomite and Shannon Formation). The Arumbera Sandstone and the youngest unit of the Pertaoorrta Group (Goyder Formation) show only minor facies changes across the central zone.

The sandstone, shale, and minor carbonate rocks of the Cambro-Ordovician Larapinta Group were deposited conformably on the Pertaoorrta Group in the north and disconformably in the south, during several marine regressions and transgressions in a widespread epeiric sea. Sedimentation was initially restricted to the northern half of the basin, but the Ordovician sea gradually spread to the south and the younger formations transgressed over the Pertaoorrta Group, and in the south and west they rest unconformably on Proterozoic rocks. During the Ordovician, the Amadeus Basin probably accounted for only a small area of the shallow epeiric sea, and the epicontinental deposits of this period extended well north of the present margin of the basin. The maximum thickness of the Larapinta Group on the northern margin of the basin is about 8000 feet. Larapinta Group sedimentation was brought to a close by the deposition of a regressive body of deltaic or estuarine sand which may represent the last marine deposits in the Amadeus Basin.

During the ensuing epeirogenic movement (Rodingan Movement), which probably started in the Silurian, the Amadeus Basin was subjected to broad vertical movements; in the northeast 5000 to 10,000 feet of sediments were eroded and the area was reduced to a peneplain. The resulting large deserts around and probably within the Siluro-Devonian basin provided the sediments of the Mereenie Sandstone, whose origin is partly shallow marine and partly fluviatile and aeolian: the shallow seas transgressed from the west. The Mereenie Sandstone is probably mostly Devonian. As a result of the constantly moving strandline there was probably no distinct form to the Amadeus Basin.

The first of several diastrophic events, the Pertnjara Movement, occurred in the late Devonian. The events resulted in successive uplifts of a large block of Precambrian rocks and the overlying Proterozoic and Palaeozoic sediments on the northern margin of the basin.

The uplifted rocks were deeply eroded, and a thick wedge of molasse, the Pertnjara Group, was deposited on the southern flanks of the mountain chains. The lacustrine siltstone at the base of the group followed conformably on the Mereenie Sandstone, and after further uplift of the provenance area coarser sediments were deposited. Along the northern margin of the basin younger formations of the Pertnjara Group successively overlap older formations; the youngest formation, the Brewer Conglomerate, overlies rocks as old as Proterozoic. The Pertnjara Group has a maximum thickness of about 12,000 feet, on the northern margin of the basin.

The molasse deposits of the Pertnjara Group become thinner and finer in grain to the south. In the southeast, the Finke Group was deposited penecontemporaneously in a basin which was probably separated from the Amadeus Basin by a peninsula.

Most of the folding took place during the Alice Springs Orogeny; but crestal thinning of the Palaeozoic sediments over anticlines shows that several of the structures were being formed during Lower Palaeozoic sedimentation; in places the folds were subsequently obscured by the Pertnjara Group. Some of the structures are the result of the formation of diapirs in the Bitter Springs Formation, and others may be caused by thrusting or a combination of thrusting and diapirism. Some were probably initiated during the Petermann Ranges Orogeny, others during the diastrophic movements in the Devonian and Carboniferous.

During the Alice Springs Orogeny the Heavitree Quartzite and Bitter Springs Formation in the north were infolded and overthrust with the crystalline basement to form structures such as the Blatherskite Nappe and the Arltunga and Ormiston Nappe Complexes.

Décollement surfaces in the Bitter Springs Formation and Cambrian sediments in the northeast, with a thrust plane connecting the upper and lower surfaces, facilitated the complex folding; the overlying sediments were strongly and isoclinally folded, but in places they yielded by thrusting. Another interpretation of the structure in the northeast suggests that the large blocks of sediments uplifted over the Arltunga Nappe Complex were possibly moved to the south by gravity sliding on the two décollement surfaces. In this way large complex thrust nappes were formed and most of the sediments are preserved in allochthonous blocks.

The Amadeus Basin was stabilized after the deposition of the Pertnjara Group and the deformation associated with the Alice Springs Orogeny. During the ensuing long period of erosion the weathered detritus was transported outside the basin and incorporated in the fluvioglacial and paludal Permian Buck and Crown Point Formations, and in the deltaic Jurassic(?) sands and the shallow-marine Cretaceous deposits in the Simpson and Gibson Deserts.

Middle Tertiary fluvial and lacustrine sediments were deposited in ancestral river valleys both before and after lateritization. The pre-silcrete sediments are Eocene to Miocene and the post-silcrete sediments are Miocene or younger. The Tertiary deposits indicate a wetter climate; tilting and minor faulting are associated with minor crustal movements.

The deposits in the Tertiary valleys were eroded, and the valleys were breached by headwater erosion by streams which were initiated by crustal downwarping in the Lake Eyre Basin. During an arid phase which followed the system of internal drainage was initiated and sand dunes formed. Resumption of drainage incision suggests an amelioration of the climate with the development of alluvial fans and flood-plains.

The two most important occurrences of hydrocarbons in the basin are wet gas accumulations in the Mereenie and Palm Valley Anticlines; in both production comes chiefly from the upper part of the Cambro-Ordovician Pacoota Sandstone. Small amounts of oil have been obtained from the Mereenie Anticline, but the permeability in the reservoir rocks is insufficient to give significant oil production. The Ordovician and Cambrian sediments offer the best prospects for petroleum accumulation, and the Missionary Syncline, in the northern part of the basin, is probably the most prospective area. Several large structures with closure in Ordovician and Cambrian rocks, have been mapped under the Devono-Carboniferous cover rocks. One of the main problems is to outline areas where there is permeability in the Ordovician reservoir sands. The transition zone where sandstone siltstone and carbonate rocks interfinger in the Pertaoorrta Group is probably worth further investigation. Reservoir rocks may be developed in the central northerly trending zone, where there was probably structural growth during deposition and which roughly corresponds to the transition zone in the Cambrian sediments. The existence of a submarine ridge during sedimentation suggests an area of reworked clean sands suitable for hydrocarbon traps.

Groundwater is important to the pastoral industry. Aquifers which usually yield large supplies of good-quality water are available in the Quaternary and Tertiary deposits, the Crown Point Formation, the De Souza Sandstone, parts of the Finke Group, the Mereenie Sandstone, and the Pacoota Sandstone and Goyder Formation.

Of a wide variety of metallic and non-metallic minerals recorded, only materials used in the building industry and semi-precious mineral and rock samples have any value at present.

Part VIII

NEW ZEALAND

Editor's Comments on Papers 29, 30, and 31

29 SCHOLZ
Excerpts from *Transform Fault Systems of California and New Zealand: Similarities in Their Tectonic and Seismic Styles*

30 HOCHSTEIN and NIXON
Excerpts from *Geophysical Study of the Hauraki Depression, North Island, New Zealand*

31 HOCHSTEIN and TEARNEY
Excerpt from *Revised Geophysical-Geological Model of the Hauraki Rift* (Abstract)

The Alpine fault of New Zealand is a rift in the sense in which the term is used for the Great Glen fault, the Dead Sea (Western Arabian) rift, and the San Andreas fault (or rift). These are among the great strike-slip, transform faults that traverse and have rifted continental lithosphere (see Introduction).

General accounts of New Zealand geology are provided by Lillie (1980) and by Suggate, Stevens, and Te Punga (1978). The standard account of the Alpine fault is that by Suggate (1963). Wellman and Willett followed out the southern part through difficult terrain (1942). Characteristic landforms of strike-slip faulting were recognized (Wellman, 1952) and Wellman (1955) established the shift as about 450 km. The faults that branch off in a northeasterly direction were matched with faults in the North Island, the most probable main fault being the Wairau (Lensen, 1958). A clear extension of a single course for the Alpine fault through the North Island is not yet established and may not exist (Wellman, 1955). The problem lies west of the fault in the abrupt change in the geology from South to North Island, where the basement structure is masked by the very great thickness of Tertiary marine sediments in the Wanganui basin (Anderton, 1981) and by the volcanics of the Central volcanic region, Tongariro and Taupo basins (Healy, 1964). However, geophysical information about it is building up (Anderton, 1981; Hunt, 1980; Hatherton, 1970; Hatherton and Hunt, 1968; Walcott, 1978). The tectonics of New Zealand and the accommodation of a mobile belt have always been of demanding interest (Lillie, 1980, 90–100; Quennell, 1965; Grindley, 1974; Sporli, 1980; Walcott and Cresswell, 1979; Wellman 1970, 1975). The Alpine fault is clearly a transform fault system of the trench-trench class, the Hikurangi trench in the north and the Puyseger trench in the south.

Paper 29, by Christopher Scholz (of Columbia University), is a brief account of the Alpine fault; his treatment of the San Andreas system must regrettably be omitted from this volume. The seismicity is given chief attention, but our interest lies mainly in the mirror-image relationship between the two systems (see also Hatherton, 1968).

Paper 30, by M. P. Hochstein and I. M. Nixon (of the Geothermal Institute, University of Aukland) describes the Hauraki Depression, a graben or rift valley of moderate size. It probably formed during the movements accompanying the oroclinal bending or its restoration (Wellman, 1975; Quennell, 1965), and perhaps as an adjustment to the collision of the Indian and Pacific plates. Paper 31 supplements the information in Paper 30 by a revision of the model originally proposed.

REFERENCES

Anderton, P. W., 1981, Structure and evolution of the South Wanganui Basin, *New Zealand Jour. Geology and Geophysics* **24:**39–63.

Grindley, G. W., 1961, Mesozoic orogenies in New Zealand, *Proc. Pacific Sci. Congr. 9, Geology and Geophysics* **12:**71–75.

Grindley, G. W., 1974, New Zealand, in *Mesozic-Cenozoic Orogenic Belts,* A. M. Spencer, ed., Geol. Soc. London Spec. Pub. 4, pp. 387–416.

Hatherton, T. 1968, Through the looking glass: a comparative study of New Zealand and California, *Nature* **220:**660–663.

Hatherton, T., 1970, Gravity, seismicity and tectonics of the North Island, New Zealand, *New Zealand Jour. Geology and Geophysics* **13:**126–144.

Hatherton, T., and T. M. Hunt, 1968, Gravity profiles across the Alpine fault, *New Zealand. Jour. Geophys. Research* **73:**5343–5352.

Healy, J., 1964, Volcanic mechanisms in the Taupo volcanic zone, *New Zealand Jour. Geology and Geophysics* **7:**6–23.

Hunt, T. M., 1980, Basement structure of the Wanganui Basin, onshore, interpretated from gravity data, *New Zealand Jour. Geology and Geophysics* **28:**1–16.

Lensen, G. J., 1958, Note on fault correlations across Cook Strait, *New Zealand Jour. Geology and Geophysics* **1:**263–268.

Lillie, A. R., 1980, *Strata and Structure in New Zealand,* Tohunga Press, Auckland, 441p.

Norris, R. J., and R. M. Carter, 1980, Offshore sedimentary basins at the southern end of the Alpine Fault, New Zealand, in *Sedimentation in Oblique-Slip Mobile Zones,* P. F. Ballance and H. G. Reading, eds., Internat. Assoc. Sedimentologists Spec. Pub. 4, pp. 237–365.

Quennell, A. M., 1965, New Zealand palaeogeography and petrology occurrence, *Commonwealth Mining and Metallurg. Congress Australia and New Zealand, 8th, Proc. 7-N. Z. Sect. Paper* 212, pp. 1–14.

Sporli, K. B., 1980, New Zealand and oblique-slip margins: tectonic development up to and during the Cainozoic, in *Sedimentation in Oblique-Slip Mobile Zones,* P. F. Ballance and H. G. Reading, eds., Internat. Assoc. Sedimentologists Spec. Pub. 4, pp. 147–170.

Suggate, R. P., 1963, The Alpine fault, *Royal Soc. New Zealand Trans.* **2:** 105–129.

Suggate, R. P., G. R. Stevens and M. T. Te Punga, eds., 1978, *The Geology of New Zealand:* vol. 1, 343p.; vol. 2, 475p., Government Printer, Wellington.

Walcott, R. I., 1978, Present tectonics and late Cenozoic evolution of New Zealand, *Royal Astron. Soc. Geophys. Jour.* **52:**137–164.

Walcott, R. I., and M. M. Cresswell, eds., 1979, The origin of the Southern Alps, *Royal Soc. New Zealand Bull.* **18**.

Wellman, H. W., 1952, The Alpine Fault in detail: River terrace displacement at Maruia River, *New Zealand Jour. Sci. Tech.* **B33:**409–414.

Wellman, H. W., 1955, New Zealand Quaternary tectonics, *Geol. Rundschau* **43:**248–257.

Wellman, H. W., 1970, Age of Alpine Fault, New Zealand, *Internat. Geol. Congr. Rept. Pt. 4,* pp. 148–162.

Wellman, H. W., 1975, New Zealand sixty million years ago, *Australian Soc. Explor. Geophys. Bull.* **6:**55–56.

Wellman, H. W., and R. W. Willett, 1942, The geology of the West Coast from about Abut Head to Milford Sound, *Royal Soc. New Zealand Trans.* **71:**282–306.

29

Reprinted by permission from pages 215-216 and 221-225 of *Geol. Soc. London Jour.* **133**:215-229 (1977)

Transform fault systems of California and New Zealand: similarities in their tectonic and seismic styles

C. H. SCHOLZ

SUMMARY

When the San Andreas fault system of California and the Alpine fault system of New Zealand are compared within the reference frame of their regional slip vectors, their major tectonic elements are seen to occupy similar positions. This argues that a genetic relationship exists between the development of these complexities.

Furthermore, three distinctive modes of slip, or 'seismic style', can be recognized for different segments of these major transform fault systems. These are: (1) slip during great ($M \geq 8$) earthquakes separated by long periods of quiescence; (2) slip during more frequent large ($6{\cdot}5 \leq M \leq 7{\cdot}5$) earthquakes, and (3) aseismic slip (creep). These different seismic styles occur in distinctive tectonic settings and it is argued that they represent fundamental variations in the mechanical properties of the faults. I suggest that style 1 occurs along sections of the fault orientated such that a component of tectonic convergence occurs, leading to a high effective normal stress and hence high frictional strength. Style 2 occurs along sections of faults which strike close to the regional slip vector, and hence have a lower normal stress. Aseismic slip occurs only along sections of faults in central California where anomalously high pore pressures reduce the effective normal stress to a very low value and the fault is entirely within the stable sliding frictional field.

IN THE IDEAL CASE, a transform fault lies along that part of a plate boundary parallel to the local vector of relative motion between the two plates. Oceanic transform faults in general obey this geometric rule, but transform faults which transect continental lithosphere often exhibit complexities which result in marked departures from the ideal. For example, in the two cases discussed here, the San Andreas fault system of California and the Alpine fault system of New Zealand, major sections of the faults strike obliquely to the slip vector, resulting in continental convergence. Other sections of these fault systems consist of a number of subparallel, active faults which together contribute to plate motion distributed over a several hundred kilometre wide plate boundary.

Furthermore, different sections of these faults exhibit distinctively different frictional properties which result in the several modes of strain release which may be called 'seismic styles'. Three seismic styles can be recognized for the San Andreas fault system: aseismic slip, slip by infrequent great earthquakes separated by long periods of quiescence, and slip by more frequent, large earthquakes. Aseismic slip, sometimes wrongly called 'creep', is accompanied by very frequent, small to moderate earthquakes ($M < 6$) which do not significantly contribute to fault slip. For the purpose of this discussion great earthquakes are defined as those of $M \geq 8$ which produce metres of slip over hundreds of kilometres of fault length, and large earthquakes as those events in the magnitude range $6{\cdot}5 \leq M \leq 7{\cdot}5$ which produce tens of centimetres of slip over tens of kilometres of fault length.

There is clearly a continuum of earthquakes spanning the entire magnitude range but these three size ranges seem to typify these three seismic styles; hence the definitions are not entirely arbitrary and are convenient for this discussion.

One must ask the question whether these seismic styles are permanent features of those sections of the fault zone on which they have been observed during the scant historic record. That is, do those regions in which slip has occurred during great earthquakes always rupture by great earthquakes? Allen (1968) has argued that this is the case by showing that sections of the faults which exhibit different seismic styles are also in distinctly different geological environments. Thus he notes that those areas of the San Andreas system which have had great earthquakes are those sections where the fault undergoes a major bend which may serve as a 'locking' device; those sections which slip aseismically have abundant serpentinite bodies along the fault zone; those areas characterized by more frequent large earthquakes are those in which several, subparallel splay faults exist.

Two fault systems are described below: the San Andreas fault system of California and the Alpine fault system of New Zealand. The paper attempts to evaluate the tectonic environment of zones of differing seismic style and points to the way in which these faults have developed the complex geometries which they now possess. The physical basis for the different seismic styles is interpreted in the light of laboratory studies of rock friction, with results that differ in some major respects from Allen's (1968) interpretation.

Both fault systems are boundaries between major lithosphere plates; hence the motion on them must reflect the motion of the plates. The San Andreas system is a transform fault of the ridge-ridge type, shown schematically in Fig. 1a. It takes up the motion between the northern end of the East Pacific rise spreading in the Gulf of California, and the Gorda and Juan de Fuca ridges to the north. The Alpine fault, on the other hand, is a trench-trench transform fault (Fig. 1b) lying between the Hikurangi trench to the north, along which the Pacific plate is subducted beneath the Indian plate, and the Puyseger trench to the south, which has oppositely directed subduction. Large, right-lateral displacements have occurred on both faults: as much as 400 km on the San Andreas (Dickenson & Grantz 1968) and 450 km on the Alpine fault (Wellman 1955). Rates of motion from marine magnetic data are similar for the two faults: 6 cm/yr for the San Andreas fault (Larson *et al.* 1968) and 5 cm/yr for the Alpine fault (Hayes & Talwani 1972), but geological data indicate rates slower by about a factor of two (Dickenson & Grantz 1968, Suggate 1963). Whichever data set one favours, the rates of motion on these two fault systems are comparable, and since the historical records from the two areas are of similar length, a direct comparison of the seismic histories of the two fault systems is possible.

[*Editor's Note:* Material has been omitted at this point.]

The aseismic slip is primarily of the type known in friction studies as 'episodic stable sliding', in that most of the motion occurs in episodes during which a few millimetres of slip occurs within a few days with little slip between episodes. Only one section of fault some 10 to 20 km long, indicated as A in Fig. 3, slips by continuous stable sliding (Yamashita & Burford 1973, Nason *et al.* 1974).

Aseismic slip in this region is accompanied by a high level of moderate to small earthquakes ($M < 6$), but the sum of their moments indicates that they contribute only a small fraction of the slip rate which is produced aseismically. Minor seismicity in this region, in contrast to that of southern California, is almost entirely limited to the major mappable faults, and delineates them clearly (Lee *et al.* 1972).

There is strong evidence that much of the length of the Hayward fault was ruptured in the two large earthquakes of 1836 and 1868 (Tocher 1959). This fault is now slipping aseismically, and this shows that, in at least some areas, the seismic style may not be a permanent characteristic of a given section of fault.

(B) PARKFIELD–CHOLAME SECTOR

A short segment about 40 km long between the northern end of the 1857 fault break at Cholame and the region of aseismic slip appears to exhibit a transitional seismic style. This segment has slipped during three moderately sized earthquakes in 1922, 1934, and 1966, and hence appears to behave in the San Jacinto style. The 1966 earthquake, however, was unusual in that it was followed by an anomalously large amount of aftercreep. It appears that slip during the earthquake was primarily at depth, and that the upper few kilometres of the fault began to slip aseismically after the earthquake at a rate that decreased exponentially with time (Scholz *et al.* 1969). The upper part of the fault thus appears to slip primarily aseismically, while the lower part slips seismically.

(C) SAN JUAN BATTISTA NORTH

Near the point where the Hayward-Calaveras fault branches off, the San Andreas fault undergoes a bend to the west. The rate of aseismic slip decreases north of that point, and is no longer recorded north of the town of San Juan Battista. The point where aseismic slip stops marks the southern end of the fault break of the great 1906 earthquake which ruptured the fault over a distance of about 500 km from the northern end of the fault at Cape Mendocino to San Juan Battista, and produced an average slip of nearly 4 m (Tocher 1959). Since 1906 this section of the fault has been seismically almost totally quiet and has shown no evidence of aseismic slip.

The section of the San Andreas fault north of San Juan Battista thus has behaved similarly to that of the 1857 fault break in the big bend area to the south. As Allen (1968) has pointed out, both areas lie where the fault zone is a single trace and where major bends exist which imply convergence along the fault and may serve to "lock" the fault.

3. The Alpine fault system

The historical earthquake record for the South Island of New Zealand is complete for large events from about 1830 (Eiby 1968) and hence is of similar length to that of California. For much of this period, however, the rugged terrain affected by earthquakes had a population density much less than that of California, so that information on early earthquakes is generally more sparse. Observations of fault breakage are much less complete in New Zealand, largely because of the thick bush which covers most of the area traversed by the Alpine fault system. For example, the Inangahua earthquake of 1968 was a shallow event of magnitude 7·1, and, although it had a 45 km long aftershock zone (Adams & Lowrey 1971), primary surface breakage could be identified at only two isolated localities where the fault crossed cleared land (Lensen & Otway 1971). In cases of this type, the extent of faulting is inferred from the aftershock distribution or from the extent of major secondary effects, principally landslides.

Although the Alpine fault exhibits a right lateral offset of 450 km (Wellman 1955) the present slip vector is parallel to the Hope fault (Fig. 4) and oblique to the Alpine fault (Scholz *et al.* 1972). To the south of Arthur's Pass, the Alpine fault has a component of thrusting which results in the uplift of the Southern Alps

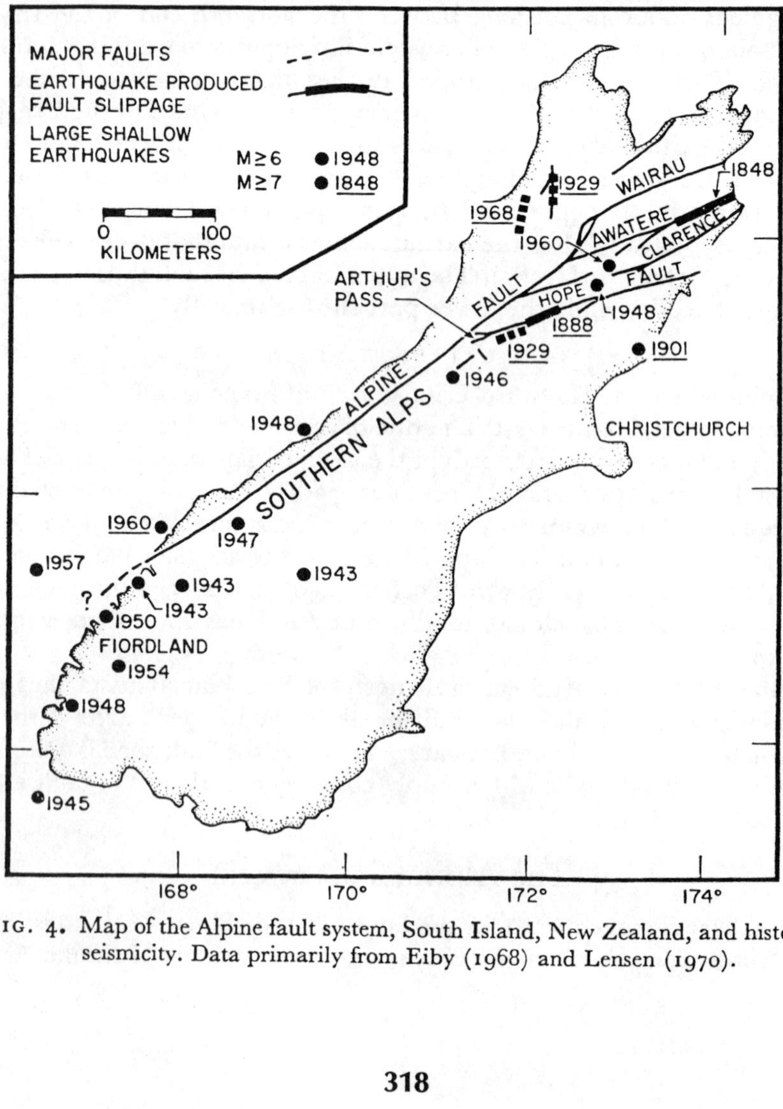

FIG. 4. Map of the Alpine fault system, South Island, New Zealand, and historical seismicity. Data primarily from Eiby (1968) and Lensen (1970).

(A) THE MARLBOROUGH–WEST NELSON FAULTS

North of Arthur's Pass, a series of subparallel faults splay off eastward from the Alpine fault. These faults, of which the most prominent are the Hope, Awatere and Wairau, are known collectively as the Marlborough–West Nelson faults. Large earthquakes occur frequently in this region, many of which are probably associated with these faults. The Amuri earthquake of 1888 produced right-lateral, strike-slip motion of as much as several metres along the Hope fault (McKay 1890), and an earthquake in 1848 is reported to have broken a 100 km length of the Awatere fault (Lensen 1970).

None of these earthquakes can be classified as great events. The frequent occurrence of earthquakes in the magnitude range $6{\cdot}5 < M < 7{\cdot}5$ and the lack of any evidence of aseismic slip on these faults (Lensen 1970) strongly indicate that their style of slip is akin to the San Jacinto type: seismic slip by relatively frequent large events.

Since all of the Marlborough–West Nelson faults must be regarded as active on geological and seismic grounds, this part of the Alpine fault zone is similar to the San Andreas system south of the big bend: the plate motion occurs as shear distributed over this entire group of strike-slip faults. A further similarity is that micro-earthquakes occur scattered over this region and are not restricted to the major faults (Scholz *et al.* 1972).

The earthquakes of 1929 and 1968 which occurred west of this fault system (Fig. 4) were high-angle, reverse-faulting events on northward striking faults (Lensen & Otway 1971, Fyfe 1929). Their mechanisms are consistent with the regional NW–SE compression and their tectonic rôle is analogous to that of the large southern California earthquakes of 1952 and 1971.

(B) ALPINE FAULT

South of the intersection of the Hope fault with the Alpine fault at Arthur's Pass· the Alpine fault defines the plate boundary as a single, relatively simple strand, It is along this part of the fault that evidence for thrusting is most prevalent (Pl. 1B).

No large or even moderate size earthquakes have occurred within the historical record that can be assigned to the Alpine fault in this section (Evison 1971). This quiescence extends to the micro-earthquake level (Scholz *et al.* 1972). This long period of total quiescence is very similar to that observed on those sections of the San Andreas fault which have had great earthquakes and suggests that the Alpine fault shares the seismic style of the 'big bend' area of the San Andreas fault.

4. Conclusion

(A) REGIONAL COMPARISONS

In the foregoing I have noted a variety of similarities of the tectonic elements of the San Andreas and Alpine fault systems and have discussed similarities of the seismic style of these elements. The two fault systems, however, can only be rationally compared in the context of the plate motions which take place across them, since it is those motions which determine the state of stress that acts on these faults. Fig. 5 shows maps of the northern part of the Alpine fault system and the southern part of the San Andreas fault system, drawn to the same scale but rotated so that the slip vectors for the two areas coincide. In California the slip vector is assumed to lie parallel to the straight section of the San Andreas fault north of the big bend which is known to be slipping by pure strike-slip motion. In New Zealand it is assumed to lie parallel to the Hope fault. The latter direction is based upon tectonics and earthquake focal mechanisms, since the South Island of New Zealand lies too close to the Indian-Pacific pole of rotation for the slip vector to be determined from sea-floor spreading data (Scholz *et al.* 1972).

This projection shows that, within the reference frame of the plate motions, the tectonic elements in these two areas occupy the same position. The main trends of the Alpine fault and the San Andreas fault are oblique to the slip vector. This implies downslip (in the direction opposite to that of the slip vector) convergence which is accomplished by uplift of the Southern Alps in New Zealand and of the Transverse Ranges in California. The mechanism of uplift is somewhat different in the two areas, being produced by thrusting on the Alpine fault in New Zealand while in California the uplift is distributed among the Transverse Range faults.

Further downslip, the plate boundary is a broad belt of shear distributed over a series of strike-slip faults nearly parallel to the slip vector. In both cases, the faults with the greatest geologically recorded offset, the Wairau and Banning–Mission Creek faults, occupy the same position. The other faults are younger and have much smaller total displacements. The onset of oblique convergence on the Alpine fault can reasonably be correlated with the mid- to late-Miocene Kaikura

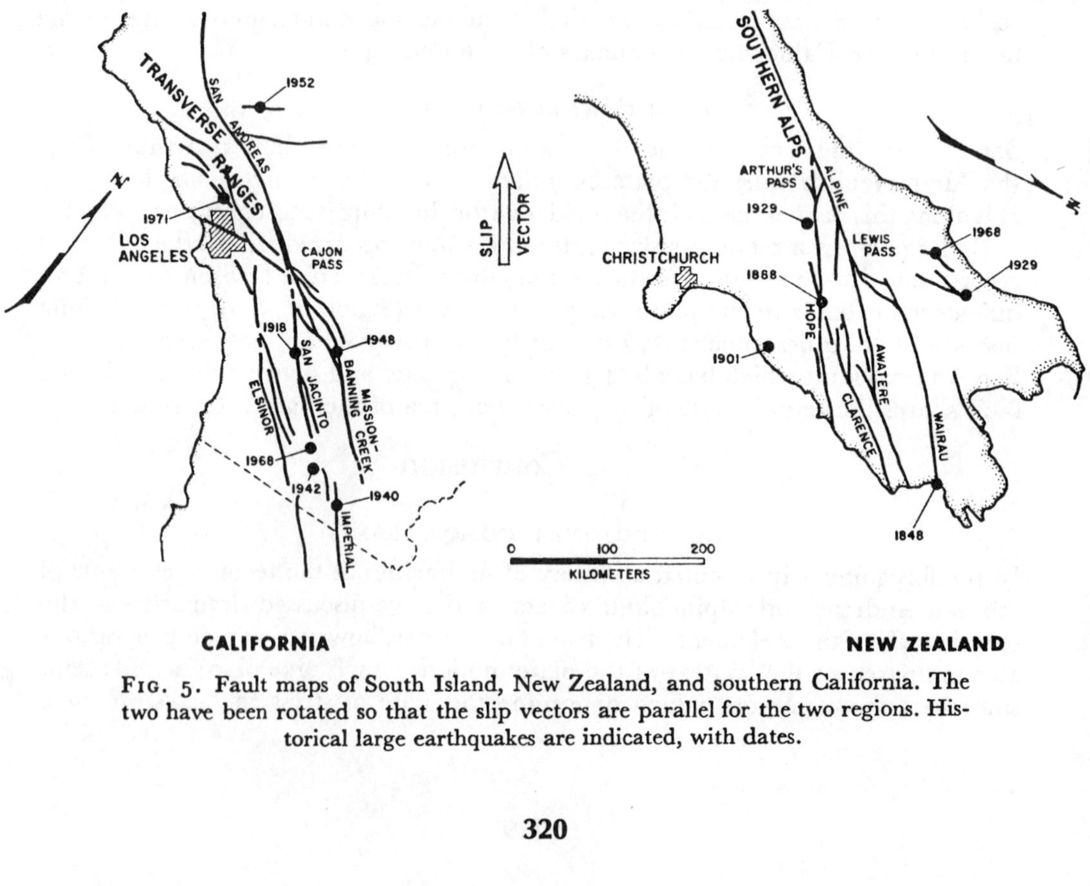

FIG. 5. Fault maps of South Island, New Zealand, and southern California. The two have been rotated so that the slip vectors are parallel for the two regions. Historical large earthquakes are indicated, with dates.

Orogeny (Suggate 1963) which was marked by uplift of the Southern Alps and formation of the Marlborough–West Nelson faults. The rôle of these faults lies in taking up some of the motion on the Alpine fault thus minimizing the amount of continental convergence.

The big bend of the San Andreas fault also appears to have developed in the Miocene, the time of the beginning of Transverse Range deformation. The bend is thought to have developed as a response to expansion in the Basin and Range Province of Nevada and Utah, which resulted in the westerly translation of that portion of the San Andreas fault north of the big bend (Scholz *et al.* 1971). The San Jacinto and Elsinor faults presumably play a similar rôle to the Marlborough–West Nelson faults in New Zealand, that of distributing the shear over a broader area, thus minimizing the continental convergence required.

[*Editor's Note:* Material has been omitted at this point and only the references cited in the preceding excerpts are reproduced here. In the original, Larson et al. (1968) and Hayes and Talwani (1972) were omitted.]

REFERENCES

ADAMS, R. & LOWREY, M. 1971. The Inangahua earthquake sequence. *In, Recent Crustal Movements,* Proceedings of Symposium. Wellington 129-35.

ALLEN, C. R. 1968. The tectonic environments of seismically active and inactive areas along the San Andreas fault. Proceedings of a conference *Geological Problems of the San Andreas fault, Stanford Univ. Publs. geol. Sci.* **11,** 70-83.

DICKENSON, W. & GRANTZ, A. (eds.) 1968. Proceedings of a conference *Geological Problems of the San Andreas fault system. Stanford Univ. Publs. geol. Sci.* **11,** 374 pp.

EIBY, G. 1968. An annotated list of New Zealand Earthquakes, 1460-1965. *N. Z. J. Geol. Geophys.* **11,** 630-47.

EVISION, F., 1971, Seismicity of the Alpine fault, New Zealand, *In, Recent Crustal Movements.* Proceedings of Symposium, Wellington. 161-5.

FYFE, H. 1929. Movements of White Creek fault, New Zealand, during the Murchison earthquake of 17 June 1929. *N. Z. Jl. Sci. Technol.* **11,** 192-7.

LEE, W., MEAGHER, K., BENNET, R. & MATAMORIS, E. 1972. Catalogue of earthquakes along the San Andreas fault system in central California. *U. S. geol. Surv. Open. File Report.* 67 pp.

LENSEN, G. 1970, Elastic and non-elastic surface deformation in New Zealand, *Bull. N. Z. Soc. Earthquake Engineers* **3,** 131-42.

LENSEN G. & OTWAY, P. 1971. Earthshift and post-earthshift deformation associated with the May 1968 Inangahua earthquake, New Zealand, *In, Recent Crustal Movements.* Proceedings of Symposium Wellington. 107-16.

MCKAY, A. 1890. On earthquakes of Sept. 1888, in the Amuri and Marlborough districts of the South Island. *Rep. geol. Surv. N. Z. Geological Exploration* 1888-89 **20,** 1-16.

NASON, R., PHILLIPSBORN, F. & YAMASHITA, P. 1974. Catalogue of creepmeter measurements in California 1968-1972. *U. S. geol. Surv. Open. File Report* 74-31, 287 pp.

SCHOLZ, C., WYSS, M. & SMITH, S. 1969. Seismic and aseismic slip on the San Andreas fault. *J. geophys. Res.* **74,** 2049–69.

SCHOLZ, C., BARAZANGI, M. & SBAR, M. 1971, Late Cenozoic evolution of the Great Basin, Western United States, as an ensialic interarc basin. *Bull. geol. Soc. Am.* **82,** 2979–90.

SCHOLZ, C., MOLNAR, P. & JOHNSON, T., 1972, Detailed studies of frictional sliding of granite and implications for the earthquake mechanism. *J. geophys. Res.* **77,** 6392–406.

SUGGATE, R. 1963. The Alpine fault. *Trans. R. Soc. N. Z.* **2,** 105–29.

TOCHER, D. 1959. Seismic history of the San Francisco Bay region. *San Francisco earthquakes of March,* 1956. *Calif. Div. Mines. Spec. Rept.* **57,** 39–48.

WELLMAN, H. 1955. New Zealand quaternary tectonics. *Geol. Rundsh.* **43,** 248–57.

YAMASHITA, P. & BURFORD, R. 1973. Catalogue of preliminary results from an 18 station creepmeter network, central California, 1969–1973, *U.S. geol. Surv. Open File Report.* 215 pp.

30

Reprinted from pages 1-3 and 15-17 of *New Zealand Jour. Geol. Geophys.*
22(1):1-19 (1979)

Geophysical study of the Hauraki Depression, North Island, New Zealand

M. P. HOCHSTEIN AND I. M. NIXON

Department of Geology, University of Auckland, New Zealand

ABSTRACT

Geophysical data show that the southern part of the Hauraki Depression is a rift structure. In W–E cross section it is made up of a fault angle depression, a median horst, and a graben. A maximum thickness of about 3 km of Quaternary and Tertiary sediments fill the two depressions. The fault angle depression and the graben are bounded at the eastern side by major normal faults which dip about 70 ± 10° to the west, namely, the recently discovered **Kerepehi Fault** in the centre and the Hauraki Fault which forms the eastern boundary of the depression. A minor hinge fault (Firth of Thames Fault) probably runs along the western boundary. Transverse faults cross the depression causing horizontal offsets of the major normal faults of up to 3 km. Recent rifting of the southernmost part is indicated by active faulting and by shallow earthquake activity beneath the depression (focal depth ⩽ 12 km). The chemistry of numerous hot springs implies that high temperatures (250–300°c) occur at depths of less than 5 km; these springs occur over the boundary faults and over the median basement ridge.

The Hauraki Depression continues north up the Hauraki Gulf, as indicated by the parallel course of the Firth of Thames shore lines. Sea gravity data across the Hauraki Gulf, and sea seismic data between Leigh and the Moko Hinau Islands suggest a basement structure similar to that of the southern part of the depression. The Hauraki Depression is, therefore, a large tectonic feature about 20–30 km wide and at least 220 km long. A comparison between major geological and geophysical features of the Hauraki Depression and those of other well-studied continental rifts shows striking similarities. It is inferred that the Hauraki Depression is an active rift which was formed after the collision of the Indian and the Pacific plates.

INTRODUCTION

Various geophysical measurements have been made across the Hauraki Depression to obtain information about its gross tectonic structure. The Hauraki Depression is a broad valley, about 25 km wide, extending from Matamata to the Hauraki Gulf over a distance of at least 150 km (Fig. 1). It is bounded on the western side by Jurassic metagreywackes (Torlesse Supergroup) which form the Hunua Range and its southern continuation, the Hapuakohe Range, with summit heights of up to 700 m. These greywackes constitute the basement of the whole region and are referred to here as "basement rocks". Tertiary volcanics of the Coromandel Range in the north and of the Kaimai Range in the south, with summit heights of 500–800 m, lie on the eastern side of the valley. These volcanics overlie the Jurassic greywackes, which outcrop at sea level near Thames township and in the northern part of the Coromandel Peninsula. No greywacke rocks are exposed south of Thames, although it can be inferred from gravity data that the greywacke basement lies at shallow depths of a few hundred metres beneath the western foothills of the Coromandel Range shown in Fig. 1. Upper Tertiary eruptive rocks, the Kiwitahi Volcanics, are also present on the western side as isolated extrusive bodies which line the western boundary of the depression. The geology of the northern part of the area shown in Fig. 1 has been summarised by Schofield (1967), and that of the southern part, by Healy *et al.* (1964).

The shallow Firth of Thames, with water depths up to 35 m, occupies the northern part of the Hauraki Depression, and the Hauraki Plains occupy the southern part. The Hauraki Plains, between the Firth of Thames foreshore and Waitoa, are a floodplain which stands only a few metres above sea level. Until the beginning of this century it was regularly flooded by streams and, in part, by high tides which produced large swamps, some of which exist today. South of Waitoa, the mean surface elevation increases to a few tens of metres and the plain changes to a partly eroded depositional surface, the southern end of which is covered by terminal flows of a Pleistocene ignimbrite sheet.

The overall parallel boundaries of the Hauraki Depression imply some fault control, although fault planes at these boundaries have not been observed. Schofield (1967) depicted these inferred faults as steeply dipping and reversed, on the basis of hot water fissures in volcanics at Te Aroha and a pyritised shear zone in basement rocks near Torehape. The inferred surface traces of the faults were shown as irregular lines following, in general, contacts between Pleistocene terraces and the plains. Arguments against the inference that the boundary faults were reversed have been given by Ballance (1967). The traces of the boundary faults in the map of Healy *et al.* (1964) are not continuous and diverge in the southernmost part of the depression. A complex tectonic structure is suggested in the south by segmented active fault traces in the central part of the depression (Healy *et al.* 1964) and also by the location of some of the widespread hot springs along the median

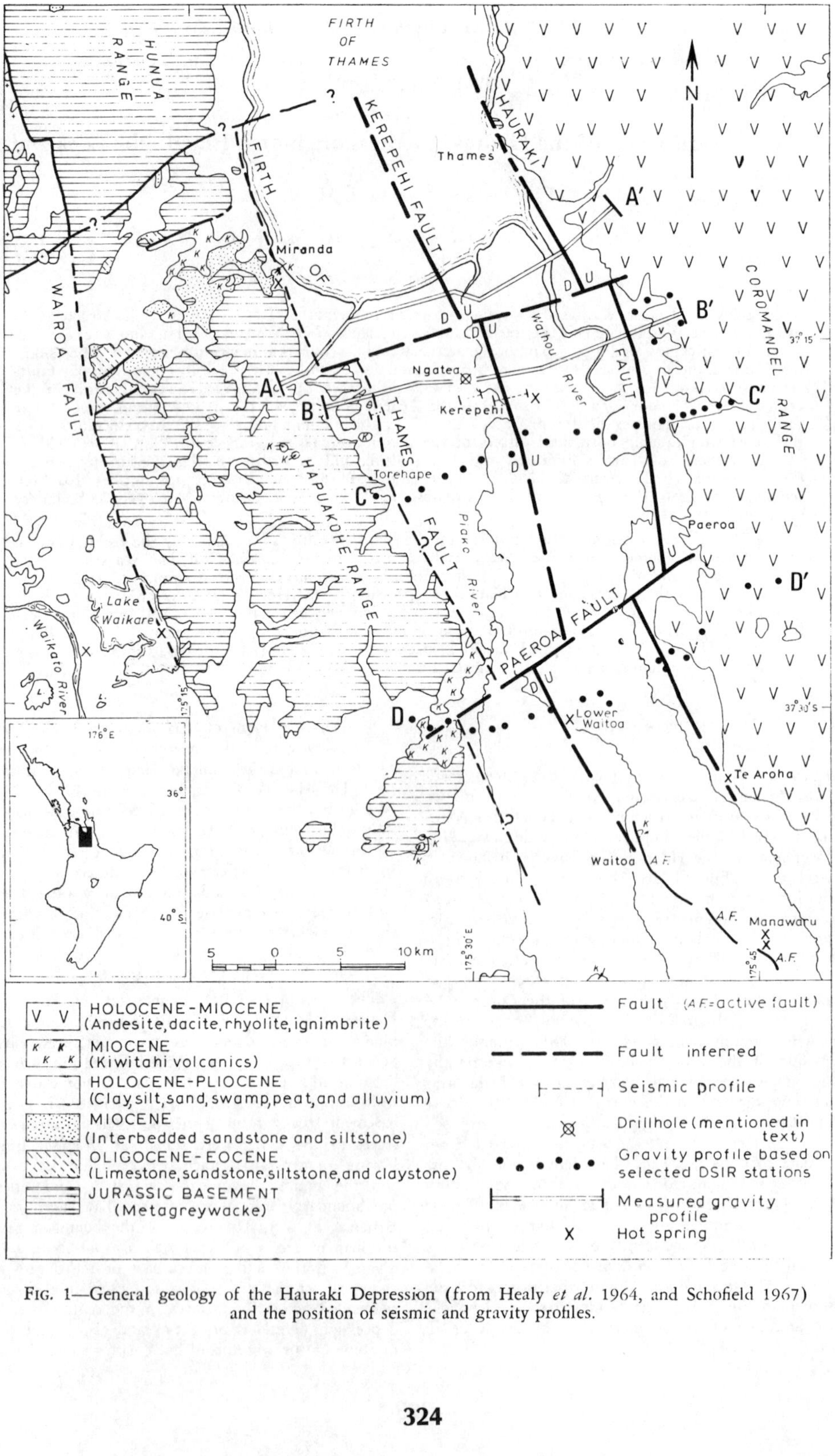

FIG. 1—General geology of the Hauraki Depression (from Healy *et al.* 1964, and Schofield 1967) and the position of seismic and gravity profiles.

part of the depression. Other hot springs are found in the vicinity of the eastern and western margins (see Fig. 1). Some tectonic activity beneath the depression is indicated by crustal earthquakes. Schofield (1968) inferred that the Hauraki Depression is a collapse feature caused by the extrusion of extensive crustal melts since mid-Tertiary times.

Little was known before our studies about the thickness of sediments which have filled the depression. Fraser (1910) recorded 340 m of Pleistocene sediments, in a bore on the foreshore at Thames, which did not reach the base of the deposit. Unconsolidated sediments were also encountered throughout a 350 m deep hole at Ngatea (R. M. Kennedy pers. comm.). A much greater thickness is indicated by the magnitude of gravity anomalies shown in maps by Woodward (1971) and Woodward & Ferry (1973), reaching values of about -25 to -30 mgal ($0 \cdot 1$ mgal $= 10^{-6}$ m/s^2) over the central part of the depression. However, the pattern of the published gravity anomalies is rather irregular and shows numerous gravity lows and gravity highs. It was found later during the interpretation of our data that some of the gravity structures over the Hauraki Plains, shown in these maps, are significantly distorted, especially in areas with low station density; the distortions are probably the by-product of the smoothing and interpolation procedures used in constructing the gravity maps (Reilly 1972).

[*Editor's Note:* Material has been omitted at this point.]

SUMMARY OF IMPORTANT FEATURES OF THE HAURAKI DEPRESSION

Interpretation of the geophysical and geochemical data indicates that the southern part of the Hauraki Depression is a rift structure. This is shown, in part, by the presence of principal normal faults which have an average dip of about $70 \pm 10°$, similar to the theoretical dip of normal faults brought about by tensional forces (Illies 1969); further evidence for rifting is given by transverse faults, which might represent incipient transform faults and which dissect the depression. The total horizontal separation is at least 2 km, given by the dip of the major faults and the average thickness ($2 \cdot 5$–$3 \cdot 5$ km) of the downfaulted sediments.

Present day rifting is indicated directly by active faulting in the south along the southern extension of the Kerepehi Fault and indirectly by the shallow earthquake activity. The chemistry of hot springs implies that high temperatures beneath the Hauraki Depression are most likely caused by anomalously hot rocks in the lower crust or at the crust-upper mantle boundary.

The Hauraki Depression extends into the Hauraki Gulf, as indicated by the parallel course of the Firth of Thames shore line. This has already been suggested by Henderson & Bartrum (1913) and also more recently by Schofield (1968). Marine geophysical data are now available to confirm this: sea gravity data across the Hauraki Gulf near $36 \cdot 5°$ S (T. Hatherton pers. comm.) show a pattern which is similar to that of our profiles A and B, although the anomalies are smaller over the western fault angle depression (-20 mgal) and the eastern graben (-25 mgal). It is likely that the depression continues even further north, since a similar basement structure has been found off shore by sea seismic measurements along a profile between Leigh and the Moko Hinau Islands (Fergusson & Hochstein 1974); this profile lies about 50 km to the north of the Hauraki Gulf profile. Between Leigh and the Moko Hinau Islands, the western fault angle depression is still recognisable, and sediments attain a maximum thickness of about 2 km. Sediments in the eastern depression are not much thicker than $0 \cdot 5$ km and lie nearly horizontal.

These data show that the Hauraki Depression is a large tectonic feature which extends probably over a distance of at least 220 km; its width increases from about 20 km in the southernmost part to about 30 km in the north.

Another important feature of the Hauraki Depression is its setting over an active subduction zone which, as part of the Indian-Pacific plate boundary, follows the Tonga-Kermadec arc into the North Island and then continues without any apparent underthrust through southern New Zealand along the Alpine Fault. As inferred from the focal depth of north-west-dipping deep earthquakes (Benioff Zone), the subduction surface lies about 200 km beneath the southernmost part of the Hauraki Depression. The angle between the axes of the inferred plate boundary and the Hauraki Depression is about 70°.

COMPARISON OF HAURAKI DEPRESSION WITH OTHER RIFTS

Various references have been made in this paper which indicate that many features of the Hauraki Depression are similar to those of continental rifts (i.e., Rhine graben, Baikal rift). However, the global tectonic setting of the Hauraki Depression, which lies over an active subduction zone, is significantly different from that of continental rifts which occur in the central part of large plates with a continental type of crust (Le Pichon *et al.* 1973). For the question whether the Hauraki Depression can be compared with other rifts, models will be discussed which have been proposed for rifts near active margins.

Arching and large scale extension of the crust can be induced by the rise of hot "upper mantle diapirs" when two plates with an oceanic type of crust collide. The resulting structures at the sea bottom are the "marginal basins" (Karig 1971) which run parallel to the plate boundary, but which are usually much wider (>100 km) than continental rifts. So far, no marginal basins have been described which strike at high angles to active margins and which occur in a plate with a continental type of crust. For these reasons, but also because of the large dimensions of these basins, we believe that the Hauraki Depression cannot be interpreted as a marginal basin.

While all authors agree that continental rifts are caused by extensional forces within the crust, some believe that these forces are brought about by doming or arching of the crust, which, in turn, is caused by some secular upwelling of deeper upper mantle material (Cloos 1939; Burke & Dewey 1973); others have suggested that rifts develop as a result of a tensional stress field (Artemjev & Artyushkov 1971) and that these rifts grow because of horizontal crack propagation (Molnar & Tapponnier 1975). In the latter case, some uplift of the crust can occur as a result of the response of a deformed deeper layer. Both the doming and the horizontal crack propagation hypotheses have been used to model continental rifts which strike at high angles to an active margin.

In the case of the doming model, high angle rifts have been interpreted by Burke & Dewey (1973) as "failed arms" of three-armed rift systems, in a setting where a plate with a continental type of crust breaks up. Most recently, Sengoer *et al.* (1978) have used the model to explain high angle rifts, also in terms of a "failed arm", in a setting where two plates with a continental type of crust collide; the authors cite the Rhine graben as an example. If one uses the doming model to explain the Hauraki Depression as a "failed arm" rift, the Taupo Volcanic Zone could be interpreted as a linked-up rift structure which represents the other arms. If both structures were caused by doming of the crust, excessive doming should occur where both structures intersect. Since there is no evidence for any large scale doming in this area, but also because it is unlikely that deep rift-valleys can result from the collapse of the central parts of large crustal arches (Artemjev & Artyushkov 1971), we believe that the Hauraki Depression cannot be explained as a "failed arm" rift.

The horizontal crack propagation model has been proposed to explain high angle rifts which have developed, for example, during the India-Eurasia collision as crustal tension cracks in an existing tensional stress field near the ends, and oblique to, extensive shear zones which run about parallel to the collision front. Molnar & Tapponnier (1975) cite the Baikal rift and the Shansi graben system as examples for such active high angle rifts and infer that the same mechanism might also apply for the Rhine graben. We believe that the horizontal crack propagation model is suitable for explaining both the formation and the position of the Hauraki Depression, since the depression also lies at the north-western end of, and strikes oblique to, an extensive shear zone which continues from Marlborough in the South Island to the eastern part of the North Island (Lensen 1975); the shear zone runs almost parallel to the still active collision front of the Indian and Pacific plate. Since other continental rifts can also be explained by the Molnar & Tapponnier model, a comparison between characteristic features of these rifts, and those of the Hauraki Depression, seems to be justified.

For this comparison, the summary of characteristic features of continental rifts by Illies (1969) can be used as a checking list. It was found that eight out of eleven features listed by Illies correspond more-or-less with those of the Hauraki Depression, namely: parallel course and steep dip of the major faults; some uplift of the flanks; a thick cover of unconsolidated sediments in the depression; segmentation of the depression with lateral offsets of the graben axis; present day subsidence and uplift of the depression and flanks, respectively; seismic activity beneath the depression with shallow foci, high heat flow, and occurrence of thermal springs. Two of the remaining three points in Illies' list have not been found yet in our case, namely, that the depression follows an old shear zone and the widening of individual segments.

Although many features of the Hauraki Depression are similar to those of other continental rifts, there are also differences.

1. In Illies' list, the characteristic tectonics of continental rifts is symmetrical antithetic normal faulting

in the centre of a much broader topographic high. In the Hauraki Depression, the established major faults dip westward and the antithetic set is poorly developed. Such asymmetry, however, could be caused by an asymmetric tensional stress field (J. H. Illies pers. comm.); the active Baikal rift is also asymmetric (Florensov 1969).

2. We have no direct evidence which shows that the Hauraki Depression is still subsiding; however, the fact that a significant portion of the depression lies at sea level, and that the unconsolidated sediments within the depression are rather thick, can be used as indirect evidence for present day subsidence.

3. Doming and present day uplift of the flanks are also important features of continental rifts, although, for the Molnar & Tapponnier model of high angle rifts, doming is not such an important feature as for the mantle plume model of Burke & Dewey. However, some uplift of the flanks, as a response to the deformation of deeper, less viscous layers (Fuchs 1974), can also be postulated for the Molnar & Tapponnier model. There is some evidence for uplift of the flanks along the Hauraki Depression, as indicated by the absence of Tertiary sediments, except for some thin pockets on top of the western and eastern basement ranges. These sediments attain a thickness of more than 2 km about 75 km to the west of the centre of the depression (Hochstein & Nunns 1976).

4. The chemical composition of the mainly alkaline flood basalts, which can be found over the flanks of, and inside, other continental rifts, is significantly different from that of the suite of only mildly alkaline basalts, andesites, and rhyolites which occur over the flanks of the Hauraki Depression. If one considers the global tectonic setting of the Hauraki Depression, this discrepancy is not as significant as it appears, since basaltic and acidic rocks are usually found together in the vicinity of active subduction zones.

In view of the similarities between active continental rifts and the Hauraki Depression, we propose that the Hauraki Depression is an active rift, to which we refer, from now on, as the **Hauraki Rift.**

GEOLOGICAL AGE OF HAURAKI DEPRESSION

If some uplift of the basement is assumed to have occurred in the southern part of the depression, the rift cannot be older than Lower Miocene to Upper Oligocene, as indicated by the widespread pockets of such sediments on the western and north-eastern basement ranges. In the case that the Molnar & Tapponnier model holds, one can also use the age of about 15×10^6 y. as a lower limit for the age of the Hauraki Rift; according to Walcott (1978), subduction of the Pacific plate in the New Zealand region, and, hence, collision of the Indian and Pacific plates, commenced at this time.

Assuming that the Kiwitahi Volcanics (Fig. 1) near the western boundary are directly associated with the tectonic development of the depression, an upper Miocene age as a lower limit for the rift is obtained, since the oldest radiometric age of these rocks is about 10×10^6 y. (Stipp 1968). Schofield (1968) also inferred that the Hauraki Depression started to develop in the Middle Tertiary.

It is possible that the rift might be significantly younger than 10×10^6 y. A tentative argument for a more recent age could be based on a comment made by Battey (1949) that boulders of banded rhyolites in an early Quaternary deposit to the west of Miranda (Fig. 1) might have come from the Coromandel Range and were transported across the rift prior to the formation of the Hauraki Depression. These boulders, however, could also have come from a yet unknown source lying to the south.

CONCLUSION

Despite various discrepancies, similarities between major geological and geophysical features of well-studied continental rifts and those of the Hauraki Depression are rather striking, especially in relation to the list of important features of continental rifts given by Illies (1969). In view of these similarities, we believe that the Hauraki Depression is an active rift, which may have developed as a high angle rift after the collision of the Indian and Pacific plates. Hence, the rift is one of the major tectonic structures in the North Island.

The hypothesis that a significant part of the volcanic activity on each side of the Hauraki Depression is related to the inferred upwelling of upper mantle material, which also caused some uplift of the flanks, is at least as attractive as the alternative hypothesis that these volcanics might be the result of a rather complex rotation, segmentation, and drifting of northern New Zealand over the Indian-Pacific plate boundary (Ballance 1976). Our model of a north-south-trending rift, underlain by an upper mantle swell, provides a more satisfactory explanation for the distribution of the volcanics, for major tectonic features within the rift, and for tectonic processes which have affected the whole region since the Miocene.

ACKNOWLEDGMENTS

Gravity and magnetic measurements over the Hauraki Depression along profile A were made by students of the Geology and Physics Departments of the University of Auckland, who participated in the 1974 Geophysics Field Camp. Professor R. F. Keam (Physics Department, University of Auckland) assisted with the analysis of the geodetic data. Gravity measurements at DSIR stations cited in the paper were provided by Dr W. I. Reilly (Geophysics Division, DSIR, Wellington). M. Bertram and D. C. Lawton (Geology Department, University of Auckland) helped with the seismic and resistivity measurements. Thermal waters from the Hauraki Plains were analysed by T. H. Wilson (Geology Department, University of Auckland). The geothermal studies were sponsored by the University of Auckland Grants Committee (AURC 141 Geol 61). Dr R. Adams, Dr P. Ballance, Professor R. N. Brothers, Professor A. Lillie, Dr W. I. Reilly, Dr K. B. Spörli, and Mr J. Schofield contributed valuable criticism at the draft stage of the paper. Professor J. H. Illies (Geology Department, University of Karlsruhe, Germany) made helpful comments during the revision of the manuscript.

[*Editor's Note:* Only the references cited in the preceding excerpts are reprinted here.]

REFERENCES

Artemjev, M. E.; Artyushkov, E. V. 1971: Structure and isostasy of the Baikal rift and the mechanism of rifting. *Journal of Geophysical Research 76 (5):* 1197-211.

Ballance, P. E. 1967: Review, Sheet 3 Auckland, Geological Map of New Zealand 1 : 250,000. *Geological Society of New Zealand Newsletter 24:* 17-19.

Ballance, P. E. 1976: Evolution of the Upper Cenozoic magmatic arc and plate boundary in northern New Zealand. *Earth and Planetary Science Letters 28:* 356-70.

Battey, M. H. 1949: The Geology of the Tuakau-Mercer Area, Auckland. *Transactions of the Royal Society of New Zealand 77 (3):* 429-55.

Burke, K.; Dewey, J. F. 1973: Plume generated triple junctions: key indicators in applying plate tectonics to old rocks. *Journal of Geology 81:* 406-33.

Cloos, H. 1939: Hebung, Spaltung, Vulkanismus. *Geologische Rundschau 30:* 405-525.

Fergusson, S. R.; Hochstein, M. P. 1974: Seismic structure of the sediments and basement between Leigh and the Mokohinau Islands. *Nelcon '74 Conference, paper 7 (B).* (Mimeographed. Published by Centre for Continuing Education, The University of Auckland, Auckland, New Zealand). 6 p.

Florensov, N. A. 1969: Rifts of the Baikal Mountain Region. *Tectonophysics 8:* 443-56.

Frazer, C. 1910: The geology of the Thames Subdivision, Hauraki, Auckland. *Bulletin New Zealand Geological Survey No. 10.* 136 p.

Fuchs, K. 1974: Geophysical contributions to taphrogenesis. In: "Approaches to Taphrogenesis". *Inter-Union Commission on Geodynamics, Scientific Report No. 8.* Schweizerbart, Stuttgart. 460 p.

Healy, J.; Schofield, J. C.; Thompson, B. N. 1964: Sheet 5 Rotorua. "Geological Map of New Zealand 1 : 250,000". New Zealand Department of Scientific and Industrial Research, Wellington.

Henderson, J.; Bartrum, J. A. 1913: The geology of the Te Aroha Subdivision, Hauraki, Auckland. *Bulletin New Zealand Geological Survey No. 16.* 127 p.

Hochstein, M. P.; Nunns, A. G. 1976: Gravity measurements across the Waikato Fault, North Island, New Zealand. *N. Z. Journal of Geology and Geophysics* 19: 347-58.

Illies, J. H. 1969: An intercontinental belt of the world rift system. *Tectonophysics 8:* 5-29.

Karig, D. E. 1971: Origin and development of marginal basins in the western Pacific. *Journal of Geophysical Research 76:* 2542-60.

Lensen, G. J. 1975: Earth-deformation studies in New Zealand. *Tectonophysics 29:* 541-51.

Le Pichon, X.; Francheteau, J.; Bonnin, J. 1973: "Plate Tectonics". Elsevier Scientific Publishing Company, Amsterdam.

Molnar, P.; Tapponnier, P. 1975: Cenozoic tectonics of Asia: effects of a continental collision. *Science 189 (4201):* 419-26.

Reilly, W. I. 1972: New Zealand gravity map series. *N. Z. Journal of Geology and Geophysics 15:* 3-15.

Schofield, J. C. 1967: Sheet 3 Auckland. "Geological Map of New Zealand 1 : 250,000". New Zealand Department of Scientific and Industrial Research, Wellington.

Schofield, J. C. 1968: Regional aspects of Cainozoic volcanology in the North Island of New Zealand—Crustal fusion produces intermediate magma. *N. Z. Journal of Geology and Geophysics 11:* 277-90.

Sengoer, A. M. C.; Burke, K.; Dewey, J. F. 1978: Rifts at high angles to orogenic belts: tests for their origin and the Upper Rhine Graben as an example *American Journal of Science 278:* 24-40.

Stipp, J. J. 1968: The geochronology and petrogenesis of the Cenozoic Volcanics of North Island, New Zealand. (Unpublished Ph.D. thesis lodged in the Library, Australian National University, Canberra.)

Walcott, R. I. 1978: Present tectonics and Late Cenozoic evolution of New Zealand. *Geophysical Journal of the Royal Astronomical Society 52:* 137-64.

Woodward, D. J. 1971: Sheet 3 Auckland. "Gravity Map of New Zealand 1 : 250,000". New Zealand Department of Scientific and Industrial Research, Wellington.

Woodward, D. J.; Ferry, L. M. 1973: Sheet 5 Rotorua. "Gravity Map of New Zealand 1 : 250,000". New Zealand Department of Scientific and Industrial Research, Wellington.

31

Reprinted from page 49 of *Proceedings of the Geological Society of New Zealand, 1981 Conference, Hamilton, November 23–26, 1981.*

REVISED GEOPHYSICAL-GEOLOGICAL MODEL OF THE HAURAKI RIFT

M. P. Hochstein and K. W. Tearney
Geology Department, University of Auckland

The first geophysical model of the Hauraki Rift (Hochstein and Nixon, 1979) was based mainly on the interpretation of negative residual Bouguer anomalies (up to 30 mgal) over the Hauraki Depression between the Firth of Thames and the Morrinsville-Te Aroha Region. The crucial assumptions for this model were that the thickness of the Quaternary layer did not exceed 600 m throughout the rift and that the remaining part of the negative gravity anomaly was caused by a 1.5–2.5 km thick sequence of Tertiary rocks.

The model had to be revised recently, as an interpretation of marine seismic and gravity data over the northern extension of the rift beneath the Hauraki Gulf has shown that Tertiary rocks are absent within the rift. The same result also applies to an interpretation of seismic reflection and gravity data along a profile crossing the depression in the Morrinsville-Te Aroha region. For the new model the density structure of the sediments in filling the rift was based on siesmic velocities. The main features of the new model are:

1. The maximum thickness of the unconsolidated (Quaternary) sediments in the rift lies between 1000 to 1500 m. These sediments are underlain by greywacke 'basement'.
2. The average dip of the major, axial, normal faults in the rift is only about 40°, pointing to concealed eroded fault scarps.
3. The en-echelon offset of the active median fault in the Morrinsville-Te Aroha region is reflected by an en-echelon basin structure of the western fault-angle depression.
4. A much younger tectonic history of the rift is indicated by these results than that postulated previously.

BIBLIOGRAPHY

1. *Upper Mantle Committee/UNESCO Seminar on the East African Rift System, Nairobi, 1965.* Sect. I. Report of the UMC/UNESCO Seminar on the East African Rift System (145 p.) Sect. II. Report on the geology and geophysics of the East African Rift System (116 p.), University College, Nairobi.
2. Irvine, T. N., ed., 1965, The World Rift System, Report of the Upper Mantle Committee Symposium on the World Rift System, *Geol. Surv. Canada Paper 66-14,* 471p.
3. Rothe, J. P., and K. Sauer, eds., 1967, *The Rhinegraben Progress Report, 1967,* International Rhinegraben Research Group, Abh. Geol. Landesamt Baden-Wurttemberg, Heft 6 *et* Serv. Carte geol. Als. Lorr., n° hors série, 146p.
4. Knopoff, L., B. C. Heezen, and G. J. F. MacDonald, eds., 1969, Upper Mantle Committee Symposium on the World Rift System, 1967, Zurich, Switzerland, *Tectonophysics* **8**(No. 4-6): 265-586.
5. Falcon, N. L., I. G. Gass, R. W. Girdler, and A. S. Laughton, eds., 1970, Structure and evolution of the Red Sea and the nature of the Red Sea, Gulf of Aden and Ethiopian Rift Junction, *Royal Soc. London Philos. Trans.* **A267:**1-417p.
6. Illies, J. H., and St. Mueller, eds., 1970, *Graben Problems,* Proceedings of an International Rift Symposium, Karlsruhe, 1968, E. Schweizerbart'sche Verlagsbuchhandlung, Stuttgart, 316p.
7. Girdler, R. W., ed., 1972, Symposium on African Rifts, Upper Mantle Committee, Moscow, 1971, *Tectonophysics* (special issue) **15** (No. 1-2):1-353.
8. Illies, J. H., and K. Fuchs, eds., 1974, *Approaches to Taphrogenesis,* Proceedings of an International Rift Symposium, Karlsruhe, 1972, E. Schweizerbart'sche Verlagsbuchhandlung, Stuttgart, 460p.
9. Pilger, A., and A. Rösler, eds., 1975, *Afar Depression of Ethiopia,* Proceedings of an International Symposium on the Afar Region and Related Rift Problems, vol. 1, Bad Bergzabern, West Germany, 1974, I. Schweizerbart'sche Verlagsbuhhandlung, Stuttgart, 416p.
10. Pilger, A., and A. Rösler, eds., 1976, *Afar between Continental and Oceanic Rifting,* Proceedings of an International Symposium on the Afar Region and Related Rift Problems, vol. 2, Bad Bergzabern, West Germany, 1974, E. Schweizerbart'sche Verlagsbuchhandlung, Stuttgart, 216p.
11. Logatchev, N. A., and P. Mohr, 1977, *Geodynamics of the Baikal Rift*

Zone, Symposium on the Rift Zones of the Earth, Irkutsk, 1975, USSR, *Tectonophysics* (special issue) **45**(No. 1):1–106.

12. Illies, J. H., ed., 1981, *Mechanism of Graben Formation,* ICG Symposium, Canberra, Australia, 1979, *Tectonophysics* (special issue) **73**(No. 1-3): 1–266.
13. Smith, R. B., and G. P. Eaton, eds., 1978, *Cenozoic Tectonics and Regional Geophysics of the Western Cordillera,* Geol. Soc. America Mem. 152, 385p.
14. E.-R. Neumann, and I. B. Ramberg, eds., 1978, *Petrology and Geochemistry of Continental Rifts,* Proceedings of the NATO Advanced Study Inst. Paleorift Systems with emphasis on the Permian Oslo Rift, Oslo, Norway, July 27–Aug. 5, 1977, vol. 1, D. Reidel, Dordrecht, 290p.
15. Ramberg, I. B., and E.-R. Neumann, eds., 1978, *Tectonics and Geophysics of Continental Rifts,* Proceedings of the NATO Advanced Study Inst. Paleorift Systems with emphasis on the Permian Oslo Rift, Oslo, Norway, July 27–Aug. 5, 1977, vol. 2, D. Reidel, Dordrecht, 444p.
16. Riecher, R. E., ed., 1979, *Rio Grande Rift: Tectonics and Magmatism,* selected papers from International Symposium on Rio Grande Rift, Santa Fe, New Mexico, Oct. 1978 (Working Group 4, Inter-Union Commission on Geodynamics), American Geophysical Union, Washington, D.C., 445p.
17. *Geodynamic Evolution of the Afro-Arabian Rift System,* 1980, International Meeting organized by the Accademia Nazionale dei Lincei and the Consiglio Nazionale delle Ricerche, Rome, 18–20 April, 1979, Accademia Nazionale dei Lincei, Rome, 705p.
18. Freund, R., and Z. Garfunkel, eds., 1981, *The Dead Sea Rift,* International Symposium on Rift Zones of the Earth, Jerusalem, Sept., 1979, *Tectonophysics* (special issue) **80:**1–30A.
19. Quennell, A. M., ed., 1982, *Rift Valleys: Afro-Arabian,* Benchmark Papers in Geology, vol. 60, Hutchinson Ross, Stroudsburg, Pa., 419p.
20. Palmason, G., ed., 1981, *Continental and Oceanic rifts,* Geodynamics Series, vol. 8 (Final Report Working Group 4, Inter-Union Commission on Geodynamics), American Geophysical Union/Geological Society of America, Washington D.C., 309p.
21. Morgan P., and B. H. Baker, eds., 1983, *Processes of Continental Rifting,* Lunar and Planetary Inst. Topical Conf. on Processes of Planetary Rifting, St. Helena, California, Dec. 1981, (Developments in Geotectonics, vol. 19) *Tectonophysics* **94:**1–680.
22. Ballance, P. F., and H. G. Reading, eds., 1980, *Sedimentation in Oblique-Slip Mobile Zones,* Int. Assoc. Sedimentologists, Spec. Pub. 4, Blackwell, Oxford, 265p.

AUTHOR CITATION INDEX

SUBJECT INDEX

About the Editor

A. M. QUENNELL is a research associate in geology at the University of Bristol, England. He retired in 1974 from an editing post with the Institute of Geological Sciences, London. He was graduated with the B.Sc. in physics and the M.Sc. in geology in the 1930s from Otago and Victoria universities, New Zealand. He is an associate of the Otago School of Mines. He served before and after World War II with the New Zealand Geological Survey and the Shell Company. He was assistant director of the Trans-Jordan Lands and Surveys Department. He made the first definitive geological survey of Trans-Jordan, which included the eastern flank of the Dead Sea Rift. He was with the Tanganyika (now Tanzania) Geological Survey until 1960, and for the last years he was director. He then practiced as consultant geologist in New Zealand and, in the later 1960s, was manager of United Nations projects in Nigeria and the Sudan.

In the course of his service in the Middle East and Africa, Mr. Quennell had unusual opportunities for the study of the Afro-Arabian Rift System along most of its length. On the basis of comparative geology, he further advanced the strike-slip hypothesis of earlier writers for the formation of the Dead Sea Rift and the opening of the Red Sea. In the course of this, he became the first to recognize in 1956 and publish (1959) the location of what is now known as a pole of rotation—that of the Arabian plate in relation to Sinai-Palestine.

Mr. Quennell is the author and editor of publications on the stratigraphy of New Zealand; on the geology, especially of the Precambrian, of eastern Africa, Jordan, and the CENTO countries.

He is a fellow of the Geological Society, London, and of the Institute of Mining and Metallurgy, and is a member of the Institution of Civil Engineers.

Benchmark Papers in Geology

Series Editor: Rhodes W. Fairbridge, Columbia University

ENVIRONMENTAL GEOMORPHOLOGY AND LANDSCAPE CONSERVATION, Volume I: Prior to 1900 / *Donald R. Coates*
RIVER MORPHOLOGY / *Stanley A. Schumm*
SPITS AND BARS / *Maurice L. Schwartz*
TEKTITES / *Virgil E. Barnes and Mildred A. Barnes*
GEOCHRONOLOGY: Radiometric Dating of Rocks and Minerals / *C. T. Harper*
SLOPE MORPHOLOGY / *Stanley A. Schumm and M. Paul Mosely*
MARINE EVAPORITES: Origin, Diagenesis, and Geochemistry / *Douglas W. Kirkland and Robert Evans*
ENVIRONMENTAL GEOMORPHOLOGY AND LANDSCAPE CONSERVATION, Volume III: Non-Urban Areas / *Donald R. Coates*
BARRIER ISLANDS / *Maurice L. Schwartz*
GLACIAL ISOSTASY / *John T. Andrews*
GEOCHEMISTRY OF GERMANIUM / *Jon N. Weber*
ENVIRONMENTAL GEOMORPHOLOGY AND LANDSCAPE CONSERVATION, Volume II: Urban Areas / *Donald R. Coates*
PHILOSOPHY OF GEOHISTORY: 1785–1970 / *Claude C. Albritton, Jr.*
GEOCHEMISTRY AND THE ORIGIN OF LIFE / *Keith A. Kvenvolden*
SEDIMENTARY ROCKS: Concepts and History / *Albert V. Carozzi*
GEOCHEMISTRY OF WATER / *Yasushi Kitano*
METAMORPHISM AND PLATE TECTONIC REGIMES / *W. G. Ernst*
GEOCHEMISTRY OF IRON / *Henry Lepp*
SUBDUCTION ZONE METAMORPHISM / *W. G. Ernst*
PLAYAS AND DRIED LAKES: Occurrence and Development / *James T. Neal*
GLACIAL DEPOSITS / *Richard P. Goldthwait*
PLANATION SURFACES: Peneplains, Pediplains, and Etchplains / *George F. Adams*
GEOCHEMISTRY OF BORON / *C. T. Walker*
SUBMARINE CANYONS AND DEEP-SEA FANS: Modern and Ancient / *J. H. McD. Whitaker*
ENVIRONMENTAL GEOLOGY / *Frederick Betz, Jr.*
LOESS: Lithology and Genesis / *Ian J. Smalley*
PERIGLACIAL PROCESSES / *Cuchlaine A. M. King*
LANDFORMS AND GEOMORPHOLOGY: Concepts and History / *Cuchlaine A. M. King*
METALLOGENY AND GLOBAL TECTONICS / *Wilfred Walker*
HOLOCENE TIDAL SEDIMENTATION / *George deVries Klein*
PALEOBIOGEOGRAPHY / *Charles A. Ross*
MECHANICS OF THRUST FAULTS AND DÉCOLLEMENT / *Barry Voight*

WEST INDIES ISLAND ARCS / *Peter H. Mattson*
CRYSTAL FORM AND STRUCTURE / *Cecil J. Schneer*
OCEANOGRAPHY: Concepts and History / *Margaret B. Deacon*
METEORITE CRATERS / *G. J. H. McCall*
STATISTICAL ANALYSIS IN GEOLOGY / *John M. Cubitt and Stephen Henley*
AIR PHOTOGRAPHY AND COASTAL PROBLEMS / *Mohamed T. El-Ashry*
BEACH PROCESSES AND COASTAL HYDRODYNAMICS / *John S. Fisher and Robert Dolan*
DIAGENESIS OF DEEP-SEA BIOGENIC SEDIMENTS / *Gerrit J. van der Lingen*
DRAINAGE BASIN MORPHOLOGY / *Stanley A. Schumm*
COASTAL SEDIMENTATION / *Donald J. P. Swift and Harold D. Palmer*
ANCIENT CONTINENTAL DEPOSITS / *Franklyn B. Van Houten*
MINERAL DEPOSITS, CONTINENTAL DRIFT AND PLATE TECTONICS / *J. B. Wright*
SEA WATER: Cycles of the Major Elements / *James I. Drever*
PALYNOLOGY, PART I: Spores and Pollen / *Marjorie D. Muir and William A. S. Sarjeant*
PALYNOLOGY, PART II: Dinoflagellates, Acritarchs, and Other Microfossils / *Marjorie D. Muir and William A. S. Sarjeant*
GEOLOGY OF THE PLANET MARS / *Vivien Gornitz*
GEOCHEMISTRY OF BISMUTH / *Ernest E. Angino and David T. Long*
ASTROBLEMES—CRYPTOEXPLOSION STRUCTURES / *G. J. H. McCall*
NORTH AMERICAN GEOLOGY: Early Writings / *Robert Hazen*
GEOCHEMISTRY OF ORGANIC MOLECULES / *Keith A. Kvenvolden*
TETHYS: The Ancestral Mediterranean / *Peter Sonnenfeld*
MAGNETIC STRATIGRAPHY OF SEDIMENTS / *James P. Kennett*
CATASTROPHIC FLOODING: The Origin of the Channeled Scabland / *Victor R. Baker*
SEAFLOOR SPREADING CENTERS: Hydrothermal Systems / *Peter A. Rona and Robert P. Lowell*
MEGACYCLES: Long-Term Episodicity in Earth and Planetary History / *G. E. Williams*
OVERWASH PROCESSES / *Stephen P. Leatherman*
KARST GEOMORPHOLOGY / *M. M. Sweeting*
RIFT VALLEYS: Afro-Arabian / *A. M. Quennell*
MODERN CONCEPTS OF OCEANOGRAPHY / *G. E. R. Deacon and Margaret B. Deacon*
OROGENY / *John G. Dennis*
EROSION AND SEDIMENT YIELD / *J. B. Laronne and M. P. Mosley*
GEOSYNCLINES: Concept and Place Within Plate Tectonics / *F. L. Schwab*
DOLOMITIZATION / *Donald H. Zenger and S. J. Mazzullo*
OPHIOLITIC AND RELATED MELANGES / *G. J. H. McCall*
ECONOMIC EVALUATION OF MINERAL PROPERTY / *Sam L. VanLandingham*
SUNSPOT CYCLES / *D. Justin Schove*
MINING GEOLOGY / *Willard C. Lacy*
MINERAL EXPLORATION / *Willard C. Lacy*

BASALTS / *Paul C. Ragland and John J. W. Rogers*
PHYSICAL HYDROGEOLOGY / *R. Allan Freeze and William Back*
CHEMICAL HYDROGEOLOGY / *William Back and R. Allan Freeze*
MODERN CARBONATE ENVIRONMENTS / *Ajit Bhattacharyya and G. M. Friedman*
FABRIC OF DUCTILE STRAIN / *Mel Stauffer*
TERRESTRIAL TRACE-FOSSILS / *William A. S. Sarjeant*
GEOLOGY OF COAL / *Charles Ross and June R. P. Ross*
NANNOFOSSIL BIOSTRATIGRAPHY / *Bilal U. Haq*
CALCAREOUS NANNOPLANKTON / *Bilal U. Haq*
RIVER NETWORKS / *Richard S. Jarvis and Michael J. Woldenberg*
STABILITY OF HEAVY MINERALS IN SEDIMENTS / *Gretchen Luepke*
STRATIGRAPHY: Foundations and Concepts / *Barbara M. Conkin and James E. Conkin*
VERTEBRATE PALEONTOLOGY / *Robert Schoch*
MANTLE FLOW AND PLATE THEORY / *Zvi Garfunkel*
EARTH TIDES / *J. C. Harrison*
ECONOMIC ANALYSIS OF HEAVY MINERALS IN SEDIMENTS / *Gretchen Luepke*
MODERN AND ANCIENT ALLUVIAL FAN DEPOSITS / *Tor H. Nilsen*
CONTINENTAL DRIFT / *James H. Shea*
PLATE TECTONICS / *James H. Shea*
CONTINENTAL RIFTS / *A. M. Quennell*